KVAERNER PROCESS
HVAC ENGINEERING

9/17/01

KVAERNER PROCESS
HVAC ENGINEERING

MEANS
MECHANICAL
ESTIMATING

Standards and Procedures

Second Edition

Senior Editors
John J. Moylan/Melville Mossman

Illustrator
Carl W. Linde

RS**Means**
CMDGROUP

MEANS MECHANICAL ESTIMATING

Standards and Procedures

Second Edition

RSMeans
CMDGROUP

Copyright 1992

R.S. Means Company, Inc.
Construction Publishers & Consultants
Construction Plaza
63 Smiths Lane
Kingston, MA 02364-0800
(781) 585-7880

The editors for this book were John Moylan, Melville Mossman, and Julia Willard. The managing editor was Mary Greene. Composition was supervised by Joan Marshman. The book and cover were designed by Norman R. Forgit. Illustrations by Carl Linde.

Printed in the United States of America

10 9 8

Library of Congress Catalog Number 92-187016

ISBN 0-87629-213-9

TABLE OF CONTENTS

ACKNOWLEDGMENTS

The editors of *Means Mechanical Estimating* would like to express their appreciation to the following estimators and contractors who contributed to the preparation of this book by sharing their experience and knowledge: Martin E. Ahlstrom, Insulation; Gene E. Corrigan, Sheet Metal; John B. Gwynn, Temperature Control; Fred Thornley, Mechanical Estimator and George H. Twomey, Master Plumber.

FOREWORD

For over 50 years, R.S. Means Co., Inc. has researched and published cost data for the building construction industry. *Means Mechanical Estimating* combines that valuable experience with the senior editor's 40-year involvement with mechanical contracting. This book was written for construction professionals and beginners who desire an understanding of the estimating processes used in the mechanical industry, and is one of a series of estimating reference books authored and published by R.S. Means Co., Inc.

Means Mechanical Estimating contains estimating standards and procedures for piping, plumbing, heating, cooling, ventilating, and fire protection. These are dynamic systems, which are constantly in motion serving the structure and its occupants.

Placing boilers, insulating pipes, and other functions accomplished by the mechanical contractor may involve the services of other trades. For this reason, the "total mechanical contractor," much like a general contractor, may employ or subcontract all of the individual mechanical trades, as well as riggers, temperature control workers, electricians, balancing contractors, or excavators. Although there may be thousands of total mechanical contractors, these numbers are far outweighed by the numbers of individual plumbing shops, HVAC firms, sprinkler contractors, and sheet metal contractors accomplishing the bulk of mechanical work today. This book has been designed both to be read in full by the total mechanical contractor, and to be used as a reference tool by the individual trades.

This book is presented in three parts. The first part, "The Estimating Process," defines the types and components of estimates and outlines the progression of each type as a building project advances from an idea to a set of contract plans and specifications. Bidding, purchasing, scheduling, and project management techniques round off this section.

The second part of this book, "Components of Mechanical Systems," describes the individual components used in some or all mechanical systems. This part begins with Chapter 13, "Piping," since pipe is the common material used in all mechanical systems. A description of each component is followed by labor and material pricing and takeoff procedures. Cost adjustment factors are given when appropriate, as well as cost-saving "field expedients."

The last part, "Sample Estimates," utilizes a reduced scale set of mechanical building plans as a model for the takeoff and estimating procedures previously discussed, using Means forms and methods to complete square foot, systems, and unit price mechanical estimates. Subcontractor quotes, pricing materials and equipment, and an alternate method of unit pricing are included in "Sample Estimating."

An Appendix of reference information, tables, graphic symbols, and abbreviations concludes *Means Mechanical Estimating*.

Part 1
THE ESTIMATING PROCESS

Chapter 1
GUIDELINES FOR ESTIMATING

The term "estimating accuracy" is a relative concept. What is the correct or accurate cost of a given construction project? Is it the total price paid to the contractor? Could another reputable contractor perform the same work for a different cost, whether higher or lower? There is no one correct estimated cost for a given project. There are too many variables in construction. At best, the estimator can determine a very close approximation of what his final costs will be. The resulting accuracy of this approximation is directly affected by the amount of detail provided and the amount of time spent on the estimate.

Every cost estimate requires three basic components. The first is the establishment of standard *units of measure*. The second component of an estimate is the determination of the *quantity* of units for each component, an actual counting process: how many linear feet of pipe, how many water closets, pounds of sheet metal, etc. The third component, and perhaps the most difficult to obtain, is the determination of a reasonable cost for each unit to purchase and install.

The first element, the designation of measurement units, is the step which determines and defines the level of detail, and thus the degree of accuracy of a cost estimate. In mechanical construction, such units could be as all-encompassing as the HVAC (heating, ventilation, and air-conditioning) system per square foot of floor area, or as detailed as a pipe fitting.

Depending on the estimator's intended use, the designation of the "unit" may describe a complete system, or it may imply only an isolated entity. The choice and detail of units also determines the time required to complete an estimate.

The second component of every estimate, the determination of quantity, is more than the counting of units. In construction, this process is called the "material takeoff." In order to perform this function successfully, the estimator should have a working knowledge of the materials, methods, and codes used in mechanical construction. An understanding of the design specification intent is particularly important. This knowledge helps to assure that each quantity is correctly tabulated and that essential items are not forgotten or omitted. The estimator with a thorough knowledge of construction is also more likely to account for all requirements in the estimate whether or not they are called for on the plans or in the specifications. Many of the items to be taken off may not involve any material but, rather, entail labor costs only. Testing is an example of a labor-only item. Experience is, therefore, invaluable to ensure a complete estimate.

The third component is the determination of a reasonable cost for each unit. This aspect of the estimate is significantly responsible for variations in estimating. Rarely do two estimators arrive at exactly the same material cost for a project. However, even if total material costs for an installation are similar for competing contractors, the labor cost estimate for installing that material will account for a variation of bids. Labor costs may vary, due to productivity as well as the pay scales in different geographical areas. The use of specialized equipment can decrease installation time and, therefore, cost. Finally, material prices do fluctuate within the market. These cost differences occur from city to city and even from supplier to supplier in the same town. It is the experienced and well prepared estimator who can keep track of these variations and fluctuations and use them to the best advantage when preparing accurate estimates. Preferential and quantity discounts play a significant part in determining material costs.

This third component of the estimate, the determination of costs, can be defined in three different ways by the estimator. With one approach, the estimator uses a predetermined cost which includes all the elements (i.e., material, installation, overhead, and profit) as one number expressed in dollars per unit. A variation of this approach is to use a unit cost which combines total material and installation only, adding a percent markup for overhead and profit to the "bottom line."

A second method is to use unit costs—in dollars—for material and for installation. This is done separately for each item, without markups. These are called "bare costs." Different profit and overhead markups are applied to each before the material and installation prices are added; the result is the total "selling" price.

A third method of unit pricing uses unit costs for materials, and man-hours, man-days, or crew-days as the units of labor. Again, these figures are totaled separately; one represents the value for materials (in dollars), the other shows the total man-days for installation. The average cost per day of craft labor is determined by allowing for the expected ratios of foremen, journeymen, and apprentices. This is called the "composite labor rate." This rate is multiplied by the total man-days or crew-days to get the total labor cost of installation. Different overhead and profit markups can then be applied to each, material and labor, and the results added to get the total selling price.

The word *unit* is used in many ways, as can be seen in the above definitions. Keeping the concepts of units clearly defined is vital to achieving an accurate, professional estimate. For the purposes of this book the following references to different types of units are used:

- **Unit of Measure**: The standard by which the quantities are counted, such as *linear feet* of pipe, *number* of fixtures, *pounds* of ductwork, *square feet* of blanket insulation.
- **Material Unit Cost**: The cost to purchase each unit of measure: this cost includes any ancillary items that will not be taken off and priced separately, such as pipe nipples, test caps, sheet metal screws, etc.
- **Labor Unit**: The man-hours required to install or fabricate and install a unit of measure. (Note: Labor units multiplied by the labor rate equals the installation unit cost in dollars.)

Chapter 2
TYPES OF ESTIMATES

Building construction estimators use four basic types of estimates. These types may be referred to by different names and may not be recognized by all as definitive. Most estimators, however, will agree that each type has its place in the construction estimating process. The four types of estimates are described below.

Order of Magnitude Estimates: Order of magnitude costs are defined in relation to the usable units that have been designed for a facility. If, for example, a hospital administrator is planning to enlarge a hospital, he needs to know the projected cost per bed. If an estimator knows the quantity of beds in a proposed hospital (or the number of apartments in an apartment building, or the tons of air conditioning), the cost of the project can be estimated. Accuracy may be plus or minus 20%.

Square Foot and Cubic Foot Estimates: This type is most often useful when only the proposed size and use of a planned building is known. This method can be completed within an hour or two. Depending on the source of cost information an accuracy of plus or minus 15% can be expected.

Assemblies (or Systems) Estimates: An Assemblies Estimate is best used as a budgetary tool in the planning stages of a project when some parameters have been decided (e.g., size of the space, owner's requirements, etc.). This type of estimate could require as much as one day to complete. Because more specific information is known about the project, a plus or minus 10% accuracy can be attained from this estimating method.

Unit Price Estimates: Working drawings and full specifications are required to complete a unit price estimate. It is the most accurate of the four types but is also the most time-consuming. Used primarily for bidding purposes, the accuracy of a unit price estimate can be plus or minus 5%.

Figure 2.1 graphically demonstrates the relative relationship of required time versus resultant accuracy of a complete building estimate for each of these four basic estimate types. It should be recognized that, as an estimator *and* his company gain repetitive experience on similar or identical projects, the accuracy of all four types of estimates should improve dramatically. In fact, given enough experience, Square Foot and Assemblies estimates may closely approach the accuracy of Unit Price Estimates.

Order of Magnitude Estimates

The Order of Magnitude Estimate, also known as a Conceptual Estimate, can be completed with only a minimum of information. The "units," as described in the first chapter of this book, can be very general for this type and need not be well-defined. For example: "The mechanical work for the office building of a small service company in a suburban industrial park will cost about $20,000." This type of statement (or estimate) can be made after a few minutes of thought used to draw upon experience and to make comparisons with similar projects from the past. While this rough figure might be appropriate for a project in one region of the country, substantial adjustments may be required for a change of geographic location due to climate and for cost changes over time, due to, for example, changes of materials, inflation, or code changes.

Figure 2.2, from *Means Mechanical Cost Data*, includes data for a refined approach to the Order of Magnitude Estimate. This format is based on unit of use. Please note at the bottom of the category "Nursing Homes," for example, that costs are given "per bed" or "per person." The proposed use and magnitude of the planned structure–such as the number of beds for a hospital building or the number of apartments in a complex may be the only parameters known at the time the Order of Magnitude Estimate is done. The data given in Figure 2.2 does not require that details of the proposed project be known in order to determine rough costs; the only required information is the intended use and capacity of the building. What is lacking in accuracy (plus or minus 20%) is more than compensated by the minimal time required to complete the Order of Magnitude Estimate–a matter of minutes. Some mechanical units which are frequently used are "tons of air-conditioning," "number of radiators," "number of plumbing fixtures," "CFM of air" to be handled, "number of sprinkler heads, " etc.

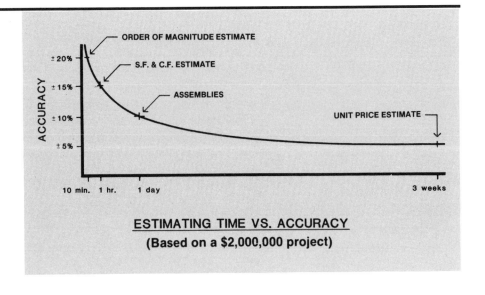

ESTIMATING TIME VS. ACCURACY
(Based on a $2,000,000 project)

Figure 2.1

171 000		S.F. & C.F. Costs	UNIT	UNIT COSTS			% OF TOTAL			
				¼	MEDIAN	¾	¼	MEDIAN	¾	
570	0010	**MEDICAL OFFICES**	S.F.	57.20	71.90	87.45				570
	0020	Total project costs	C.F.	4.44	5.95	7.90				
	2720	Plumbing	S.F.	3.49	5.20	7.10	5.70%	6.90%	9.20%	
	2770	Heating, ventilating, air conditioning		4.04	6.10	7.90	6.50%	8%	10.40%	
	2900	Electrical		4.83	6.95	9.30	7.60%	9.70%	11.70%	
	3100	Total: Mechanical & Electrical		11.30	16.05	21.05	17.30%	22.40%	27.10%	
590	0010	**MOTELS**	▼	41.75	55.85	75.85				590
	0020	Total project costs	C.F.	3.61	5.55	8.50				
	2720	Plumbing	S.F.	3.76	4.70	5.70	9.40%	10.60%	12.50%	
	2770	Heating, ventilating, air conditioning		1.99	3.42	4.99	4.90%	5.60%	8.20%	
	2900	Electrical		3.48	4.27	5.50	7%	8.10%	10.40%	
	3100	Total: Mechanical & Electrical	▼	8.25	10.85	14.50	17.90%	23.10%	26.20%	
	9000	Per rental unit, total cost	Unit	14,900	28,400	38,800				
	9500	Total: Mechanical & Electrical	"	4,025	5,700	6,150				
600	0010	**NURSING HOMES**	S.F.	57.40	75.75	92.15				600
	0020	Total project costs	C.F.	4.58	6	8				
	2720	Plumbing	S.F.	5.10	6.25	9.10	9.30%	10.30%	13.30%	
	2770	Heating, ventilating, air conditioning		5.20	7.25	9.30	9.20%	11.40%	11.80%	
	2900	Electrical		5.75	7.25	9.45	9.70%	11%	12.80%	
	3100	Total: Mechanical & Electrical		13.40	17.90	26.95	22.30%	28.30%	33.30%	
	9000	Per bed or person, total cost	Bed	22,200	28,900	36,700				
610	0010	**OFFICES** Low-Rise (1 to 4 story)	S.F.	47.10	60.35	79.90				610
	0020	Total project costs	C.F.	3.49	4.86	6.55				
	2720	Plumbing	S.F.	1.79	2.70	3.86	3.60%	4.50%	6%	
	2770	Heating, ventilating, air conditioning		3.81	5.35	7.84	7.20%	10.40%	11.90%	
	2900	Electrical		3.98	5.50	7.55	7.40%	9.50%	11%	
	3100	Total: Mechanical & Electrical		8.25	12.25	17.90	14.90%	20.80%	26.80%	
620	0010	**OFFICES** Mid-Rise (5 to 10 story)	▼	52.95	65.05	87.25				620
	0020	Total project costs	C.F.	3.63	4.69	6.60				
	2720	Plumbing	S.F.	1.60	2.43	3.50	2.80%	3.60%	4.50%	
	2770	Heating, ventilating, air conditioning		3.94	5.65	9	7.60%	9.30%	11%	
	2900	Electrical		3.36	4.81	7.40	6.50%	8%	10%	
	3100	Total: Mechanical & Electrical		9.50	12.20	20.35	16.70%	20.50%	25.70%	
630	0010	**OFFICES** High-Rise (11 to 20 story)	▼	62.70	79.25	98.30				630
	0020	Total project costs	C.F.	4.01	5.70	8.10				
	2900	Electrical	S.F.	3.13	4.56	7	5.80%	7%	10.50%	
	3100	Total: Mechanical & Electrical		11.45	14.55	24.20	17.20%	21.40%	29.40%	
640	0010	**POLICE STATIONS**	▼	78.40	101	129				640
	0020	Total project costs	C.F.	5.75	7.45	9.90				
	2720	Plumbing	S.F.	4.28	6.25	10.50	5.60%	6.80%	10.70%	
	2770	Heating, ventilating, air conditioning		6.30	8.50	12.10	7%	10.50%	11.90%	
	2900	Electrical		7.85	12.50	15.80	9.40%	11.70%	14.70%	
	3100	Total: Mechanical & Electrical		21.25	26.55	34.70	22.60%	27.50%	33.10%	
650	0010	**POST OFFICES**	▼	63.85	76.20	99.55				650
	0020	Total project costs	C.F.	3.52	4.60	5.45				
	2720	Plumbing	S.F.	2.72	3.44	4.36	4.20%	5.30%	5.60%	
	2770	Heating, ventilating, air conditioning		3.93	5.25	8	6.60%	8%	9.80%	
	2900	Electrical		5	7.05	8.35	7.40%	9.40%	11%	
	3100	Total: Mechanical & Electrical		11	1.51	21	16.50%	21.40%	26.30%	
660	0010	**POWER PLANTS**	▼	375	555	850				660
	0020	Total project costs	C.F.	10.20	18.85	52.10				
	2900	Electrical	S.F.	22.05	57.65	93.80	9.20%	12.30%	18.40%	
	8100	Total: Mechanical & Electrical		66.55	143	322	28.80%	32.50%	52.60%	
670	0010	**RELIGIOUS EDUCATION**	▼	47.65	56.30	69.80				670
	0020	Total project costs	C.F.	2.81	3.97	5.25				
	2720	Plumbing	S.F.	2.02	2.94	4.06	3.90%	4.90%	6.90%	
	2770	Heating, ventilating, air conditioning	"	4.64	5.50	7.25	8.10%	9.90%	11.20%	

For expanded coverage of these items see *Means Square Foot Cost Data 1991*   283

(Reprinted from Means Mechanical Cost Data 1991.)

Figure 2.2

Square Foot and Cubic Foot Estimates

The use of Square Foot and Cubic Foot estimates is most appropriate prior to the preparation of plans or preliminary drawings, when budgetary parameters are being analyzed and established. Please refer again to Figure 2.2 and note that costs for each type of project are presented first as "Total Project Costs" by square foot and by cubic foot. These costs are broken down into different components, and then into the relationship of each component to the project as a whole, in terms of costs per square foot. This breakdown enables the designer, planner, or estimator to adjust certain components according to the unique requirements of the proposed project. The costs on this and other pages of *Means Mechanical Cost Data* and *Means Plumbing Cost Data* were derived from more than 11,300 projects contained in the Means data bank of construction costs. These costs include the contractor's overhead and profit but do not include architectural fees or land costs.

Historical data for square foot costs of new construction are plentiful (see *Means Mechanical Cost Data* and *Means Plumbing Cost Data*, Division 17). However, the best source of square foot costs is the estimator's own cost records for similar projects, adjusted to the parameters of the project in question. While helpful for preparing preliminary budgets, Square Foot and Cubic Foot estimates can also be useful as checks against other, more detailed estimates. While slightly more time is required than with Order of Magnitude Estimates, a greater accuracy (plus or minus 15%) is achieved due to a more specific definition of the project.

Assemblies (or Systems) Estimates

One of the primary advantages of Assemblies (or Systems) Estimating is to enable alternate construction techniques to be readily compared for budgetary purposes. Rapidly rising design and construction costs in the past have made budgeting and cost effectiveness studies increasingly important in the early stages of building projects. Never before has the estimating process had such a crucial role in the initial planning process. Unit Price Estimating, because of the time and detailed information required, is not possible as a budgetary or planning tool. A faster and more cost-effective method is needed for the planning phase of a building project; this is the Assemblies, or Systems Estimate.

The Assemblies method is a logical, sequential approach that reflects how a building is constructed. Twelve "Uniformat" divisions organize building construction into major components that can be used in Assemblies Estimates. These Uniformat divisions are listed below:

Assemblies Estimating Divisions:
Division 1 - Foundations
Division 2 - Substructures
Division 3 - Superstructure
Division 4 - Exterior Closure
Division 5 - Roofing
Division 6 - Interior Construction
Division 7 - Conveying
Division 8 - Mechanical
Division 9 - Electrical
Division 10 - General Conditions
Division 11 - Special
Division 12 - Site Work

Each division is further broken down into systems. Division 8, which covers mechanical construction, is comprised of the following major groups of systems: Plumbing, Fire Protection, Heating and Air-Conditioning, and Special Systems.

Each system incorporates several different components into an assemblage that is commonly used in construction. Figure 2.3 is an example of a typical system, in this case "Plumbing–Three Fixture Bathroom" (from *Means Plumbing Cost Data*).

A great advantage of the Assemblies Estimate is that the estimator/designer is able to substitute one system for another during design development and can quickly determine the relative cost differential. The owner can then anticipate budget requirements before the final details and dimensions are established.

The Assemblies method does not require the degree of final design detail needed for a Unit Price Estimate, but estimators who use this approach must have a solid background knowledge of construction materials and methods, code requirements, design options, and budget considerations.

The Assemblies Estimate should not be used as a substitute for the Unit Price Estimate. While the Assemblies approach can be an invaluable tool in the planning stages of a project, it should be supported by Unit Price Estimating when greater accuracy is required.

Unit Price Estimates

The Unit Price Estimate is the most accurate and detailed of the four estimate types and therefore takes the most time to complete. Detailed working drawings and specifications must be available to the unit price estimator. All decisions regarding the project's design, materials, and methods must have been made in order to complete this type of estimate. Because there are fewer variables, the estimate can be more accurate. Working drawings and specifications are used to determine the quantities of materials, equipment, and labor. Current and accurate unit costs for these items are a necessity. These costs can come from different sources: actual quotations for the particular job from normal sources of supply; manufacturers' price sheets; or prices from an up-to-date industry source book such as *Means Mechanical Cost Data* and *Means Plumbing Cost Data*.

Because of the detail involved and the need for accuracy, completion of a Unit Price Estimate entails a great deal of time and expense. Unit Price Estimating is best suited for construction bidding. It can also be an effective method for determining certain detailed costs in a conceptual budget or during design development.

Most construction specification manuals and cost reference books, such as *Means Mechanical* and *Plumbing Cost Data*, compile and present unit price information into the 16 divisions of the MASTERFORMAT of the Construction Specifications Institute, Inc.

CSI Masterformat Divisions:
Division 1 - General Requirements
Division 2 - Site Work
Division 3 - Concrete
Division 4 - Masonry
Division 5 - Metals
Division 6 - Wood & Plastics
Division 7 - Moisture-Thermal Control
Division 8 - Doors, Windows & Glass
Division 9 - Finishes
Division 10 - Specialties
Division 11 - Equipment
Division 12 - Furnishings

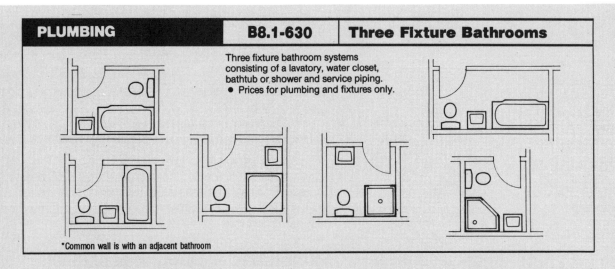

PLUMBING	B8.1-630	Three Fixture Bathrooms

Three fixture bathroom systems consisting of a lavatory, water closet, bathtub or shower and service piping.
● Prices for plumbing and fixtures only.

*Common wall is with an adjacent bathroom

System Components	QUANTITY	UNIT	COST EACH		
			MAT.	INST.	TOTAL
SYSTEM 08.1-630-1170					
BATHROOM, LAVATORY, WATER CLOSET & BATHTUB					
ONE WALL PLUMBING, STAND ALONE					
Wtr closet, 2 pc close cpld vit china flr mntd w/seat supply & stop	1.000	Ea.	151.80	103.20	255
Water closet, rough-in waste & vent	1.000	Set	94.22	280.78	375
Lavatory w/ftngs, wall hung, white, PE on CI, 20″ x 18″	1.000	Ea.	121	69	190
Lavatory, rough-in waste & vent	1.000	Set	143.11	331.89	475
Bathtub, white PE on CI, w/ftgs, mat bottom, recessed, 5′ long	1.000	Ea.	309.10	125.90	435
Baths, rough-in waste and vent	1.000	Set	96.22	236.78	333
TOTAL			915.45	1,147.55	2,063

8.1-630	Three Fixture Bathroom, One Wall Plumbing	COST EACH		
		MAT.	INST.	TOTAL
1150	Bathroom, three fixture, one wall plumbing			
1160	Lavatory, water closet & bathtub			
1170	Stand alone	915	1,150	2,065
	Share common plumbing wall *	790	855	1,645

8.1-630	Three Fixture Bathroom, Two Wall Plumbing	COST EACH		
		MAT.	INST.	TOTAL
2130	Bathroom, three fixture, two wall plumbing			
2140	Lavatory, water closet & bathtub			
2160	Stand alone	920	1,150	2,070
2180	Long plumbing wall common *	830	945	1,775
3610	Lavatory, bathtub & water closet			
3620	Stand alone	990	1,300	2,290
3640	Long plumbing wall common *	935	1,200	2,135
4660	Water closet, corner bathtub & lavatory			
4680	Stand alone	1,700	1,175	2,875
4700	Long plumbing wall common *	1,600	915	2,515
6100	Water closet, stall shower & lavatory			
6120	Stand alone	1,000	1,475	2,475
6140	Long plumbing wall common *	960	1,375	2,335
7060	Lavatory, corner stall shower & water closet			
7080	Stand alone	1,225	1,350	2,575
7100	Short plumbing wall common *	1,125	1,000	2,125

264

(Reprinted from Means Plumbing Cost Data 1991.)

Figure 2.3

Division 13 - Special Construction
Division 14 - Conveying Systems
Division 15 - Mechanical
Division 16 - Electrical

Division 15, Mechanical, is further divided into the following subdivisions in *Means Mechanical Cost Data* and *Means Plumbing Cost Data* books:

151 Pipe and Fittings
152 Plumbing Fixtures
153 Plumbing Appliances
154 Fire Extinguishing Systems
155 Heating
156 HVAC Piping Specialties
157 Air-Conditioning and Ventilating

This method of organizing the various components provides a standard of uniformity that is widely used by construction industry professionals: contractors, material suppliers, engineers, and architects. A sample unit price page from the 1991 edition of *Means Plumbing Cost Data* is shown in Figure 2.4. This page lists various types of elements. (Please note that the heading "151 Pipe and Fittings" denotes the subdivision classification for these items in Division 15.) Each page contains a wealth of information useful in Unit Price Estimating. The type of work to be performed is described in detail: typical crew makeups, daily outputs, units of measure, and separate costs for material and installation. Total costs are extended to include the installing contractor's overhead and profit.

Pipe and fittings are installed by plumbers, pipefitters, sprinkler fitters, steamfitters, and gas fitters. Rather than duplicating subdivision 151 for each trade, only one trade is shown in the crew makeup in question. Many of the materials in 151 are used in each of the mechanical trades.

A fifth type of estimate warrants mention, a refinement of the unit price method. This is the "Scheduling Estimate," which involves the application of realistic manpower allocations. A complete Unit Price Estimate is a prerequisite for the preparation of a Scheduling Estimate. The purpose of the Scheduling Estimate is to determine costs based upon actual working conditions. This is done using data obtained from the Unit Price Estimate. A "human factor" can be applied. For example, if a task requires 7.5 hours to complete, based on unit price data, a tradesman will most likely take 8 hours to complete the work, and will, in any event, be paid for 8 hours work. Costs can be adjusted accordingly. A thorough discussion of scheduling and scheduling estimating is beyond the scope of this book but a brief overview is included in Chapter 9; this subject is covered in more detail in *Means Scheduling Manual* by F. William Horsley.

		151 100	Miscellaneous Fittings	CREW	DAILY OUTPUT	MAN-HOURS	UNIT	BARE COSTS				TOTAL INCL O&P	
								MAT.	LABOR	EQUIP.	TOTAL		
101	0010	AVERAGE Square foot and percent of total											101
	0100	job cost for plumbing, see division 171											
105	0010	BACKFLOW PREVENTER Includes gate valves,											105
	0020	and four test cocks, corrosion resistant, automatic operation											
	1000	Double check principle											
	1080	Threaded 											
	1100	¾" pipe size	1 Plum	16	.500	Ea.	135	12.75		147.75	165	
	1120	1" pipe size		14	.571		155	14.55		169.55	190	
	1140	1-½" pipe size		10	.800		210	20		230	260	
	1160	2" pipe size		7	1.140		270	29		299	340	
	1300	Flanged										
	1380	3" pipe size	Q-1	4.50	3.560	Ea.	1,200	81		1,281	1,450	
	1400	4" pipe size	"	3	5.330		1,725	120		1,845	2,075	
	1420	6" pipe size	Q-2	3	8		2,900	190		3,090	3,475	
	4000	Reduced pressure principle										
	4100	Threaded										
	4120	¾" pipe size	1 Plum	16	.500	Ea.	215	12.75		227.75	255	
	4140	1" pipe size		14	.571		270	14.55		284.55	320	
	4150	1-¼" pipe size		12	.667		345	16.95		361.95	405	
	4160	1-½" pipe size		10	.800		400	20		420	470	
	4180	2" pipe size		7	1.140		480	29		509	570	
	5000	Flanged, bronze										
	5060	2-½" pipe size 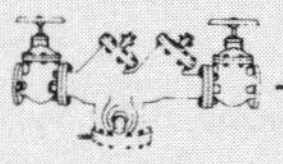	Q-1	5	3.200	Ea.	1,725	73		1,798	2,000	
	5080	3" pipe size		4.50	3.560		1,900	81		1,981	2,200	
	5100	4" pipe size		3	5.330		3,450	120		3,570	3,975	
	5120	6" pipe size	Q-2	3	8		6,150	190		6,340	7,050	
	5600	Flanged, iron										
	5660	2-½" pipe size	Q-1	5	3.200	Ea.	1,550	73		1,623	1,825	
	5680	3" pipe size		4.50	3.560		1,750	81		1,831	2,050	
	5700	4" pipe size		3	5.330		2,425	120		2,545	2,850	
	5720	6" pipe size	Q-2	3	8		3,725	190		3,915	4,375	
	5740	8" pipe size		2	12		8,000	285		8,285	9,225	
	5760	10" pipe size		1	24		10,600	570		11,170	12,500	
110	0010	CLEANOUTS 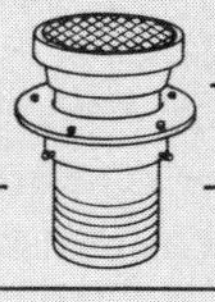											110
	0060	Floor type										
	0080	Round or square, scoriated nickel bronze top										
	0100	2" pipe size	1 Plum	10	.800	Ea.	46	20		66	81	
	0120	3" pipe size		8	1		49	25		74	92	
	0140	4" pipe size		6	1.330		65	34		99	120	
	0160	5" pipe size		4	2		105	51		156	190	
	0180	6" pipe size	Q-1	6	2.670		105	61		166	205	
	0200	8" pipe size	"	4	4		120	92		212	270	
	0340	Recessed for tile, same price										
	0980	Round top, recessed for terrazzo										
	1000	2" pipe size	1 Plum	9	.889	Ea.	78	23		101	120	
	1080	3" pipe size		6	1.330		83	34		117	140	
	1100	4" pipe size		4	2		91	51		142	175	
	1120	5" pipe size	Q-1	6	2.670		145	61		206	250	
	1140	6" pipe size		5	3.200		150	73		223	275	
	1160	8" pipe size		4	4		190	92		282	345	
	2000	Round scoriated nickel bronze top, extra heavy duty										
	2060	2" pipe size	1 Plum	9	.889	Ea.	59	23		82	99	
	2080	3" pipe size		6	1.330		64	34		98	120	
	2100	4" pipe size		4	2		79	51		130	165	
	2120	5" pipe size	Q-1	6	2.670		125	61		186	230	
	2140	6" pipe size		5	3.200		140	73		213	265	
	2160	8" pipe size		4	4		210	92		302	370	

45

(Reprinted from Means Plumbing Cost Data 1991.)

Figure 2.4

Chapter 3

BEFORE STARTING THE ESTIMATE

Selecting a project to bid on can be done in numerous ways. A mechanical contractor may be invited to bid by an architect, general contractor, or an owner. The contractor may hear of a project from a construction publication or an advertisement for bids from a local newspaper. Whichever source is used, the process may involve weeks of hard work with only a chance of bidding success. It can also mean the opportunity to obtain a contract for a successful and lucrative project. It is not uncommon for the contractor to bid ten or more jobs in order to win just one. The successful bid depends heavily upon estimating accuracy and thus, the preparation, organization, and care that go into the estimating process. (Conversely, if a job is won due to omissions in the estimate, it is likely *not* to be a "successful" bid.)

The first step before starting the estimate is to obtain copies of the plans and specifications in *sufficient quantities*. Most estimators mark up plans with colored pencils and make numerous notes and references. For one estimator to work from plans that have been used by another is difficult at best and may easily lead to errors. Most often, only one complete set is provided to the mechanical contractor. Additional sets must be purchased, if needed. When more than one estimator works on a particular project, especially if they are working from the same plans, careful coordination is required to prevent omissions as well as duplications. Color coding of the various piping circuits, when neatly done, can be used to mark the takeoff progress and will allow for another estimator to finish a takeoff, if necessary.

The estimator should be aware of and note any instructions to bidders, which may be included in the specifications. To avoid future confusion, the bid's due date, time, and place should be clearly stated and understood upon receipt of the construction documents (plans and specifications). The due date should be marked on a calendar to avoid conflicting bid dates and closing times. Completion of the takeoff should be made as soon as possible, not two days prior to the bid deadline. Utility companies should be contacted to obtain the service charges, and the local municipal building department contacted for inspection charges, permit fees, and building service charges.

If bid security, or a bid bond, is required, it should be arranged for at this time, especially if the bonding capability or capacity of a contractor is limited or has not previously been established. If a bond or bid security is to be based on a percentage of the bid, then a preliminary square foot or systems estimate must be prepared to figure the amount of the bid security. This allows the bonding company to analyze the project up front, and thus prevents the contractor from estimating and bidding on a job that cannot be bonded.

The estimator should attend any prebid meetings with the owner or architect, preferably *after* reviewing the plans and specifications. Important points are often brought up at such meetings and details clarified. Attendance is important, not only to show the owner your interest in the project, but also to assure equal and competitive bidding. For many projects, attendance is required or the bid will not be accepted. It is to the estimator's advantage to examine and review the plans and specifications before any such meetings and before the initial site visit. It is important to become familiar with the project as soon as possible.

All contract documents should be read thoroughly. They exist to protect all parties involved in the construction process. The contract documents are written so that the contractors will be bidding equally and competitively, and to ensure that all items in a project are included. The contract documents protect the designer (the architect or engineer) by ensuring that all work is supplied and installed as specified. The owner also benefits from thorough and complete construction documents by being assured of a quality job and a complete, functional project. Finally, the contractor benefits from good contract documents because the scope of work is well-defined, eliminating the gray areas of what is implied but not stated. "Extras" are more readily avoided. Change orders, if required, are accepted with less argument if the original contract documents are complete, well stated, and most importantly, read by all concerned parties. The Appendix of this book contains a reproduction of the two mechanical pages from SPEC-AID, an R.S. Means Company, Inc. publication from *Means Forms for Building Construction Professionals*. This SPEC-AID was created not only to assist designers and planners when developing project specifications, but also as an aid to the contractor's estimator. The mechanical section of the SPEC-AID lists some typical mechanical components and variables that are normally included in building construction. The estimator can use this form as a means of outlining a project's requirements, and as a checklist to be sure that all items have been included. A preprinted summary sheet (shown in Figure 3.1) serves both as the SPEC-AID and as a pricing estimate sheet for plumbing.

During the first review of the specifications, any items to be priced should be identified and noted. All work to be subcontracted should be examined for "related work" required from other trades or contractors. Such work is usually referenced and described in a thorough project specification. "Work by others" or "Not in Contract" should be clearly defined both on the drawings and in the specifications. Certain materials are sometimes specified by the designer and purchased by the owner to be installed (labor only) by the contractor. These items should be noted and the responsibilities of each party clearly understood. An example of such an item might involve allocating responsibility for receiving, temporary storage, and protection of restaurant fixtures furnished by others but installed by plumbers.

The *General Conditions*, *Supplemental Conditions*, and *Special Conditions* sections of the specifications should be examined carefully by *all* parties involved in the project. These sections describe the items that have a direct bearing on the proposed project, but may not be part of the actual, physical installation. Temporary heat and water are examples of these kinds of items. Also included in these sections is information regarding completion dates, payment schedules (e.g. retainage), submittal requirements, allowances, alternates, and other important project requirements. Each of these conditions can have a significant bearing on the ultimate cost of the project. They must be read and understood prior to performing the estimate. The most popular General Conditions will

		1		2	
Labor w/Payroll Taxes, Welfare					
Travel					
Fixtures					
Drains, Carriers, etc.					
Sanitary - Interior					
Storm - Interior					
Water - Interior					
Valves - Interior					
Water Exterior - Street Connection					
Water Exterior - Extension to Building					
Sanitary & Storm Exterior - Street Connection					
Sanitary & Storm Exterior - Extension to Building					
Water Meter					
Gas Service					
Gas Interior					
Accessories					
Flashings for: Vents, Floor & Roof Drains, Pans					
Hot Water Tanks: Automatic w/trim					
Hot Water Tanks: Storage, w/stand and trim					
Hot Water Circ. Pumps, Aquastats, Mag. Starters					
Sump Pumps and/or Ejectors; Controls, Mag. Starters					
Flue Piping					
Rigging, Painting, Excavation and Backfill					
Therm. Gauges, P.R.V. Controls, Specialties, etc.					
Record Dwgs., Tags and Charts					
Fire Extinguishers, Equipment, and/or Standpipes					
Permits					
Sleeves, Inserts and Hangers					
Staging					
Insulation					
Acid Waste System					
Sales Tax					
Trucking and Cartage					

Name of Job: ___ Due Date: ___

Location: ___ Bid Sent To: ___

Architect: ___ Engineer: ___

Figure 3.1

soon be memorized by an estimator, who will then need only to skim through them looking for any deletions or changes. *Study* the supplemental or special conditions, as these are prepared for each particular job.

While analyzing the plans and specifications, the estimator should evaluate the different portions of the project to determine which areas warrant the most attention. The estimator should focus first on those items which represent the largest cost centers of the project or which entail the greatest risk. These cost centers are not always the portions of the job that require the most time to estimate but are those items that will have the most significant impact on the estimate.

Other sections of the specifications should be read because items such as electric motor controls may be duplicated under the mechanical and electrical specifications.

Figure 3.2 from Division 17 of *Means Mechanical Cost Data* is a chart showing typical percentages of mechanical and electrical work relative to projects as a whole. These charts have been developed to represent the overall average percentages for new construction. Commonly used building types are included.

When the overall scope of the work has been identified, the drawings should be examined to confirm the information in the specifications. This is the time to clarify details while reviewing the general content. The estimator should note which sections, elevations, and detail drawings are for which plans. At this point and throughout the whole estimating process, the estimator should note and list any discrepancies between the plans and specifications, as well as any possible omissions. It is often stated in bid documents that bidders are obliged to notify the owner or architect/engineer of any such discrepancies. When so notified, the designer will most often issue an addendum to the contract documents in order to properly notify all parties concerned and to assure equal and competitive bidding. Competition can be fair only if the same information is provided for all bidders. Notifying services such as Construction Market Data (CMD) or F.W. Dodge, of intent to bid will increase receipt of sub-bids and suppliers' material quotations.

Once familiar with the contract documents, the estimator should solicit bids by notifying appropriate subcontractors, manufacturers, and vendors. Those whose work may be affected by the site conditions should accompany the estimator on a job site visit (especially in cases of renovation and remodeling, where existing conditions can have a significant effect on the cost of a project). This notification is usually done by postcard or telephone call. It may also be beneficial to inform these vendors of other locations where the bid documents can be reviewed, to avoid tie-up of your plans or plan room.

During a site visit, the estimator should take notes, and possibly photographs, of any unusual situations pertinent to the construction and, thus, to the project estimate. If unusual site conditions exist, or if questions arise during the takeoff, a second site visit is recommended.

In some areas, questions are likely to arise that cannot be answered clearly by the plans and specifications. It is crucial that the owner or responsible party be notified quickly, preferably in writing, so that these questions may be resolved before unnecessary problems arise. Often such items involve more than one contractor and can only be resolved by the owner or architect/engineer. A proper estimate cannot be completed until all such questions are answered.

171 | S.F., C.F. and % of Total Costs

		171 000 S.F. & C.F. Costs	UNIT	UNIT COSTS			% OF TOTAL			
				¼	MEDIAN	¾	¼	MEDIAN	¾	
570	0010	**MEDICAL OFFICES**	S.F.	57.20	71.90	87.45				570
	0020	Total project costs	C.F.	4.44	5.95	7.90				
	2720	Plumbing	S.F.	3.49	5.20	7.10	5.70%	6.90%	9.20%	
	2770	Heating, ventilating, air conditioning		4.04	6.10	7.90	6.50%	8%	10.40%	
	2900	Electrical		4.83	6.95	9.30	7.60%	9.70%	11.70%	
	3100	Total: Mechanical & Electrical		11.30	16.05	21.05	17.30%	22.40%	27.10%	
590	0010	**MOTELS**		41.75	55.85	75.85				590
	0020	Total project costs	C.F.	3.61	5.55	8.50				
	2720	Plumbing	S.F.	3.76	4.70	5.70	9.40%	10.60%	12.50%	
	2770	Heating, ventilating, air conditioning		1.99	3.42	4.99	4.90%	5.60%	8.20%	
	2900	Electrical		3.48	4.27	5.50	7%	8.10%	10.40%	
	3100	Total: Mechanical & Electrical		8.25	10.85	14.50	17.90%	23.10%	26.20%	
	9000	Per rental unit, total cost	Unit	14,900	28,400	38,800				
	9500	Total: Mechanical & Electrical	"	4,025	5,700	6,150				
600	0010	**NURSING HOMES**	S.F.	57.40	75.75	92.15				600
	0020	Total project costs	C.F.	4.58	6	8				
	2720	Plumbing	S.F.	5.10	6.25	9.10	9.30%	10.30%	13.30%	
	2770	Heating, ventilating, air conditioning		5.20	7.25	9.30	9.20%	11.40%	11.80%	
	2900	Electrical		5.75	7.25	9.45	9.70%	11%	12.80%	
	3100	Total: Mechanical & Electrical		13.40	17.90	26.95	22.30%	28.30%	33.30%	
	9000	Per bed or person, total cost	Bed	22,200	28,900	36,700				
610	0010	**OFFICES Low-Rise (1 to 4 story)**	S.F.	47.10	60.35	79.90				610
	0020	Total project costs	C.F.	3.49	4.86	6.55				
	2720	Plumbing	S.F.	1.79	2.70	3.86	3.60%	4.50%	6%	
	2770	Heating, ventilating, air conditioning		3.81	5.35	7.84	7.20%	10.40%	11.90%	
	2900	Electrical		3.98	5.50	7.55	7.40%	9.50%	11%	
	3100	Total: Mechanical & Electrical		8.25	12.25	17.90	14.90%	20.80%	26.80%	
620	0010	**OFFICES Mid-Rise (5 to 10 story)**		52.95	65.05	87.25				620
	0020	Total project costs	C.F.	3.63	4.69	6.60				
	2720	Plumbing	S.F.	1.60	2.43	3.50	2.80%	3.60%	4.50%	
	2770	Heating, ventilating, air conditioning		3.94	5.65	9	7.60%	9.30%	11%	
	2900	Electrical		3.36	4.81	7.40	6.50%	8%	10%	
	3100	Total: Mechanical & Electrical		9.50	12.20	20.35	16.70%	20.50%	25.70%	
630	0010	**OFFICES High-Rise (11 to 20 story)**		62.70	79.25	98.30				630
	0020	Total project costs	C.F.	4.01	5.70	8.10				
	2900	Electrical	S.F.	3.13	4.56	7	5.80%	7%	10.50%	
	3100	Total: Mechanical & Electrical		11.45	14.55	24.20	17.20%	21.40%	29.40%	
640	0010	**POLICE STATIONS**		78.40	101	129				640
	0020	Total project costs	C.F.	5.75	7.45	9.90				
	2720	Plumbing	S.F.	4.28	6.25	10.50	5.60%	6.80%	10.70%	
	2770	Heating, ventilating, air conditioning		6.30	8.50	12.10	7%	10.50%	11.90%	
	2900	Electrical		7.85	12.50	15.80	9.40%	11.70%	14.70%	
	3100	Total: Mechanical & Electrical		21.25	26.55	34.70	22.60%	27.50%	33.10%	
650	0010	**POST OFFICES**		63.85	76.20	99.55				650
	0020	Total project costs	C.F.	3.52	4.60	5.45				
	2720	Plumbing	S.F.	2.72	3.44	4.36	4.20%	5.30%	5.60%	
	2770	Heating, ventilating, air conditioning		3.93	5.25	8	6.60%	8%	9.80%	
	2900	Electrical		5	7.05	8.35	7.40%	9.40%	11%	
	3100	Total: Mechanical & Electrical		11	1.51	21	16.50%	21.40%	26.30%	
660	0010	**POWER PLANTS**		375	555	850				660
	0020	Total project costs	C.F.	10.20	18.85	52.10				
	2900	Electrical	S.F.	22.05	57.65	93.80	9.20%	12.30%	18.40%	
	8100	Total: Mechanical & Electrical		66.55	143	322	28.80%	32.50%	52.60%	
670	0010	**RELIGIOUS EDUCATION**		47.65	56.30	69.80				670
	0020	Total project costs	C.F.	2.81	3.97	5.25				
	2720	Plumbing	S.F.	2.02	2.94	4.06	3.90%	4.90%	6.90%	
	2770	Heating, ventilating, air conditioning	"	4.64	5.50	7.25	8.10%	9.90%	11.20%	

For expanded coverage of these items see *Means Square Foot Cost Data 1991*

283

Figure 3.2

Chapter 4
THE QUANTITY TAKEOFF

The quantity takeoff is the cornerstone of construction bidding, therefore it should be organized so that the information gathered can be used to future advantage. Purchasing and scheduling can be made easier if items are taken off and listed by system, material type, construction phase, or floor. Material purchasing will similarly benefit.

Much of the methods and materials utilized by the mechanical trades are "understood" or standardized, and not usually indicated on the drawings. Thus, in the quantity takeoff, the mechanical estimator must count not only the items shown on the drawings but must envision the completed design and include the proper number of fittings, hangers, shields, inserts, bolts, nuts, and gaskets, to name but a few. The estimator should also be familiar with or have access to all codes and regulations. To effectively cover all aspects of the project, certain steps should be followed.

Units for each item should be used consistently throughout the whole project–from takeoff to cost control. If the units in a takeoff are consistent, the original estimate can be equitably compared to progress reports and final cost reports (see Chapter 11, "Cost Control"). Part Two of this book is devoted to descriptions of over 80 mechanical components. In that section, a takeoff procedure is suggested for the components. Typical material and labor units are also given for component installation. Each material and labor unit consists of a list of items that are generally included in the cost of the component. Also given are the units by which the component is measured or counted.

Quantities should be taken off by one person if the project is not too large and if time allows. For larger projects, the plans are often split and the work assigned to two or more estimators. In this case, a project leader ought to be assigned to coordinate and assemble the estimate.

Traditionally, quantities are taken off from the drawings in the same sequence as they are erected or installed. Recently, however, the sequence of takeoff and estimating is more often based on the systems or type of work. In this case, preference in the estimating process is given to certain items or components based on the relative costs.

When working with the plans during the quantity takeoff, consistency is a very important consideration. If each job is approached in the same manner, a pattern will develop, such as moving from the lower floors to the top, clockwise or counterclockwise. The choice of method is not important, but consistency is. The purpose of being consistent is to avoid duplications as well as omissions and errors. Preprinted forms provide an excellent means for developing consistent patterns. Figures 4.1 and 4.2 are two examples of such forms. These Quantity Sheets are designed purely for quantity accumulation. Parts Two and Three of this book contain examples of the use of quantity sheets. Figure 4.3, a Consolidated Estimate sheet, is designed to be used for both quantity takeoff and pricing on one form. There are many other variations.

Every contractor could benefit from designing custom company forms. If employees of a company use the same types of forms, communications and coordination of the estimating process will proceed more smoothly. One estimator will be able to more easily understand the work of another. R.S. Means has published two books completely devoted to forms and their use, *Means Forms for Contractors* and *Means Forms for Building Construction Professionals*. Scores of forms, examples, and instructions for use are included.

Appropriate and easy-to-use forms are the first, and most important, of the "tools of the trade" for estimators. Other tools useful to the estimator include tapes, scales, rotometers, mechanical counters, and colored pencils.

A number of other shortcuts can be used for the quantity takeoff. If approached logically and systematically, these techniques help to save time without sacrificing accuracy. Consistent use of accepted abbreviations saves the time of writing things out. An abbreviations list similar to those that appear in the Appendix of this book might be posted in a conspicuous place for each estimator to provide a consistent pattern of definitions for use within an office.

All dimensions–whether printed, measured, or calculated–that can be used for determining quantities of more than one item should be listed on a separate sheet and posted for easy reference. Posted gross dimensions can also be used to quickly check for order of magnitude errors.

Rounding off, or decreasing the number of significant digits, should be done only when it will not statistically affect the resulting product. The estimator must use good judgement to determine instances when rounding is appropriate. An overall 2-3% variation in a competitive market can often be the difference between getting or losing a job, or between profit or no profit. The estimator should establish rules for rounding to achieve a consistent level of precision. In general, it is best not to round numbers until the final summary of quantities. Pennies should be rounded off, and, when appropriate, dollars can be rounded off to the closest five or ten.

The final summary is also the time to convert units of measure into standards for practical use (linear feet of copper tube to twenty-foot divisions, for example). This is done to keep the numerical value of the unit equitable to what will be purchased and handled.

Be sure to quantify (count) and include "labor only" items that are not shown on the plans. Such items may or may not be indicated in the specifications and might include cleanup, special labor for handling materials, testing, code inspectors, etc.

The following list summarizes the aforementioned suggestions plus a few more guidelines which will be helpful during the quantity takeoff:

- Use preprinted forms.
- Use only the front side of each piece of paper.
- Transfer carefully when copying numbers from one sheet to the next.
- List dimensions (width, length) in a consistent order.
- Verify the scale of drawings before using them as a basis for measurement. A good check is to scale off a bathtub, (usually 5'), a light fixture (2'-4'), or a doorway (3').
- Mark drawings neatly and consistently as quantities are counted.

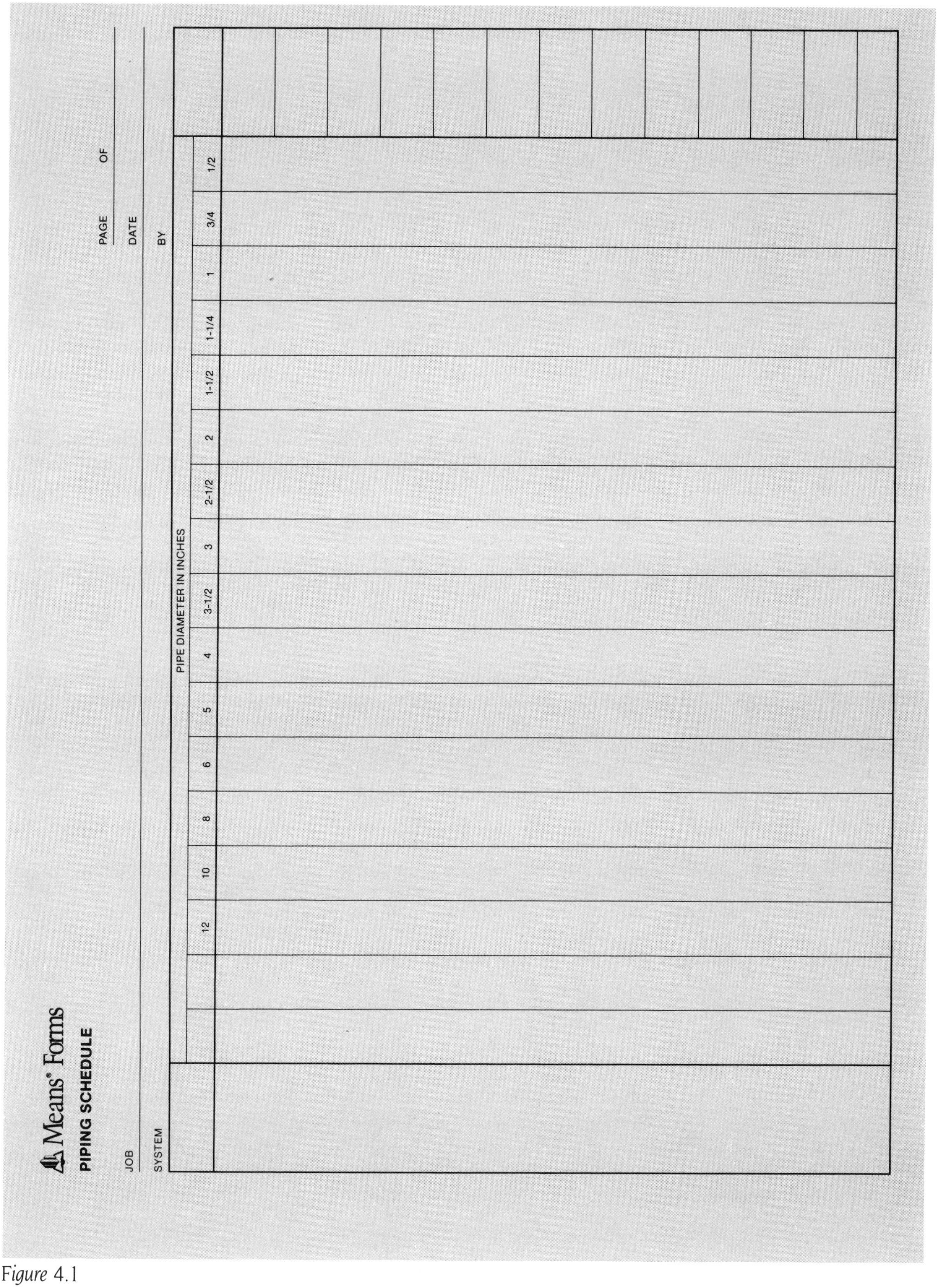

Means® Forms

PIPING SCHEDULE

JOB

SYSTEM

PAGE ____ OF ____

DATE ____

BY ____

PIPE DIAMETER IN INCHES

| 12 | 10 | 8 | 6 | 5 | 4 | 3-1/2 | 3 | 2-1/2 | 2 | 1-1/2 | 1-1/4 | 1 | 3/4 | 1/2 |

Figure 4.1

▲ Means® Forms

DUCTWORK SCHEDULE

PAGE	OF	
JOB		DATE
SYSTEM		BY
MATERIAL		DRAWING NO.

DUCT SIZE	LINING	INSUL.	GAUGE	LENGTH	TOTAL LENGTH	LBS./ FOOT	TOTAL POUNDS
X							
X							
X							
X							
X							
X							
X							
X							
X							
X							
X							
X							
X							
X							
X							
X							
X							
X							
X							
X							
X							
X							
X							
X							
X							
X							
X							
X							
X							
X							
X							
X							
X							
X							
X							
X							
X							
X							
X							
X							

Figure 4.2

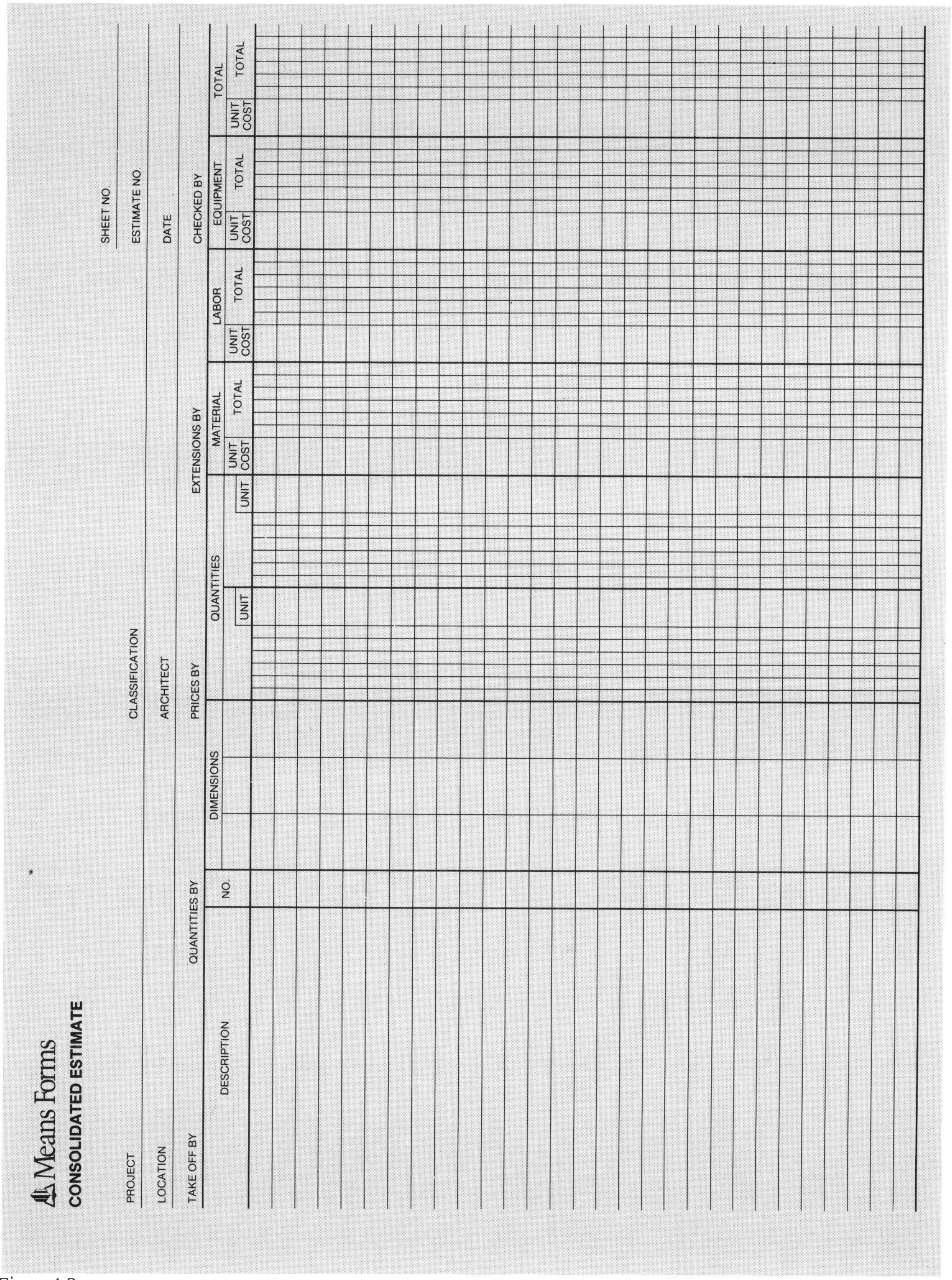

Means Forms

CONSOLIDATED ESTIMATE

PROJECT _____ CLASSIFICATION _____ SHEET NO. _____

LOCATION _____ ARCHITECT _____ ESTIMATE NO. _____

TAKE OFF BY _____ QUANTITIES BY _____ PRICES BY _____ EXTENSIONS BY _____ DATE _____

CHECKED BY _____

DESCRIPTION	NO.	DIMENSIONS	QUANTITIES		EXTENSIONS								
					MATERIAL		LABOR		EQUIPMENT		TOTAL	
			UNIT		UNIT COST	TOTAL	UNIT COST	TOTAL	UNIT COST	TOTAL	UNIT COST	TOTAL

Figure 4.3

23

- Be alert for changes in scale, or notes such as "N.T.S." (not to scale). Sometimes these drawings have been photographically reduced.
- Include required items which may not appear in the plans and specs.
- Be alert for discrepancies between the plans and the specifications.

And perhaps the four most important points:

- Print legibly.
- Be organized.
- Use common sense.
- Be consistent.

Chapter 5

PRICING THE ESTIMATE

When the quantities have been counted, values, in the form of unit costs, must be applied and project burdens (overhead and profit) added in order to determine the total selling price (the quote). Depending upon the chosen estimating method (based on the degree of accuracy required) and the level of detail, these unit costs may be direct or *bare*, or may include overhead, profit, or contingencies. In Unit Price Estimating, the unit costs most commonly used are *bare*, or *unburdened*. Most extend labor quantities as man-hours or man-days and price material quantities in dollars. When the total man-days are determined, it is necessary to multiply these figures by the appropriate daily rate, before adding markups. Items such as overhead and profit are usually added to the total direct costs at the estimate summary.

Sources of Cost Information

One of the most difficult aspects of the estimator's job is determining accurate and reliable bare cost data. Sources for such data are varied, but can be categorized in terms of their relative reliability. The most reliable of any cost information is direct quotations from usual sources of supply for this particular job, and accurate, up-to-date, well-kept records of completed work by the estimator's own company. There is no better cost for a particular construction item than the *actual* cost to the contractor of that item from another recent job, modified (if necessary) to meet the requirements of the project being estimated.

Bids from responsible subcontractors are the next most reliable source of cost data. Any estimating inaccuracies are essentially absorbed by the subcontractor. A subcontract bid is a known, fixed cost, prior to the project. Whether the price is "right" or "wrong" does not matter (as long as it is a responsible competitive bid with no apparent errors). The bid is what the appropriate portion of work will cost. No estimating is required of the prime contractor, except for possible verification of the quote and comparison with other subcontractors' quotes.

Quotations by vendors for material costs are, for the same reasons, as reliable as subcontract bids. In this case, however, the estimator must apply estimated labor costs. Thus, the "installed" price for a particular item may be more variable.

Whenever possible, all price quotations from vendors or subcontractors should be confirmed in writing. Qualifications and exclusions should be clearly stated. The items quoted should be checked to be sure that they are complete and as specified. One way to assure these requirements is to prepare a form on which all subcontractors and vendors must submit their quotations. This form, generally called a "request for quote," can suggest all of the appropriate questions. It also provides a format to organize the information needed by the estimator and, later, the purchasing agent. This technique can be especially useful to the estimator in a smaller organization, since he must often act as purchasing agent as well.

The above procedures are ideal, but in the realistic haste of estimating and bidding, quotations are often received verbally, either in person or by telephone. The importance of gathering all pertinent information is heightened because omissions are more likely. A preprinted form, such as the one shown in Figure 5.1, is essential to assure that all required information and qualifications are obtained and understood. How often has the subcontractor or vendor said, "I didn't know that I was supposed to include that?" With the help of such forms, the appropriate questions may be covered.

If the estimator has no cost records for a particular item and is unable to obtain a quotation, then another reliable source of price information is current unit price cost books such as *Means Mechanical Cost Data* and *Means Plumbing Cost Data*. R.S. Means presents all such data in the form of national averages; these figures can be adjusted to local conditions, a procedure that will be explained in Part Three of this book. In addition to being a source of primary costs, unit price books are useful as a reference or cross-check for verifying costs obtained from unfamiliar sources.

Lacking cost information from any of the above-mentioned sources, the estimator may have to rely on experience and personal knowledge of the field to develop costs.

No matter which source of cost information is used, the system and sequence of pricing should be the same as that used for the quantity takeoff. This consistent approach should continue through both accounting and cost control during work on the project.

Types of Costs

All costs included in a Unit Price Estimate can be divided into two types: direct and indirect. Direct costs are those dedicated solely to the physical construction of a specific project. Material, labor, equipment, and subcontract costs, as well as project overhead costs are all direct.

Types of Costs in a Construction Estimate:

Direct Costs	Indirect Costs
Material	Taxes
Labor	Insurance
Equipment	Office Overhead
Subcontractors	Profit
Project Overhead	Contingencies
Sales Tax	
Bonds	

Indirect costs are usually added to the estimate at the summary stage and are most often calculated as a percentage of the direct costs. They include such items as taxes, insurance, overhead, profit, and contingencies. The indirect costs account for great variation in estimates among different bidders.

A clear understanding of direct and indirect cost factors is a fundamental part of pricing the estimate. Chapters 6 and 7 address the components of direct and indirect costs in detail.

▲ Means° Forms

TELEPHONE QUOTATION

PROJECT	DATE _____
FIRM QUOTING	TIME _____
ADDRESS	PHONE ( ) _____
ITEM QUOTED	BY _____
	RECEIVED BY _____

WORK INCLUDED	AMOUNT OF QUOTATION

DELIVERY TIME	**TOTAL BID**	
DOES QUOTATION INCLUDE THE FOLLOWING	If ☐ NO is checked, determine the following	
STATE & LOCAL SALES TAXES	☐ YES ☐ NO	MATERIAL VALUE
DELIVERY TO THE JOB SITE	☐ YES ☐ NO	WEIGHT
COMPLETE INSTALLATION	☐ YES ☐ NO	QUANTITY
COMPLETE SECTION AS PER PLANS & SPECIFICATIONS	☐ YES ☐ NO	DESCRIBE BELOW

EXCLUSIONS AND QUALIFICATIONS	

ADDENDA ACKNOWLEDGEMENT	**TOTAL ADJUSTMENTS**	
	ADJUSTED TOTAL BID	

ALTERNATES	
ALTERNATE NO.	
ALTERNATE NO.	
ALTERNATE NO.	
ALTERNATE NO.	
ALTERNATE NO.	
ALTERNATE NO.	
ALTERNATE NO.	

Figure 5.1

Chapter 6

DIRECT COSTS

Direct costs can be defined as those necessary for the completion of the project, in other words, the hard costs. Material, labor, and equipment are among the more obvious items in this category. While subcontract costs include the overhead and profit (indirect costs) of the subcontractor, they are considered to be direct costs to the prime mechanical contractor. Also included are certain project overhead costs for items that are necessary for construction. Examples are a storage trailer, telephone, tools, and possibly temporary power and lighting. Sales tax and bonds are additional direct costs, since they are essential for the performance of the project. On most jobs, items such as temporary lighting and temporary sanitary facilities are stated in the Supplementary General Conditions as being the responsibility of the general or prime contractor.

Material

When quantities have been carefully taken off, estimates of material cost can be very accurate. For a high level of accuracy, the material unit prices must be reliable and current. The most reliable source of material costs is a quotation from a vendor for the particular job in question. Ideally, the vendor should have access to the plans and specifications for verification of quantities and specified products. Providing material quotes (and submittals for approval) is a service which most suppliers will perform.

Material pricing appears relatively simple and straightforward. There are, however, certain considerations that the estimator must address when analyzing material quotations. The reputation of the vendor is a significant factor. Can the vendor "deliver," both figuratively and literally? Estimators may choose not to rely on a "competitive" lower price from an unknown vendor, but will instead use a slightly higher price from a known, reliable vendor. Experience is the best judge for such decisions.

There are many other questions that the estimator should ask. How long is the price guaranteed? When does the guarantee begin? Is there an escalation clause? Does the price include delivery charges or sales tax, if required? Where is the point of FOB? This can be an extremely important factor. Are guarantees and warranties in compliance with the specification requirements? Will there be adequate and appropriate storage space available? If not, can staggered shipments be made? Note that most of these questions can be addressed on the form shown in Chapter 5, Figure 5.1. More information should be obtained, however, to assure that a quoted price is accurate and competitive.

The estimator must be sure that the quotation or obtained price is for the materials as per plans and specifications. Architects and engineers may write into the specifications that: a) a particular type or brand of product must be used with no substitution, b) a particular type or brand of product is specified, but alternate brands of equal quality and performance may be accepted *upon approval*, or c) no particular type or brand is specified. Depending upon the options, the estimator may be able to find an acceptable, less expensive alternative. In some cases, these substitutions can substantially lower the cost of a project. Note also that most specifications will require that "catalog cuts" or "shop drawings" be submitted for certain materials as part of the contract conditions. In this case, there is pressure on the estimator to obtain the lowest possible price on materials that he believes will meet the specified criteria and gain the architect's approval. A typical submittal or transmittal cover sheet is shown in Figure 6. 1.

When the estimator has received material quotations, there are still other considerations which should have a bearing on the final choice of a vendor. Lead time–the amount of time between order and delivery–must be determined and considered. It does not matter how competitive or low a quote is if the material cannot be delivered to the job site on time to support the schedule. If a delivery date is promised, is there a guarantee, or a penalty clause for late delivery?

Delivery and lead time problems can be substantially eliminated by having the major equipment encased and delivered immediately after approval. This equipment is then shipped to a staging area for the job. The staging area may be the contractor's premises, the job site, supply house, etc. The most efficient method for decreasing delivery and lead time, particularly for large equipment, is to ship it to the yard of the rigger who is going to put this equipment in position. Payment for this equipment can be arranged through the architect, providing insurance and ownership have been established.

The estimator should also determine if there are any unusual payment requirements. Cash flow for a company can be severely affected if a large material purchase, thought to be payable in 30 days, is delivered C.O.D. or "site draft." Truck drivers may not allow unloading until payment has been received. Such requirements must be determined during the estimating stage so that the cost of borrowing money, if necessary, can be included.

If unable to obtain the quotation of a vendor from whom the material would be purchased, the estimator has other sources for obtaining material prices. These include, in order of reliability:

1. Current price lists from manufacturers' catalogs. Be sure to check that the list indicates "contractor discounts."
2. Cost records from previous jobs. Historical costs must be updated for present market conditions.
3. Reputable and current annual unit price cost books, such as Means Mechanical Cost Data and Means Plumbing Cost Data. Such books usually represent national averages and must be factored to local markets.

No matter which price source is used, the estimator must be sure to include an allowance for any burdens, such as delivery, taxes, or finance charges, over the actual cost of the material. (The same kinds of concerns that apply to vendor quotations should be taken into consideration when using these other price sources.)

Labor

In order to determine the installation cost for each item of construction, the estimator must know two pieces of information: first, the labor rate (hourly wage or salary) of the worker, and second, how much time a worker will need to complete a given unit or task of the installation–in other words, the productivity or output. Wage rates are usually known going into a project, but productivity may be very difficult to determine. To estimators working for contractors, the construction labor rates that the contractor pays will be known, well

⚓ Means Forms

**LETTER
OF TRANSMITTAL**

FROM:

DMV Mechanical Contractors

Braintree, Mass. 02184

TO: Raymond Supply Co.

Foxboro, Mass.

DATE: Nov. 8, 1991
PROJECT: Office Building
LOCATION: Kingston, Mass.
ATTENTION: P. V. Howard
RE:

Gentlemen:

WE ARE SENDING YOU ☑ HEREWITH ☐ DELIVERED BY HAND ☐ UNDER SEPARATE COVER

VIA _____ THE FOLLOWING ITEMS:

☐ PLANS ☐ PRINTS ☑ SHOP DRAWINGS ☐ SAMPLES ☐ SPECIFICATIONS
☐ ESTIMATES ☐ COPY OF LETTER ☐ _____

COPIES	DATE OR NO.	DESCRIPTION
2	354	H. B. Jones Cast Iron Boiler
2	VF9	Maco Pump Submittals

THESE ARE TRANSMITTED AS INDICATED BELOW

☐ FOR YOUR USE ☐ APPROVED AS NOTED ☐ RETURN _____ CORRECTED PRINTS
☐ FOR APPROVAL ☑ APPROVED FOR CONSTRUCTION ☐ SUBMIT _____ COPIES FOR_____
☐ AS REQUESTED ☐ RETURNED FOR CORRECTIONS ☐ RESUBMIT_____ COPIES FOR_____
☐ FOR REVIEW AND COMMENT ☐ RETURNED AFTER LOAN TO US ☐ FOR BIDS DUE_____
☐ _____

REMARKS: Approved as submitted — Release for
ASAP shipment to riggers yard as per
our P.O. # 0895

IF ENCLOSURES ARE NOT AS INDICATED,
PLEASE NOTIFY US AT ONCE.

SIGNED: John J. Moylan

Figure 6.1

documented, and constantly updated. Projected labor rates for construction jobs of long duration should also be contended with. Estimators for owners, architects, or engineers must determine labor rates from outside sources. Unit price data books, such as *Means Mechanical Cost Data* and *Means Plumbing Cost Data*, provide national average labor wage rates, with location factors for 162 major U.S. and Canadian cities. The unit costs for labor are based on these averages. Figure 6.2 shows national average union rates for the construction industry based on January 1, 1991. Figure 6.3 lists national average open shop rates, again based on January 1, 1991.

Note the column entitled "Workers' Compensation." Plumbers, pipe fitters, sprinkler installers, and sheet metal workers have among the lowest rates. These are average numbers which vary from state to state and between trades (see Figure 6.4). Workers' Compensation insurance is often considered a direct cost when applied to field labor.

If more accurate union labor rates are required, the estimator has different options. Union locals can provide rates (as well as negotiated increases) for a particular location. Employer bargaining groups can usually provide labor cost data as well. R.S. Means Company, Inc. publishes *Means Labor Rates for the Construction Industry* on an annual basis. This book lists the union labor rates by trade for over 300 U.S. and Canadian cities.

Determination of nonunion, or "open shop" rates is much more difficult. In larger cities, employer organizations often exist to represent open shop (nonunion) contractors. These organizations may have records of local pay scales, but ultimately, wage rates are determined by individual contractors.

Labor units (or man-hours) are the least predictable of all factors for building projects. It is important to determine, as accurately as possible, prevailing productivity. The best source of labor productivity or labor units (and therefore labor costs) is the estimator's well-kept records from previous projects. If there are no company records for productivity, cost data books, such as *Means Mechanical Cost Data* and *Means Plumbing Cost Data*, and productivity reference books, such as *Means Man-Hour Standards for Construction* can be invaluable. Included with the listing for each individual construction item is the designation of a suggested crew makeup (usually 2-4 workers). The crew is the minimum grouping of workers which can be expected to accomplish the task efficiently. Figure 6.5, a typical page from *Means Mechanical Cost Data*, includes this data and indicates the man-hours as well as the average daily output of the designated crew for each construction item. When more than one or two tradespersons are designated, refer to the crew listings in the Foreword of *Means Mechanical Cost Data* and *Means Plumbing Cost Data*.

The estimator who has neither company records nor the sources described above must put together the appropriate crews and determine the expected output or productivity. This type of estimating should only be attempted based upon strong experience and considerable exposure to construction methods and practices. There are rare occasions when this approach is necessary to estimate a particular item or a new technique. Even then, the new labor units are often extrapolated from existing figures for similar work, rather than being created from scratch.

Equipment

In recent years, construction equipment has become more important, not only because of the incentive to reduce labor costs, but also as a response to new, high technology construction methods and materials. As a result, these costs represent an increasing percentage of total project costs in construction.

Installing Contractor's Overhead & Profit

Below are the **average** installing contractor's percentage mark-ups applied to base labor rates to arrive at typical billing rates.

Column A: Labor rates are based on union wages averaged for 30 major U.S. cities. Base rates including fringe benefits are listed hourly and daily. These figures are the sum of the wage rate and employer-paid fringe benefits such as vacation pay, employer-paid health and welfare costs, pension costs, plus appropriate training and industry advancement funds costs.

Column B: Workers' Compensation rates are the national average of state rates established for each trade.

Column C: Column C lists average fixed overhead figures for all trades. Included are Federal and State Unemployment costs set at 6.2%; Social Security Taxes (FICA) set at 7.65%; Builder's Risk Insurance costs set at 0.34%; and Public Liability costs set at 1.55%. All the percentages except those for Social Security Taxes vary from state to state as well as from company to company.

Column D and E: Percentages in Columns D and E are based on the presumption that the installing contractor has annual billing of $500,000 and up. Overhead percentages may increase with smaller annual billing. The overhead percentages for any given contractor may vary greatly and depend on a number of factors, such as the contractor's annual volume, engineering and logistical support costs, and staff requirements. The figures for overhead and profit will also vary depending on the type of job, the job location, and the prevailing economic conditions. All factors should be examined very carefully for each job.

Column F: Column F lists the total of columns B, C, D, and E.

Column G: Column G is Column A (hourly base labor rate) multiplied by the percentage in Column F (O&P percentage).

Column H: Column H is the total of Column A (hourly base labor rate) plus Column G (Total O&P).

Column I: Column I is Column H multiplied by eight hours.

Abbr.	Trade	A Base Rate Incl. Fringes Hourly	A Base Rate Incl. Fringes Daily	B Workers' Comp. Ins.	C Average Fixed Overhead	D Overhead	E Profit	F Total Overhead & Profit %	F Total Overhead & Profit Amount	H Rate with O&P Hourly	I Rate with O&P Daily
Skwk	Skilled Workers Average (35 trades)	$22.65	$181.20	15.1%	15.7%	12.8%	10%	53.6%	$12.15	$34.80	$278.40
	Helpers Average (5 trades)	17.10	136.80	16.2		13.0		54.9	9.40	26.50	212.00
	Foremen Average, Inside (50¢ over trade)	23.15	185.20	15.1		12.8		53.6	12.40	35.55	284.40
	Foremen Average, Outside ($2.00 over trade)	24.65	197.20	15.1		12.8		53.6	13.20	37.85	302.80
Clab	Common Building Laborers	17.50	140.00	16.6		11.0		53.3	9.35	26.85	214.80
Asbe	Asbestos Workers	24.70	197.60	13.5		16.0		55.2	13.65	38.35	306.80
Boil	Boilermakers	25.05	200.40	9.3		16.0		51.0	12.80	37.85	302.80
Bric	Bricklayers	22.75	182.00	13.7		11.0		50.4	11.45	34.20	273.60
Brhe	Bricklayer Helpers	17.65	141.20	13.7		11.0		50.4	8.90	26.55	212.40
Carp	Carpenters	22.00	176.00	16.6		11.0		53.3	11.75	33.75	270.00
Cefi	Cement Finishers	21.65	173.20	9.6		11.0		46.3	10.00	31.65	253.20
Elec	Electricians	25.15	201.20	6.0		16.0		47.7	12.00	37.15	297.20
Elev	Elevator Constructors	25.35	202.80	7.7		16.0		49.4	12.50	37.85	302.80
Eqhv	Equipment Operators, Crane or Shovel	23.30	186.40	10.4		14.0		50.1	11.65	34.95	279.60
Eqmd	Equipment Operators, Medium Equipment	22.50	180.00	10.4		14.0		50.1	11.25	33.75	270.00
Eqlt	Equipment Operators, Light Equipment	21.40	171.20	10.4		14.0		50.1	10.70	32.10	256.80
Eqol	Equipment Operators, Oilers	19.20	153.60	10.4		14.0		50.1	9.60	28.80	230.40
Eqmm	Equipment Operators, Master Mechanics	24.00	192.00	10.4		14.0		50.1	12.00	36.00	288.00
Glaz	Glaziers	22.55	180.40	11.9		11.0		48.6	10.95	33.50	268.00
Lath	Lathers	21.95	175.60	10.2		11.0		46.9	10.30	32.25	258.00
Marb	Marble Setters	22.55	180.40	13.7		11.0		50.4	11.35	33.90	271.20
Mill	Millwrights	22.95	183.60	9.9		11.0		46.6	10.70	33.65	269.20
Mstz	Mosaic and Terrazzo Workers	22.10	176.80	8.3		11.0		45.0	9.95	32.05	256.40
Pord	Painters, Ordinary	20.80	166.40	12.4		11.0		49.1	10.20	31.00	248.00
Psst	Painters, Structural Steel	21.45	171.60	42.9		11.0		79.6	17.05	38.50	308.00
Pape	Paper Hangers	20.80	166.40	12.4		11.0		49.1	10.20	31.00	248.00
Pile	Pile Drivers	22.20	177.60	25.4		16.0		67.1	14.90	37.10	296.80
Plas	Plasterers	21.80	174.40	13.4		11.0		50.1	10.90	32.70	261.60
Plah	Plasterer Helpers	17.90	143.20	13.4		11.0		50.1	8.95	26.85	214.80
Plum	Plumbers	25.45	203.60	7.5		16.0		49.2	12.50	37.95	303.60
Rodm	Rodmen (Reinforcing)	23.90	191.20	27.6		14.0		67.3	16.10	40.00	320.00
Rofc	Roofers, Composition	20.35	162.80	28.9		11.0		65.6	13.35	33.70	269.60
Rots	Roofers, Tile & Slate	20.45	163.60	28.9		11.0		65.6	13.40	33.85	270.80
Rohe	Roofer Helpers (Composition)	14.90	119.20	28.9		11.0		65.6	9.75	24.65	197.20
Shee	Sheet Metal Workers	25.00	200.00	10.2		16.0		51.9	13.00	38.00	304.00
Spri	Sprinkler Installers	26.40	211.20	7.7		16.0		49.4	13.05	39.45	315.60
Stpi	Steamfitters or Pipefitters	25.50	204.00	7.5		16.0		49.2	12.55	38.05	304.40
Ston	Stone Masons	22.65	181.20	13.7		11.0		50.4	11.40	34.05	272.40
Sswk	Structural Steel Workers	24.10	192.80	35.4		14.0		75.1	18.10	42.20	337.60
Tilf	Tile Layers (Floor)	22.15	177.20	8.3		11.0		45.0	9.95	32.10	256.80
Tilh	Tile Layer Helpers	17.45	139.60	8.3		11.0		45.0	7.85	25.30	202.40
Trlt	Truck Drivers, Light	18.10	144.80	13.5		11.0		50.2	9.10	27.20	217.60
Trhv	Truck Drivers, Heavy	18.40	147.20	13.5		11.0		50.2	9.25	27.65	221.20
Sswl	Welders, Structural Steel	24.10	192.80	35.4		14.0		75.1	18.10	42.20	337.60
Wrck	*Wrecking	17.50	140.00	35.5		11.0		72.2	12.65	30.15	241.20

*Not included in Averages.

(Reprinted from Means Mechanical Cost Data 1991.)

Figure 6.2

33

Installing Contractor's Overhead & Profit

Below are the **average** installing contractor's percentage mark-ups applied to base labor rates to arrive at typical billing rates.

Column A: Labor rates are based on average open shop wages for 7 major U.S. regions. Base rates including fringe benefits are listed hourly and daily. These figures are the sum of the wage rate and employer-paid fringe benefits such as vacation pay and employer-paid health costs.

Column B: Workers' Compensation rates are the national average of state rates established for each trade.

Column C: Column C lists average fixed overhead figures for all trades. Included are Federal and State Unemployment costs set at 6.2%; Social Security Taxes (FICA) set at 7.65%; Builder's Risk Insurance costs set at 0.34%; and Public Liability costs set at 1.55%. All the percentages except those for Social Security Taxes vary from state to state as well as from company to company.

Column D and E: Percentages in Columns D and E are based on the presumption that the installing contractor has annual billing of $500,000 and up. Overhead percentages may increase with smaller annual billing. The overhead percentages for any given contractor may vary greatly and depend on a number of factors, such as the contractor's annual volume, engineering and logistical support costs, and staff requirements. The figures for overhead and profit will also vary depending on the type of job, the job location, and the prevailing economic conditions. All factors should be examined very carefully for each job.

Column F: Column F lists the total of columns B, C, D, and E.

Column G: Column G is Column A (hourly base labor rate) multiplied by the percentage in Column F (O&P percentage).

Column H: Column H is the total of Column A (hourly base labor rate) plus Column G (Total O&P).

Column I: Column I is Column H multiplied by eight hours.

		A		B	C	D	E	F	G	H	I
		Base Rate Incl. Fringes		Workers' Comp. Ins.	Average Fixed Overhead	Over-head	Profit	**Total Overhead & Profit**		**Rate with O & P**	
Abbr.	Trade	Hourly	Daily					%	Amount	Hourly	Daily
Skwk	Skilled Workers Average (35 trades)	$14.70	$117.60	15.1%	15.7%	26.8%	10%	67.6%	$ 9.95	$24.65	$197.20
	Helpers Average (5 trades)	11.10	88.80	16.2		27.0		68.9	7.65	18.75	150.00
	Foremen Average, Inside (50¢ over trade)	15.20	121.60	15.1		26.8		67.6	10.25	25.45	203.60
	Foremen Average, Outside ($2.00 over trade)	16.70	133.60	15.1		26.8		67.6	11.30	28.00	224.00
Clab	Common Building Laborers	11.35	90.80	16.6		25.0		67.3	7.65	19.00	152.00
Asbe	Asbestos Workers	16.05	128.40	13.5		30.0		69.2	11.10	27.15	217.20
Boil	Boilermakers	16.30	130.40	9.3		30.0		65.0	10.60	26.90	215.20
Bric	Bricklayers	14.80	118.40	13.7		25.0		64.4	9.55	24.35	194.80
Brhe	Bricklayer Helpers	11.45	91.60	13.7		25.0		64.4	7.35	18.80	150.40
Carp	Carpenters	14.30	114.40	16.6		25.0		67.3	9.60	23.90	191.20
Cefi	Cement Finishers	14.05	112.40	9.6		25.0		60.3	8.45	22.50	180.00
Elec	Electricians	16.35	130.80	6.0		30.0		61.7	10.10	26.45	211.60
Elev	Elevator Constructors	16.45	131.60	7.7		30.0		63.4	10.45	26.90	215.20
Eqhv	Equipment Operators, Crane or Shovel	15.15	121.20	10.4		28.0		64.1	9.70	24.85	198.80
Eqmd	Equipment Operators, Medium Equipment	14.60	116.80	10.4		28.0		64.1	9.35	23.95	191.60
Eqlt	Equipment Operators, Light Equipment	13.90	111.20	10.4		28.0		64.1	8.90	22.80	182.40
Eqol	Equipment Operators, Oilers	12.50	100.00	10.4		28.0		64.1	8.00	20.50	164.00
Eqmm	Equipment Operators, Master Mechanics	15.60	124.80	10.4		28.0		64.1	10.00	25.60	204.80
Glaz	Glaziers	14.65	117.20	11.9		25.0		62.6	9.15	23.80	190.40
Lath	Lathers	14.25	114.00	10.2		25.0		60.9	8.70	22.95	183.60
Marb	Marble Setters	14.65	117.20	13.7		25.0		64.4	9.45	24.10	192.80
Mill	Millwrights	14.90	119.20	9.9		25.0		60.6	9.05	23.95	191.60
Mstz	Mosaic and Terrazzo Workers	14.35	114.80	8.3		25.0		59.0	8.45	22.80	182.40
Pord	Painters, Ordinary	13.50	108.00	12.4		25.0		63.1 .	8.50	22.00	176.00
Psst	Painters, Structural Steel	13.95	111.60	42.9		25.0		93.6	13.05	27.00	216.00
Pape	Paper Hangers	13.50	108.00	12.4		25.0		63.1	8.50	22.00	176.00
Pile	Pile Drivers	14.45	115.60	25.4		30.0		81.1	11.70	26.15	209.20
Plas	Plasterers	14.15	113.20	13.4		25.0		64.1	9.05	23.20	185.60
Plah	Plasterer Helpers	11.65	93.20	13.4		25.0		64.1	7.45	19.10	152.80
Plum	Plumbers	16.55	132.40	7.5		30.0		63.2	10.45	27.00	216.00
Rodm	Rodmen (Reinforcing)	15.55	124.40	27.6		28.0		81.3	12.65	28.20	225.60
Rofc	Roofers, Composition	13.20	105.60	28.9		25.0		79.6	10.50	23.70	189.60
Rots	Roofers, Tile & Slate	13.30	106.40	28.9		25.0		79.6	10.60	23.90	191.20
Rohe	Roofer Helpers (Composition)	9.70	77.60	28.9		25.0		79.6	7.70	17.40	139.20
Shee	Sheet Metal Workers	16.25	130.00	10.2		30.0		65.9	10.70	26.95	215.60
Spri	Sprinkler Installers	17.15	137.20	7.7		30.0		63.4	10.85	28.00	224.00
Stpi	Steamfitters or Pipefitters	16.55	132.40	7.5		30.0		63.2	10.45	27.00	216.00
Ston	Stone Masons	14.70	117.60	13.7		25.0		64.4	9.45	24.15	193.20
Sswk	Structural Steel Workers	15.65	125.20	35.4		28.0		89.1	13.95	29.60	236.80
Tilf	Tile Layers (Floor)	14.40	115.20	8.3		25.0		59.0	8.50	22.90	183.20
Tilh	Tile Layer Helpers	11.35	90.80	8.3		25.0		59.0	6.70	18.05	144.40
Trlt	Truck Drivers, Light	11.75	94.00	13.5		25.0		64.2	7.55	19.30	154.40
Trhv	Truck Drivers, Heavy	11.95	95.60	13.5		25.0		64.2	7.65	19.60	156.80
Sswl	Welders, Structural Steel	15.65	125.20	35.4		28.0		89.1	13.95	29.60	236.80
Wrck	*Wrecking	11.35	90.80	35.5	↓	25.0	↓	86.2	9.80	21.15	169.20

*Not included in Averages.

(Reprinted from Means Open Shop Building Construction Cost Data 1991.)

Figure 6.3

Table 10.2-203 Workers' Compensation by Trade and State

STATE	CARPENTRY – 3 stories or less	CARPENTRY – interior cab. work	CARPENTRY – general	CONCRETE WORK–NOC	CONCRETE WORK – flat (flr., sdwk.)	ELECTRICAL WIRING – inside	EXCAVATION – earth NOC	EXCAVATION – rock	GLAZIERS	INSULATION WORK	LATHING	MASONRY	PAINTING & DECORATING	PILE DRIVING	PLASTERING	PLUMBING	ROOFING	SHEET METAL WORK (HVAC)	STEEL ERECTION – door & sash	STEEL ERECTION – inter. ornam.	STEEL ERECTION – structure	STEEL ERECTION – NOC	TILE WORK – (interior ceramic)	WATERPROOFING	WRECKING
	5651	5437	5403	5213	5221	5190	6217	6217	5462	5479	5443	5022	5474	6003	5480	5183	5551	5538	5102	5102	5040	5057	5348	9014	5701
AL	14.97	8.55	14.18	11.06	7.25	5.86	8.68	8.68	11.20	11.17	8.35	10.59	11.54	32.21	9.38	6.53	25.13	13.57	7.78	7.78	23.21	20.90	8.16	3.60	20.90
AK	16.10	11.21	14.48	18.19	10.46	9.07	12.87	12.87	18.59	19.83	13.80	12.49	10.31	41.05	21.55	12.61	29.27	14.16	19.77	19.77	74.36	56.94	10.21	6.51	74.36
AZ	19.74	7.60	23.31	12.41	10.23	9.30	7.07	7.07	13.56	17.90	8.95	16.91	11.59	27.00	19.08	6.60	24.77	10.19	12.74	12.74	35.69	17.97	6.70	7.24	17.97
AR	11.85	6.30	15.05	12.20	6.23	4.71	8.50	8.50	8.80	8.61	8.14	10.02	10.68	16.64	10.01	4.49	13.70	7.62	6.18	6.18	44.38	16.97	5.54	5.34	44.38
CA	21.96	8.01	21.96	9.33	9.33	7.82	7.55	7.55	12.94	22.76	8.13	14.05	14.90	19.03	15.92	10.36	36.32	13.11	12.18	12.18	28.27	24.76	7.68	14.90	28.35
CO	22.81	13.43	19.38	20.39	14.20	6.56	13.75	13.75	11.58	20.41	10.43	26.22	17.00	36.16	38.54	11.54	48.70	12.88	12.23	12.23	52.23	30.77	10.81	9.24	30.77
CT	18.58	20.34	25.91	26.78	15.09	8.87	12.75	12.75	21.83	25.42	15.47	28.35	17.77	42.59	21.19	10.09	51.57	15.49	16.81	16.81	50.82	26.21	9.81	4.61	50.82
DE	11.44	13.73	11.44	9.20	6.25	5.71	8.49	8.49	11.11	11.44	10.45	9.77	13.30	11.50	10.45	5.32	22.88	10.11	10.09	10.09	26.31	10.09	7.46	9.77	25.47
DC	10.32	9.62	14.81	27.57	12.88	11.76	14.68	14.68	14.11	13.79	10.72	21.41	11.84	37.37	12.32	14.82	32.92	11.25	20.59	20.59	47.36	46.44	23.24	5.77	47.36
FL	25.21	18.82	30.26	44.32	18.17	12.89	20.27	20.27	23.43	25.66	26.91	26.88	30.18	46.97	31.60	16.01	55.77	19.12	22.40	22.40	48.03	58.10	12.70	10.32	48.03
GA	15.40	10.28	17.30	11.49	9.47	5.67	11.86	11.86	9.06	12.63	13.51	11.69	10.65	27.66	11.35	6.07	24.90	9.60	8.21	8.21	18.44	26.83	5.90	7.36	26.83
HI	11.66	9.32	39.81	13.93	9.03	8.89	10.81	10.81	14.08	20.18	9.51	17.35	7.68	28.43	19.16	5.36	40.64	8.32	14.10	14.10	27.94	26.82	7.72	10.50	27.94
ID	13.17	7.20	18.71	12.78	5.84	4.67	10.20	10.20	11.13	13.14	7.93	16.25	12.51	20.62	10.67	4.53	27.80	8.95	7.45	7.45	18.09	16.83	6.59	7.26	18.09
IL	16.54	11.14	17.49	26.51	12.88	8.75	10.19	10.19	21.11	17.72	13.22	18.01	15.60	31.52	13.46	12.20	31.86	15.94	13.72	13.72	61.21	84.88	12.92	5.22	84.88
IN	9.00	4.97	6.19	6.08	3.94	2.64	5.07	5.07	6.83	5.21	3.67	5.54	4.65	13.11	5.37	2.67	10.84	3.97	3.90	3.90	8.73	14.57	3.58	3.14	14.57
IA	9.23	7.01	10.64	17.71	4.89	5.38	9.85	9.85	8.25	12.47	6.76	10.58	8.30	17.54	8.92	7.78	18.49	8.41	9.80	9.80	38.15	37.61	6.25	4.27	37.61
KS	10.35	5.61	7.94	9.82	5.58	3.66	4.70	4.70	6.25	15.84	7.81	9.75	6.57	15.96	7.75	4.59	19.70	6.82	4.74	4.74	11.50	23.34	5.08	4.75	23.34
KY	14.27	6.19	14.56	11.42	7.61	4.67	7.40	7.40	8.65	10.16	10.07	13.10	12.37	28.02	9.53	5.21	23.87	9.76	9.85	9.85	30.22	23.44	7.65	4.39	30.22
LA	16.33	10.25	16.12	8.59	7.84	5.59	9.61	9.61	12.34	8.20	7.25	7.74	13.24	37.62	8.86	6.29	20.21	7.76	7.91	7.91	24.13	13.43	6.64	6.06	24.13
ME	12.13	10.90	38.36	24.84	10.99	8.38	16.53	16.53	18.35	19.96	13.34	19.36	19.26	43.24	19.49	12.27	44.12	13.43	17.46	17.46	52.06	57.53	13.46	7.99	52.06
MD	12.59	12.33	10.70	17.28	9.77	7.64	12.85	12.85	18.78	12.60	8.29	12.40	9.07	19.45	10.26	9.40	32.69	13.58	11.27	11.27	26.18	32.54	9.59	4.30	37.63
MA	14.16	12.04	30.31	33.73	15.36	6.21	9.34	9.34	21.21	16.86	15.19	23.65	14.05	26.22	15.99	9.21	79.41	13.85	16.23	16.23	83.51	48.66	13.21	9.81	65.27
MI	11.40	7.35	12.74	12.71	9.27	5.75	12.42	12.42	14.56	15.48	9.32	16.14	13.24	26.48	14.23	7.23	26.63	8.70	11.42	11.42	29.17	20.06	8.13	NA	29.17
MN	20.86	20.86	36.23	21.03	14.74	6.95	21.20	21.20	17.38	22.76	22.29	17.84	17.92	35.90	22.29	11.77	45.20	14.86	16.41	16.41	51.17	54.54	12.08	9.34	51.17
MS	10.19	6.87	12.79	8.41	4.70	4.93	8.18	8.18	8.45	6.92	7.85	6.29	6.64	22.56	9.94	3.36	15.30	7.80	6.98	6.98	22.27	13.33	6.72	3.86	22.27
MO	7.88	5.54	7.19	6.30	5.92	3.52	6.07	6.07	5.38	9.22	5.76	7.22	4.95	16.62	6.26	3.33	17.56	4.75	5.91	5.91	15.91	13.15	4.58	3.66	13.15
MT	23.27	13.56	43.40	38.99	21.88	8.30	27.19	27.19	19.61	25.49	23.34	44.23	44.16	54.83	28.08	15.32	62.45	16.29	17.94	17.94	162.26	55.63	14.05	16.25	162.26
NE	8.22	5.45	8.84	9.53	7.35	3.72	7.11	7.11	6.84	9.78	6.05	8.01	7.17	12.60	7.58	4.53	18.79	7.89	5.42	5.42	13.54	27.21	4.76	4.92	27.21
NV	14.25	14.25	14.25	11.31	11.31	7.21	10.61	10.61	11.46	13.98	13.68	11.93	12.97	10.90	8.97	23.39	25.48	13.01	33.81	33.81	33.81	36.57	9.98	10.90	N.A.
NH	15.51	9.63	18.51	26.18	13.21	5.16	12.90	12.90	10.28	14.57	11.62	14.47	13.26	52.73	18.31	9.06	60.55	11.19	11.94	11.94	31.55	36.57	8.96	6.26	31.55
NJ	6.83	5.66	6.83	5.78	4.90	2.43	6.26	6.26	4.91	6.54	6.43	7.98	7.58	10.12	6.43	3.18	15.67	4.20	7.66	7.66	22.64	10.49	3.16	3.51	27.74
NM	14.96	11.73	19.31	24.57	9.63	5.95	10.09	10.09	16.15	14.70	11.32	16.70	11.44	49.12	16.45	10.09	37.80	12.36	18.46	18.46	21.61	25.84	10.07	7.75	21.61
NY	9.63	4.67	9.88	12.46	9.73	5.00	10.17	10.17	8.89	8.52	8.23	12.36	9.56	17.49	9.18	7.53	22.73	9.75	7.68	7.68	17.83	24.06	7.13	5.44	19.97
NC	6.06	5.70	9.29	8.46	3.63	4.32	5.94	5.94	5.09	6.14	4.15	4.01	5.05	11.79	8.46	4.46	12.70	5.31	4.73	4.73	19.37	8.00	3.27	2.59	19.37
ND	12.86	12.86	12.86	9.98	9.98	4.54	8.28	8.28	14.60	7.64	5.87	6.34	9.54	27.52	5.87	10.27	15.98	10.27	12.86	12.86	27.52	27.52	5.69	15.98	NA
OH	9.93	9.93	9.93	10.32	10.32	3.87	10.32	10.32	11.36	8.83	8.83	10.48	11.36	10.32	8.83	5.69	20.00	12.71	N.A.	N.A.	58.93	N.A.	5.03	10.32	10.32
OK	14.20	7.38	11.62	10.27	8.38	3.78	9.39	9.39	8.02	9.81	7.80	9.85	8.30	24.54	10.73	5.06	20.50	8.02	7.74	7.74	37.18	29.25	5.53	6.60	37.18
OR	34.23	14.01	33.37	26.01	19.95	8.49	20.99	20.99	15.10	28.80	13.90	24.25	24.82	53.64	26.62	11.10	60.89	14.90	16.80	16.80	48.90	36.45	23.07	17.85	48.90
PA	12.32	11.41	12.32	18.08	8.25	5.50	9.57	9.57	10.58	12.32	13.37	12.19	12.85	17.01	13.37	7.30	27.01	9.87	16.25	16.25	41.51	16.25	8.23	12.19	64.42
RI	15.52	7.13	12.32	14.83	16.49	6.22	12.89	12.89	18.05	15.23	11.26	14.94	17.87	30.71	15.05	4.69	31.27	6.80	10.13	10.13	78.01	40.30	9.28	8.00	78.01
SC	14.28	7.39	15.32	7.64	6.55	6.30	7.43	7.43	11.77	11.46	6.85	10.01	9.36	14.13	9.58	3.44	17.21	8.84	4.23	4.23	11.28	24.72	8.80	4.44	11.28
SD	8.52	5.88	13.80	10.26	5.60	4.22	7.72	7.72	8.18	11.76	6.24	7.27	8.01	17.08	8.45	7.03	27.14	6.16	6.92	6.92	17.97	13.96	5.02	3.91	17.97
TN	11.89	6.11	11.70	9.84	5.61	4.17	7.59	7.59	7.91	9.44	6.15	8.76	8.99	21.90	7.96	5.32	18.61	8.59	8.03	8.03	20.93	15.57	4.73	4.33	20.93
TX	22.09	15.58	22.09	18.49	14.60	9.81	14.91	14.91	13.54	20.74	9.35	16.27	14.06	30.47	14.57	12.00	38.41	16.85	11.50	11.50	39.93	21.52	8.82	7.96	38.04
UT	NA	NA	8.79	9.82	4.93	5.05	6.25	6.25	6.84	6.73	8.06	13.07	9.18	15.75	8.37	5.67	20.31	5.72	5.90	5.90	NA	24.80	4.77	3.07	24.80
VT	7.62	5.25	12.12	12.62	5.63	3.63	6.85	6.85	8.04	8.22	7.31	12.12	8.01	19.05	9.67	5.05	19.24	6.31	6.66	6.66	20.53	26.40	5.90	5.02	20.53
VA	7.28	5.90	9.08	9.23	6.29	3.45	5.52	5.52	8.03	8.84	7.69	8.54	9.24		5.63	4.72	25.24	7.11	5.10	5.10	21.59	15.51	3.78	3.34	21.59
WA	12.37	12.37	12.37	10.52	8.78	3.12	8.18	8.18	13.15	10.10	12.37	12.99	10.69	21.35	13.05	4.47	11.99	5.92	10.52	10.52	29.26	29.26	7.92	11.94	14.80
WV	10.80	10.80	10.80	14.90	14.90	4.00	8.48	8.48	4.13	4.13	15.81	7.03	15.81	8.40	15.81	3.52	12.91	4.13	9.94	9.93	7.96	9.93	7.03	2.73	7.96
WI	10.26	7.57	17.21	11.12	6.31	5.41	8.58	8.58	10.62	12.79	8.28	11.77	11.34	28.37	12.80	7.34	28.44	8.70	9.19	9.19	29.73	24.92	10.03	5.38	63.36
WY	5.00	5.00	5.00	5.00	5.00	5.00	5.00	5.00	5.00	5.00	5.00	5.00	5.00	5.00	5.00	5.00	5.00	5.00	5.00	5.00	5.00	5.00	5.00	5.00	5.00
AVG.	13.72	9.61	16.64	15.29	9.55	5.97	10.37	10.37	11.90	13.48	10.22	13.71	12.36	25.40	13.39	7.45	28.91	10.24	10.79	10.79	35.36	27.59	8.28	7.09	35.54

(Reprinted from Means Mechanical Cost Data 1991.)

Figure 6.4

156 200	Heat/Cool Piping Misc.	CREW	DAILY OUTPUT	MAN-HOURS	UNIT	BARE COSTS				TOTAL INCL O&P	
						MAT.	LABOR	EQUIP.	TOTAL		
201	0010	**AUTOMATIC AIR VENT**										201
	0020	Cast iron body, stainless steel internals, float type									
	0060	½" NPT inlet, 300 psi	1 Stpi	12	.667	Ea.	53.75	17		70.75	84
	0140	¾" NPT inlet, 300 psi		12	.667		53.75	17		70.75	84
	0180	½" NPT inlet, 250 psi		10	.800		174	20		194	220
	0220	¾" NPT inlet, 250 psi		10	.800		174	20		194	220
	0260	1" NPT inlet, 250 psi		10	.800		258	20		278	315
	0340	1-½" NPT inlet, 250 psi	Q-5	12	1.330		545	31		576	645
	0380	2" NPT inlet, 250 psi	"	12	1.330		545	31		576	645
	0600	Forged steel body, stainless steel internals, float type									
	0640	½" NPT inlet, 750 psi	1 Stpi	12	.667	Ea.	525	17		542	605
	0680	¾" NPT inlet, 750 psi		12	.667		525	17		542	605
	0760	¾" NPT inlet, 1000 psi		10	.800		795	20		815	905
	0800	1" NPT inlet, 1000 psi	Q-5	12	1.330		795	31		826	920
	0880	1-½" NPT inlet, 1000 psi		10	1.600		2,230	37		2,267	2,500
	0920	2" NPT inlet, 1000 psi		10	1.600		2,230	37		2,267	2,500
	1100	Formed steel body, non corrosive									
	1110	⅛" NPT inlet 150 psi	1 Stpi	32	.250	Ea.	4.20	6.40		10.60	14.15
	1120	¼" NPT inlet 150 psi		32	.250		16.45	6.40		22.85	28
	1130	¾" NPT inlet 150 psi		32	.250		16.45	6.40		22.85	28
	1300	Chrome plated brass, automatic/manual, for radiators									
	1310	⅛" NPT inlet, nickel plated brass	1 Stpi	32	.250	Ea.	2.76	6.40		9.16	12.55
205	0010	**AIR CONTROL** With strainer										205
	0040	2" diameter	Q-5	6	2.670	Ea.	381	61		442	510
	0080	2-½" diameter		5	3.200		430	73		503	585
	0100	3" diameter		4	4		660	92		752	865
	0120	4" diameter		3	5.330		950	120		1,070	1,225
	0140	6" diameter	Q-6	3.40	7.060		1,445	170		1,615	1,850
	0160	8" diameter		3	8		2,160	190		2,350	2,650
	0180	10" diameter		2.20	10.910		3,360	260		3,620	4,075
	0200	12" diameter		1.70	14.120		5,690	335		6,025	6,750
	0210	14" diameter		1.30	18.460		6,100	440		6,540	7,375
	0220	16" diameter		1	24		9,190	570		9,760	11,000
	0230	18" diameter		.80	30		12,100	715		12,815	14,400
	0240	20" diameter		.60	40		13,500	950		14,450	16,300
	0300	Without strainer									
	0310	2" diameter	Q-5	6	2.670	Ea.	295	61		356	415
	0320	2-½" diameter		5	3.200		350	73		423	495
	0330	3" diameter		4	4		495	92		587	680
	0340	4" diameter		3	5.330		775	120		895	1,025
	0350	5" diameter		2.40	6.670		1,050	155		1,205	1,375
	0360	6" diameter	Q-6	3.40	7.060		1,210	170		1,380	1,575
	0370	8" diameter		3	8		1,670	190		1,860	2,125
	0380	10" diameter		2.20	10.910		2,445	260		2,705	3,075
	0390	12" diameter		1.70	14.120		3,805	335		4,140	4,675
	0400	14" diameter		1.30	18.460		5,115	440		5,555	6,275
	0410	16" diameter		1	24		7,165	570		7,735	8,725
	0420	18" diameter		.80	30		9,345	715		10,060	11,300
	0430	20" diameter		.60	40		11,200	950		12,150	13,700
207	0010	**AIR PURGING SCOOP** with tappings										207
	0020	for air vent and expansion tank connection									
	0100	1" pipe size, threaded	1 Stpi	19	.421	Ea.	10.50	10.75		21.25	28
	0110	1-¼" pipe size, threaded		15	.533		10.60	13.60		24.20	32
	0120	1-½" pipe size, threaded		13	.615		22.70	15.70		38.40	48
	0130	2" pipe size, threaded		11	.727		25.75	18.55		44.30	56
	0140	2-½" pipe size, threaded	Q-5	15	1.070		55.70	24		79.70	98
	0150	3" pipe size, threaded	"	13	1.230		71.30	28		99.30	120

192

(Reprinted from Means Mechanical Cost Data 1991.)

Figure 6.5

Estimators must carefully address the issue of equipment and related expenses. Equipment costs can be divided into the two following categories:

- **Rental, lease or ownership costs**: These costs may be determined based on hourly, daily, weekly, monthly or annual increments. These fees or payments only buy the *right* to use the equipment (i.e., exclusive of operating costs).
- **Operating costs**: Once the *right* of use is obtained, costs are incurred for actual use or operation. These costs may include fuel, lubrication, maintenance, and parts.

Equipment costs, as described above, do not include the labor expense of operators. However, some cost books and suppliers may include the operator in the quoted price for equipment as an "operated" rental cost. In other words, the equipment is priced as if it were a subcontract cost. In mechanical construction, equipment is often procured in this way, via an arrangement with the general contractor for equipment already on site. The advantage of this approach is that it can be used on an "as needed" basis and not carried as a weekly or monthly cost.

Equipment ownership costs apply to both leased and owned equipment. The operating costs of equipment, whether rented, leased, or owned, are available from the following sources (listed in order of reliability):

1. The company's own records
2. Annual cost books containing equipment operating costs, such as *Means Mechanical Cost Data* and *Means Plumbing Cost Data*
3. Manufacturers' estimates
4. Text books dealing with equipment operating costs

These operating costs consist of fuel, lubrication, expendable parts replacement, minor maintenance, transportation, and mobilizing costs. For estimating purposes, the equipment ownership and operating costs should be listed separately. In this way, the decision to rent, subcontract, or purchase can be decided project by project.

There are two commonly used methods for including equipment costs in a construction estimate. The first is to include the equipment as a part of the construction task for which it is used. In this case, costs are included in each line item as a separate unit price. The advantage of this method is that costs are allocated to the division or task that actually incurs the expense. As a result, more accurate records can be kept for each installed component. A disadvantage of this method occurs in the pricing of equipment that may be used for many different tasks. Duplication of costs can occur in this instance. Another disadvantage is that the budget may be left short for the following reason: the estimate may only reflect two hours for a crane truck, when the minimum 'cost of the crane is usually a daily (8-hour) rental charge.

The second method for including equipment costs in the estimate is to keep all such costs separate and to include them in Division 1 as a part of Project Overhead. The advantage of this method is that all equipment costs are grouped together, and that machines used for several tasks are included (without duplication). One disadvantage is that for future estimating purposes, equipment costs will be known only on a job basis and not per installed unit.

Whichever method is used, the estimator must be consistent, and must be sure that all equipment costs are included, but not duplicated. The estimating method should be the same as that chosen for cost monitoring and accounting. In this way, the data will be available both for monitoring the project's costs and for bidding future projects.

A final word of caution about equipment is to consider its age and reliability. If an older item, such as a pick-up truck, needs frequent repair, it may cost far more in lost man-hours to the project than is reflected in its calculated cost rate.

Subcontractors

Subcontractors often account for a large percentage of the mechanical estimator's bid. When subcontractors are used, quotations should be solicited and analyzed in the same way as material quotes. A primary concern is that the bid covers the work as per plans and specifications, and that all appropriate work, alternates, and allowances, if any, are included. Any exclusions should be clearly stated and explained. If the bid is received verbally, a form such as that shown in Chapter 5 (Figure 5. 1) will help to assure that it is documented accurately. Any unique scheduling or payment requirements must be noted and evaluated prior to submission of your bid. Such requirements could affect (restrict or enhance) the normal progress of the project, and should, therefore, be known in advance.

The estimator should note how long the subcontract bid will be honored. This time period usually varies from 30 to 90 days and is usually included as a condition in complete bids.

The estimator should know or verify the bonding capability and capacity of unfamiliar subcontractors. Taking such action may be necessary when bidding in a new location. Other than word of mouth, these inquiries may be the only way to confirm subcontractor reliability.

Project Overhead

Project Overhead represents those construction costs that are usually included in Division 1–General Requirements. Site management is covered in this section. Typical items are supervisory personnel, job engineers, cleanup, and temporary heat and power. While these items may not be directly part of the physical structure, they are a part of the project. Project Overhead, like all other direct costs, can be separated into material, labor, and equipment components. Figures 6.6 and 6.7 are examples of forms that can help ensure that all appropriate costs are included.

Some may not agree that certain items (such as equipment or scaffolding) should be included in Project Overhead, and might prefer to list such items in another division. Ultimately, it is not important *where* each item is incorporated into the estimate but that *every item is included somewhere*.

Project Overhead often includes time-related items. Equipment rental, supervisory labor, and temporary utilities are examples. The cost for these items depends upon the duration of the project. A preliminary schedule should, therefore, be developed *prior* to completion of the estimate so that time-related items can be properly counted. This will be further discussed in Chapter 9, "Pre-Bid Scheduling."

Bonds

Although bonds are really a type of "direct cost," they are priced and based upon the total "bid" or "selling price." For this reason, they are generally figured after indirect costs have been added. Bonding requirements for a project will be specified in Division 1–General Requirements, and will be included in the construction contract. Various types of bonds may be required. Listed below are a few common types:

Means Forms

PROJECT
OVERHEAD SUMMARY
PROJECT

SHEET NO.

ESTIMATE NO.

LOCATION   ARCHITECT   DATE

QUANTITIES BY: PRICES BY: EXTENSIONS BY: CHECKED BY:

DESCRIPTION	QUANTITY	UNIT	MATERIAL/EQUIPMENT		LABOR		TOTAL COST	
			UNIT	TOTAL	UNIT	TOTAL	UNIT	TOTAL
Job Organization: Superintendent								
Project Manager								
Timekeeper & Material Clerk								
Clerical								
Safety, Watchman & First Aid								
Travel Expense: Superintendent								
Project Manager								
Engineering: Layout								
Inspection/Quantities								
Drawings								
CPM Schedule								
Testing: Soil								
Materials								
Structural								
Equipment: Cranes								
Concrete Pump, Conveyor, Etc.								
Elevators, Hoists								
Freight & Hauling								
Loading, Unloading, Erecting, Etc.								
Maintenance								
Pumping								
Scaffolding								
Small Power Equipment/Tools								
Field Offices: Job Office								
Architect/Owner's Office								
Temporary Telephones								
Utilities								
Temporary Toilets								
Storage Areas & Sheds								
Temporary Utilities: Heat								
Light & Power								
Water								
PAGE TOTALS								

Page 1 of 2

Figure 6.6

⚜ Means Forms

DESCRIPTION	QUANTITY	UNIT	MATERIAL/EQUIPMENT		LABOR		TOTAL COST	
			UNIT	TOTAL	UNIT	TOTAL	UNIT	TOTAL
Totals Brought Forward								
Winter Protection: Temp. Heat/Protection								
Snow Plowing								
Thawing Materials								
Temporary Roads								
Signs & Barricades: Site Sign								
Temporary Fences								
Temporary Stairs, Ladders & Floors								
Photographs								
Clean Up								
Dumpster								
Final Clean Up								
Punch List								
Permits: Building								
Misc.								
Insurance: Builders Risk								
Owner's Protective Liability								
Umbrella								
Unemployment Ins. & Social Security								
Taxes								
City Sales Tax								
State Sales Tax								
Bonds								
Performance								
Material & Equipment								
Main Office Expense								
Special Items								
TOTALS:								

Figure 6.7

- **Bid Bond**: A form of bid security executed by the bidder or principle and by a surety (bonding company) to guarantee that the bidder will enter into a contract within a specified time and furnish any required Performance or Labor and Material Payment bonds.
- **Completion Bond**: Also known as "Construction" or "Contract" bond. The guarantee by a surety that the construction contract will be completed and that it will be clear of all liens and encumbrances.
- **Labor and Material Payment Bond**: The guarantee by a surety to the owner that the contractor will pay for all labor and materials used in the performance of the contract as per the construction documents. The claimants under the bond are those having direct contracts with the contractor or any subcontractor.
- **Performance Bond**: (1) A guarantee that a contractor will perform a job according to the terms of the contracts. (2) A bond of the contractor in which a surety guarantees to the owner that the work will be performed in accordance with the contract documents. Except where prohibited by statute, the performance bond is frequently combined with the labor and material payment bond.
- **Surety Bond**: A legal instrument under which one party agrees to answer to another party for the debt, default, or failure to perform of a third party.

Sales Tax

Sales tax varies from state to state and often from city to city within a state (see Figure 6.8). Larger cities may have a sales tax in addition to the state sales tax. Some localities also impose separate sales taxes on labor and equipment.

Several contractors are renaming sales tax to a "service tax" based on the contractor's "selling" price (which includes overhead and profit).

Sales Tax Percentages on Materials by State

State	Tax	State	Tax	State	Tax	State	Tax
Alabama	4%	Illinois	5%	Montana	0%	Rhode Island	6%
Alaska	0	Indiana	5	Nebraska	4	South Carolina	5
Arizona	5	Iowa	4	Nevada	3.5	South Dakota	4
Arkansas	4	Kansas	4.25	New Hampshire	0	Tennessee	5.5
California	6	Kentucky	5	New Jersey	6	Texas	6
Colorado	3	Louisiana	4	New Mexico	4.75	Utah	6
Connecticut	7.5	Maine	5	New York	4	Vermont	4
Delaware	0	Maryland	5	North Carolina	5	Virginia	4.5
District of Columbia	6	Massachusetts	5	North Dakota	6	Washington	6.5
Florida	6	Michigan	4	Ohio	5.5	West Virginia	6
Georgia	3	Minnesota	6	Oklahoma	4	Wisconsin	5
Hawaii	4	Mississippi	6	Oregon	0	Wyoming	3
Idaho	5	Missouri	4.225	Pennsylvania	6	Average	4.44%

(Reprinted from Means Mechanical Cost Data 1991.)

Figure 6.8

When bidding takes place in unfamiliar locations, the estimator should check with local agencies regarding the amount and the method of payment of sales tax. Local authorities may require owners to withhold payments to out-of-state contractors until payment of all required sales tax has been verified. Sales tax is often taken for granted or even omitted and, as can be seen in Figure 6.8, can be as much as 7.5% of material costs. Indeed, this can represent a significant portion of the project's total cost. Conversely, some clients and/or their projects may be tax exempt. If this fact is unknown to the estimator, a large dollar amount for sales tax might needlessly be included in a bid.

Chapter 7

INDIRECT COSTS

Indirect costs are those "costs of doing business" that are incurred by the general staff. These expenses are sometimes referred to as a "burden" to the project. Indirect costs may include certain fixed, or known, expenses and percentages, as well as costs which can be variable and subjectively determined. Government authorities require payment of certain taxes and insurance, usually based upon labor costs and determined by trade. These are a type of fixed indirect cost. Office overhead, if well understood and established, can also be considered as a relatively fixed percentage. Profit and contingencies, however, are more variable and subjective. These figures are often determined based on the judgement and discretion of the person responsible for the company's growth and success.

If the direct costs for the same project have been carefully determined, they should not vary significantly from one estimator to another. It is the indirect costs that are often responsible for variations between bids. The direct costs of a project must be itemized, tabulated, and totaled before the indirect costs can be applied to the estimate. Indirect costs include:

- Taxes and Insurance
- Office or Operating Overhead (vs. Project Overhead)
- Profit
- Contingencies

Taxes and Insurance

The taxes and insurance included as indirect costs are most often related to the costs of labor and/or the type of work. This category may include Workers' Compensation, Builder's Risk, and Public Liability insurance, as well as employer-paid Social Security tax and Unemployment Insurance. By law, the employer must pay these expenses. Rates are based on the type and salary of the employees, as well as the location and/or type of business.

Office or Operating Overhead

Office overhead, or the cost of doing business, is perhaps one of the main reasons why so many contractors are unable to realize a profit, or even to stay in business. This is manifested in two ways. Either a company does not know its true overhead cost and, therefore, fails to mark up its costs enough to recover them; or management does not restrain or control overhead costs effectively and fails to remain competitive.

If a contractor does not know the costs of operating the business, then, more than likely, these costs will not be recovered. Many companies survive, and even turn a profit, by simply adding an overhead to each job, without knowing how the percentage is derived or what is included. When annual volume changes significantly, whether by increase or decrease, the previously used percentage for overhead may no longer be valid. When such a volume change occurs, the owner often finds that the company is not doing as well as before and cannot determine the reasons. Chances are, overhead costs are not being fully recovered. As an example, Figure 7.1 lists annual office costs and expenses for a "typical" mechanical contractor. It is assumed that the anticipated annual volume of the company is $1,500,000. Each of the items is described briefly below.

Owner: This includes only a reasonable base salary and does not include profits. An owner's salary is *not* a company's profit.

Engineer/Estimator: Since the owner is primarily on the road getting business, this is the person who runs the daily operation of the company and is

---

Annual Main Office Expenses

Assume: $1,500,000 Annual Volume in the Field
       30% Material, 70% Labor

Office/Operating Expenses:

Owner	$ 60,000
Engineer/Estimator	44,000
Secretary/Receptionist	18,000
Personnel Insurance & Taxes	44,030
Office Rent	10,000
Utilities	1,800
Telephone	6,000
Vehicles (2)	13,000
Office Equipment	2,400
Legal/Accounting Services	5,000
Miscellaneous:	
Advertising	1,750
Seminars	3,000
Travel & Entertainment	8,000
Uncollected Receivables	18,000
Total	$234,980

$$\frac{\text{Expenses}}{\text{Labor Volume}} = \frac{\$234,980}{\$1,050,000} = 22.4\%$$

To support this overhead, job site staffing would have to average approximately 23 workers throughout the year.

Figure 7.1

responsible for estimating. In some operations, the estimator who successfully wins a bid, then becomes the "project manager" and is responsible to the owner for its profitability.

Secretary/Receptionist: This person manages office operations and handles paperwork. A talented individual in this position can be a tremendous asset.

Office Worker Insurance & Taxes: These costs are for main office personnel only and, for this example, are calculated as 37% of the total salaries based on the following breakdown:

Workers' Compensation	6%
FICA	7%
Unemployment	4%
Medical & other insurance	10%
Profit sharing, pension, etc.	10%
	37%

Physical Plant Expenses: Whether the office, warehouse, and yard are rented or owned, roughly the same costs are incurred. Telephone and utility costs will vary depending on the size of the building and the type of business. Office equipment includes items such as the rental of a copy machine and typewriters.

Professional Services: Accountant fees are primarily for quarterly audits. Legal fees go towards collecting and contract disputes. In addition, a prudent contractor will have *every* contract read by his lawyer prior to signing.

Miscellaneous: There are many expenses that could be placed in this category. Included in the example are just a few of the possibilities. Advertising includes the Yellow Pages, promotional materials, etc.

Uncollected Receivables: This amount can vary greatly, and is often affected by the overall economic climate. Depending upon the timing of "uncollectables," cash-flow can be severely restricted and can cause serious financial problems, even for large companies. Sound cash planning and anticipation of such possibilities can help to prevent severe repercussions. While the office example used here is feasible within the industry, keep in mind that it is hypothetical and that conditions and costs vary widely from company to company.

In order for this example company to stay in business without losses (profit is not yet a factor), not only must all direct construction costs be paid, but an additional $234,980 must be recovered during the year (as a percentage of volume) in order to operate the office. The percentage may be calculated and applied in two ways:

- Office overhead applied as a percentage of labor costs only. This method requires that labor and material costs be estimated separately.
- Office overhead applied as a percentage of total project costs. This can be used whether or not material and labor costs are estimated separately.
- Remember that the anticipated volume is $1,500,000 for the year, 70% of which is expected to be labor. Office overhead costs, therefore, will be approximately 22.4% of the labor cost, or 15.7% of annual volume *for this example*. The most common method for recovering these costs is to apply this percentage to each job over the course of the year.
- The estimator must also remember that, if volume changes significantly, then the percentage for office overhead should be recalculated for current conditions. The same is true if there are changes in office staff.

Salaries are the major portion of office overhead costs. It should be noted that a percentage is commonly applied to material costs, for handling, regardless of the method of recovering office overhead costs. This percentage is more easily calculated if material costs are estimated and listed separately.

Profit

Determining a fair and reasonable percentage to be included for profit is not an easy task. This responsibility is usually left to the owner or chief estimator. Experience is crucial in anticipating what profit the market will bear. The economic climate, competition, knowledge of the project, and familiarity with the architect, engineer, or owner all affect the way in which profit is determined. Chapter 10 includes a method to mathematically determine the profit margin based on historical bidding information. As with all facets of estimating, experience is the key to success.

Contingencies

Like profit, contingencies can also be difficult to quantify. Especially appropriate in preliminary budgets, the addition of a contingency is meant to protect the contractor as well as to give the owner a realistic estimate of potential project costs.

A contingency percentage should be based on the number of "unknowns" in a project, or the level of risk involved. This percentage should be inversely proportional to the amount of planning detail that has been done for the project. If complete plans and specifications are supplied, the estimate is thorough and precise, and the market is stable, then there is little need for a contingency. Figure 7.2, from *Means Mechanical Cost Data*, lists suggested contingency percentages that may be added to an estimate based on the stage of planning and development.

If an estimate is priced and each individual item is rounded upward, or "padded," this is, in essence, adding a contingency. This method can cause problems, however, because the estimator can never be quite sure of what is the actual cost and what is the "padding," or safety margin, for each item. At the summary, the estimator cannot determine exactly how much has been included as a contingency factor for the project as a whole. A much more accurate and controllable approach is to price the estimate precisely and then add one contingency amount at the bottom line.

010 | Overhead

010 000	Overhead	CREW	DAILY OUTPUT	MAN-HOURS	UNIT	MAT.	LABOR	EQUIP.	TOTAL	TOTAL INCL O&P	
012	0011	**CONSTRUCTION COST INDEX** For 162 major U.S. and										012
	0020	Canadian cities, total cost, min. (Greensboro, NC)				%					78.60%	
	0050	Average									100%	
	0100	Maximum (Anchorage, AK)									127%	
020	0010	**CONTINGENCIES** Allowance to add at conceptual stage				Project					15%	020
	0050	Schematic stage									10%	
	0100	Preliminary working drawing stage									7%	
	0150	Final working drawing stage									2%	
022	0010	**CONTRACTOR EQUIPMENT** See division 016										022
024	0010	**CREWS** For building construction, see How To Use This Book										024
028	0010	**ENGINEERING FEES** Educational planning consultant, minimum				Project					.50%	028
	0100	Maximum				"					2.50%	
	0200	Electrical, minimum				Contrct					4.10%	
	0300	Maximum									10.10%	
	0400	Elevator & conveying systems, minimum									2.50%	
	0500	Maximum									5%	
	0600	Food service & kitchen equipment, minimum									8%	
	0700	Maximum									12%	
	1000	Mechanical (plumbing & HVAC), minimum									4.10%	
	1100	Maximum									10.10%	
032	0010	**FACTORS** To be added to construction costs for particular job										032
	0200											
	0500	Cut & patch to match existing construction, add, minimum				Costs	2%	3%				
	0550	Maximum					5%	9%				
	0800	Dust protection, add, minimum					1%	2%				
	0850	Maximum					4%	11%				
	1100	Equipment usage curtailment, add, minimum					1%	1%				
	1150	Maximum					3%	10%				
	1400	Material handling & storage limitation, add, minimum					1%	1%				
	1450	Maximum					6%	7%				
	1700	Protection of existing work, add, minimum					2%	2%				
	1750	Maximum					5%	7%				
	2000	Shift work requirements, add, minimum						5%				
	2050	Maximum						30%				
	2300	Temporary shoring and bracing, add, minimum					2%	5%				
	2350	Maximum					5%	12%				
034	0010	**FIELD OFFICE EXPENSE**										034
	0100	Office equipment rental, average				Month	135				148.50	
	0120	Office supplies, average				"	250				275	
	0125	Office trailer rental, see division 015-904										
	0140	Telephone bill; avg. bill/month incl. long dist.				Month	225				247.50	
	0160	Field office lights & HVAC				"	78				85.80	
038	0011	**HISTORICAL COST INDEXES** Back to 1947										038
040	0010	**INSURANCE** Builders risk, standard, minimum				Job					.22%	040
	0050	Maximum									.59%	
	0200	All-risk type, minimum									.25%	
	0250	Maximum									.62%	
	0400	Contractor's equipment floater, minimum				Value					.50%	
	0450	Maximum				"					1.50%	
	0600	Public liability, average				Job					1.55%	
	0800	Workers' compensation & employer's liability, average										
	0850	by trade, carpentry, general				Payroll		16.64%				
	1000	Electrical						5.97%				
	1150	Insulation						13.48%				
	1450	Plumbing						7.45%				

For expanded coverage of these items see *Means Building Construction Cost Data 1991*   1

(Reprinted from Means Mechanical Cost Data 1991.)

Figure 7.2

Chapter 8

THE ESTIMATE SUMMARY

At the pricing stage of the estimate, there is typically a large amount of paperwork that must be assembled, analyzed and tabulated. Generally, the information contained in this paperwork could be recorded on any or all of the following major categories:

- Plumbing Fixture Schedule (Figure 8. 1)
- Material supplier's written quotations (see note below)
- Equipment or material supplier's or subcontractor's quotations (Figure 8.2)
- Subcontractor's written quotations
- Equipment supplier's quotations
- Cost Analysis or Consolidated Estimate Sheets (Pricing Sheets) (Figures 8.3 and 8.4)
- Recap Summary Sheets or Estimate Summary Sheets
- Piping Schedule (Figure 8.5), Fittings & Valves Schedule (Figure 8.6), Ductwork Schedule (Figure 4.2)

    Note: Additional forms, such as Request for Quote postal cards, are often prepared by the individual contractor.

In the "real world" of estimating, many quotations, especially for large equipment and for subcontracts, are not received until the last minute before the bidding deadline. Therefore, a system is needed to efficiently handle the paperwork and to ensure that everything will get transferred once (and only once) from the quantity takeoff to the cost analysis sheets. Some general rules for this process are as follows:

- The piping, fixtures, labor, etc., should have been previously priced and entered in the estimate or summary sheet.
- Write on only one side of any document.
- Use Subcontractor's Quotation forms for uniformity in recording prices received by telephone. (See Figure 8.2)
- Document the source of every quantity and price.
- Keep each type of document in its "pile" (Quantities, Material, Subcontractors, Equipment) piled in order by classifications.
- Keep the entire estimate in one or more compartmentalized folders.
- If you are pricing your own materials, number and code each takeoff sheet and each pricing extension sheet as it is created. At the same time, keep an index list of each sheet by number. If a sheet is to be abandoned, write "VOID" on it, but do not discard it. Keep it until the bid is accepted to be able to account for all pages and sheets.

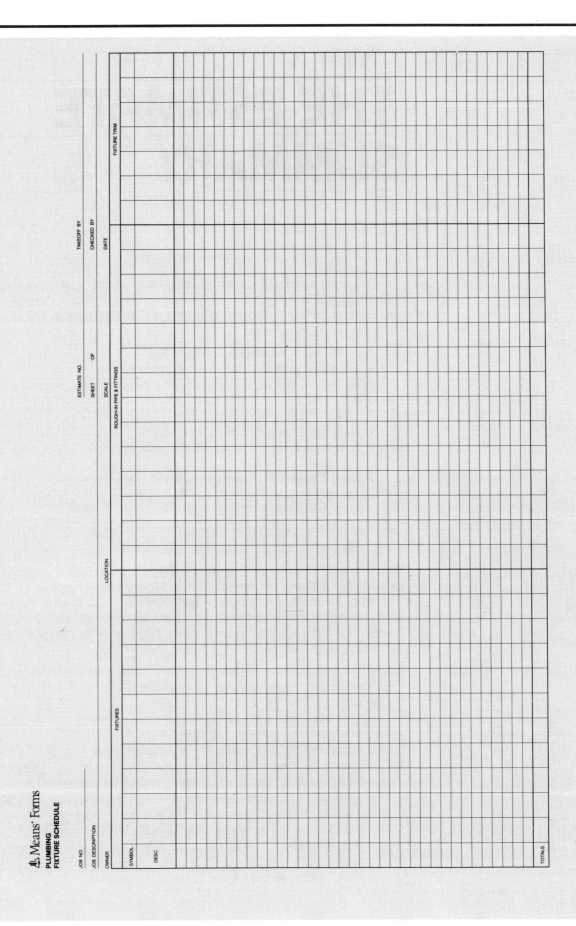

Figure 8.1

⚓ Means® Forms

MECHANICAL SUBCONTRACTORS'
QUOTES SUMMARY

PROJECT   LOCATION   DATE

SPECIALTY/CONTRACTOR	QUOTE			COMMENTS
Balancing	Air	Water	Total	
Controls	Electric	DDC	Pneumatic	
				Wiring Included?

SPECIALTY/CONTRACTOR	QUOTE				COMMENTS
Insulation	Duct	Pipe	Roof	Total	
Water Treatment	Equipment	Chemicals	Year Service	Total	

SPECIALTY/CONTRACTOR	QUOTE			COMMENTS
Sheet Metal	Duct	Miscellaneous	Total	

Figure 8.2

Means Forms

COST ANALYSIS

SHEET NO.

PROJECT _____

ESTIMATE NO.

ARCHITECT _____

DATE

| TAKE OFF BY: | QUANTITIES BY: | PRICES BY: | EXTENSIONS BY: | CHECKED BY: |

DESCRIPTION	SOURCE/DIMENSIONS			QUANTITY	UNIT	MATERIAL		LABOR		EQ./TOTAL	
						UNIT COST	TOTAL	UNIT COST	TOTAL	UNIT COST	TOTAL

Figure 8.3

52

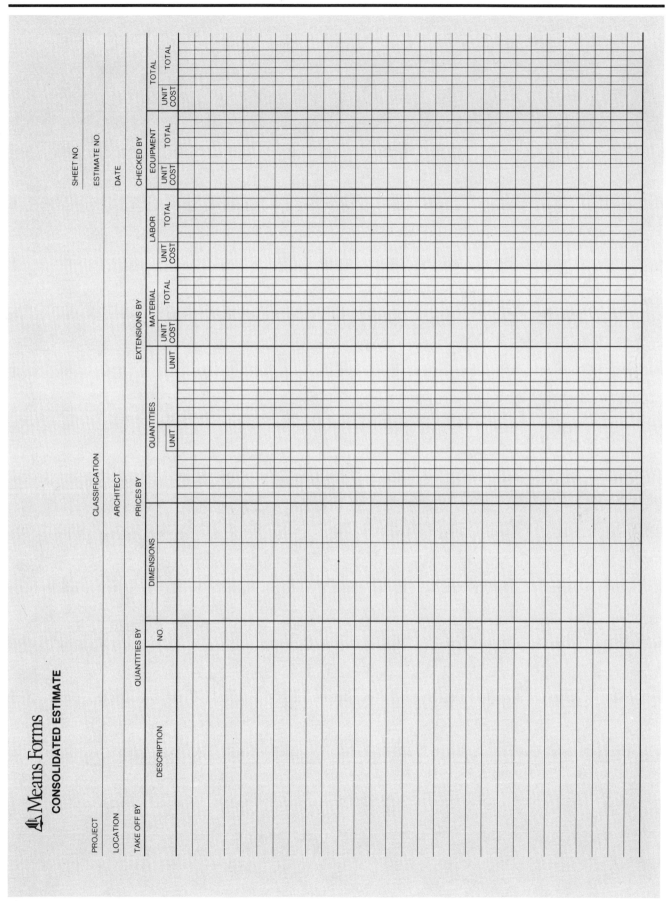

Figure 8.4

Figure 8.5

Means® Forms

FITTINGS AND VALVES SCHEDULE

JOB _____

SYSTEM _____

PAGE _____ OF _____

DATE _____

BY _____

		PIPE DIAMETER IN INCHES																	
	12	10	8	6	5	4	3-1/2	3	2-1/2	2	1-1/2	1-1/4	1	3/4	1/2				

Figure 8.6

Note: A helpful technique for organizing these sheets and forms involves the use of pastel colors to code each type of category of cost sheets. This system makes locating and revising sheets both easier and quicker.

All subcontract costs should be properly noted and listed separately. These costs contain the subcontractor's markups and may be treated differently from other direct costs when the estimator calculates the prime contractor's overhead and profit.

After all the unit prices and allowances have been entered on the pricing sheets, the costs are extended. In making the extensions, ignore the cents column and round all totals to the nearest five or ten dollars. In a column of figures, the cents will average out and will not be of consequence. Finally, each subdivision is added and the results checked, preferably by someone other than the person doing the extensions.

It is important to check the larger items for order of magnitude errors. If the total costs are divided by the building area, the resulting square foot cost figures can be used to quickly check with expected square foot costs. These cost figures should be recorded for comparison to past projects and as a resource for future estimating.

The takeoff and pricing method, as discussed, has been to utilize a Fixture Quantity Sheet for the material takeoff (see Figure 8.1), and to transfer the data to an analysis form for pricing the material, labor, and equipment items (see Figure 8.3).

An alternative to this method is a consolidation of the takeoff task and pricing on a single form. This approach works well for smaller bids and for change orders. An example, the Consolidated Estimate Form, is shown in Figure 8.4. The same sequences and recommendations used to complete the Quantity Sheet and Cost Analysis form are to be followed when using the Consolidated Estimate form to price the estimate.

When the pricing of all direct costs is complete, the estimator has two choices: 1) to make all further price changes and adjustments on the Cost Analysis or Consolidated Estimate sheets, *or* 2) to transfer the total costs for each subdivision to an Estimate Summary sheet so that all further price changes, until bid time, will be done on one sheet. Any indirect cost markups and burdens will be figured on this sheet also.

Unless the estimate has a limited number of items, it is recommended that costs be transferred to an Estimate Summary sheet. This step should be double-checked since an error of transposition may easily occur. Preprinted forms can be useful, although a plain columnar form may suffice. This summary with page numbers from each extension sheet can also serve as an index of the mechanical specifications.

A company that repeatedly uses certain standard listings can save valuable time by having a custom Estimate Summary sheet printed with these items listed. The printed division and subdivision headings may serve as another type of checklist, ensuring that all required costs are included. Appropriate column headings or categories for any estimate summary form could be as follows:

- Material
- Labor
- Equipment
- Subcontractor
- Miscellaneous
- Total

As items are listed in the proper columns, each category is added and appropriate markups applied to the total dollar values. Different percentages may be added to the sum of each column at the estimate summary. These percentages may include the following items, as discussed in Chapter 7:

- Taxes and Insurance
- Overhead
- Profit
- Contingencies

Chapter 9
PRE-BID SCHEDULING

The need for planning and scheduling is clear once the contract is signed and work commences on the project. However, some scheduling is also important during the bidding stage for the following reasons:

- To determine if the project can be completed in the allotted or specified time using normal crew sizes.
- To identify potential overtime requirements.
- To determine the time requirements for supervision.
- To anticipate possible temporary heat and power requirements.
- To price certain general requirement items and overhead costs.
- To budget for equipment usage.
- To anticipate and justify material and equipment delivery requirements. (An awareness of these requirements may, for example, justify using a more expensive vendor quote with better delivery terms.)

The schedule produced prior to bidding may be a simple bar chart or network diagram that includes overall quantities, probable delivery times, and available manpower. Network scheduling methods, such as the Critical Path Method (CPM) and the Precedence Chart simplify pre-bid scheduling because they do not require time-scaled line diagrams.

In the CPM diagram, the activity is represented by an arrow. Nodes indicate start/stop between activities. The Precedence diagram, on the other hand, shows the activity as a node with arrows used to denote precedence relationships between the activities. The precedence arrows may be used in different configurations to represent the sequential relationships between activities. Examples of CPM and Precedence diagrams are shown in Figures 9.1 and 9.2, respectively. In both systems, duration times are indicated along each path. The sequence (path) of activities requiring the most total time represents the shortest possible time (critical path) in which those activities may be completed.

For example, in both Figure 9.1 and Figure 9.2, activities A, B, and C require 20 successive days for completion before activity G can begin. Activity paths for D and E (15 days), and for F (12 days) are shorter and can easily be completed during the 20-day sequence. Therefore, this 20-day sequence is the shortest possible time (i.e., the "critical path") for the completion of these activities–before activity G can begin.

Past experience or a prepared rough schedule may suggest that the time specified in the bidding documents is insufficient to complete the required work. In such cases, a more comprehensive schedule should be produced prior to bidding; this schedule will help to determine the added overtime or premium time work costs required to meet the completion date.

A three-story office building project is used for the Sample Estimates in Part III of this book. A preliminary schedule is needed to determine the supervision and manning requirements of the job. The specifications state that the building must be completed within one year. Normally, the excavation, foundations, and superstructure can be completed in approximately one half of the construction period. Therefore, the major portion of the mechanical work is restricted to the last six months of the one year allotted.

A rough schedule for the mechanical work might be produced as shown in Figure 9.3. The man-days used to develop this schedule are derived from output figures determined in the estimate. Output can be determined based on the figures in *Means Mechanical Cost Data* or *Means Plumbing Cost Data*. Man-days can also be figured by dividing the total labor cost shown on the estimate by the cost per man-day for each appropriate tradesperson.

As shown, the preliminary schedule can be used to determine supervision requirements, to develop appropriate crew sizes, and as a basis for ordering materials. All of these factors must be considered at this preliminary stage in order to determine how to meet the required one year completion date.

A pre-bid schedule can provide much more information than simple job duration. It can be used to refine the estimate by introducing realistic manpower projections. The schedule may also help the contractor to adjust the structure

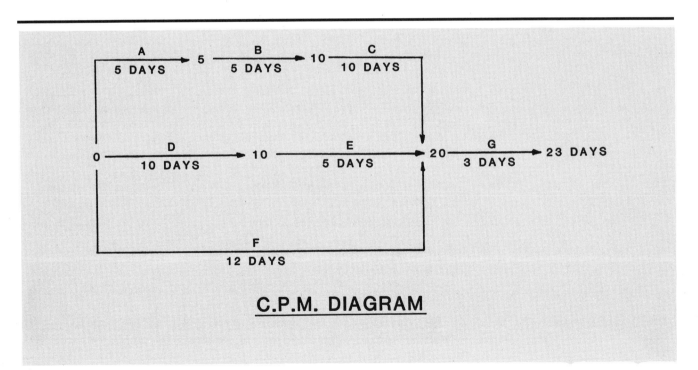

C.P.M. DIAGRAM

Figure 9.1

and size of the company based on projected requirements for months, even years, ahead. A schedule can also become an effective tool for negotiating contracts.

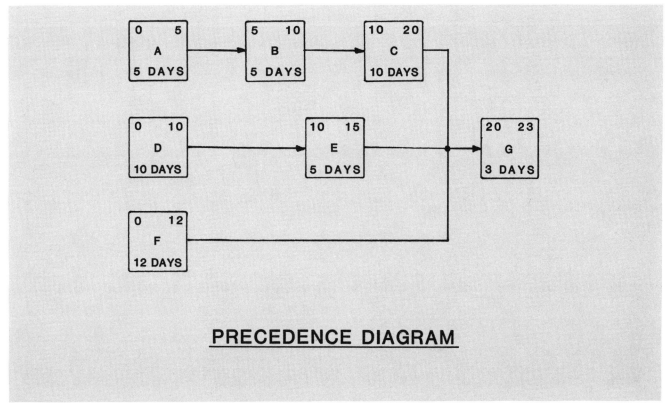

PRECEDENCE DIAGRAM

Figure 9.2

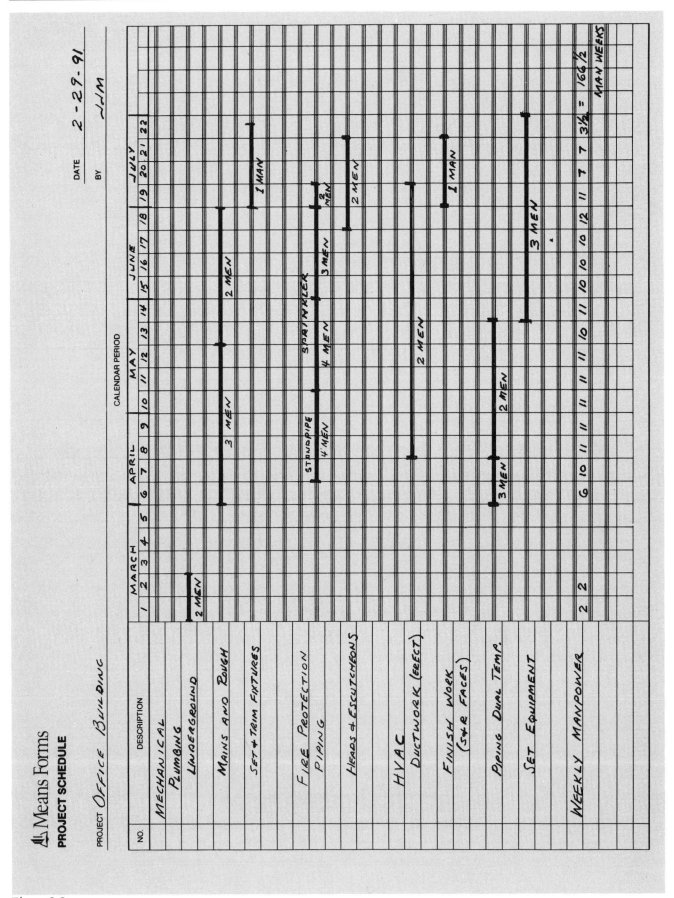

Figure 9.3

62

Chapter 10
BIDDING STRATEGIES

The goal of most contractors is to make as much money as possible on each job, but more importantly, to maximize return on investment on an annual basis. Often, this can be done by taking *fewer* jobs at a *higher* profit and by limiting bidding to the jobs which are most likely to be successful for the company.

Resource Analysis

Since most contractors cannot physically bid every job in a geographic area, a selection process must determine which projects to bid. This process should begin with an analysis of the strengths and weaknesses of the contractor. The following items must be considered as objectively as possible:

- Individual strengths of the company's top management
- Management experience with the type of construction involved, from top management to project superintendents
- Cost records adequate for the appropriate type of construction
- Bonding capability and capacity
- Size of projects with which the company is "comfortable"
- Geographic area that can be managed effectively
- Unusual corporate assets such as:
    Specialized equipment availability
    Reliable and timely cost control systems
    Strong balance sheet
    Familiarity with designer, owner, or general contractor

Market Analysis

Most contractors tend to concentrate on a few particular kinds of projects. From time to time, the company should step back and examine the portion of the industry they are serving. During this process, the following items should be carefully analyzed:

- Historical trend of the market segment
- Expected future trend of the market segment
- Geographic expectations of the market segment
- Historical and expected competition from other contractors
- Risk involved in the particular market segment
- Typical size of projects in this market
- Expected return on investment from the market segment

If several of these areas are experiencing a downturn, then it is definitely appropriate to examine an alternate market. On the other hand, many managers would feel that "bad times" are the times when they should consolidate and narrow their market. These managers are only likely to expand or broaden their market when current activities are especially strong.

Bidding Analysis

Certain steps should be taken to develop a bid strategy within a particular market. The first is to obtain the bid results of jobs in the prospective geographic area. These results should be set up on a tabular basis. This is fairly easy to do in the case of public jobs since the bid results are normally published (or at least available) from the agency responsible for the project. In private work, this step is more difficult, since the bid results are not normally divulged by the owner.

Determining Risk in a New Market Area

One way to measure success in bidding is how much money is "left on the table," the difference between the low bid and next lowest bid. The contractor who consistently takes jobs by a wide margin below the next bidder is obviously not making as much money as possible. Information on competitive public bidding is used to determine the amount of money left on the table; this information serves as the basis for fine-tuning a future bidding strategy.

For example, assume a public market where all bid prices and the total number of bidders are known. For each type of market sector, create a chart showing the percentage left on the table versus the total number of bidders. When the median figure (percent left on the table) for each number of bidders is connected with a smooth curve, the usual shape of the curve is shown in Figure 10.1.

The exact shape and magnitude of the amounts left on the table will depend on how much risk is involved with that type of work. If the percentages left on the table are high, then the work can be assumed to be very risky—with a high profit or loss potential if the award is won. If the percentages are low, the work is probably neither as risky nor as potentially profitable.

Analyzing the Bid's Risk

If a company has been bidding in a particular market, certain information should be collected and recorded as a basis for a bidding analysis. First the percentage left on the table should be tabulated (as shown in Figure 10.1), along with the number of bidders for the projects in that market on which the company was the low bidder. By probability, half the bids should be above

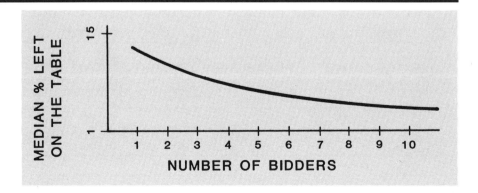

Figure 10.1

the median line and half below. If more than half are below the line, the company is doing well; if more than half are above, the bidding strategy should be examined.

Maximizing the Profit-to-Volume Ratio

Once the bidding track record for the company has been established, the next step is to reduce the historical percentage left on the table. One method is to create a chart showing, for instance, the last ten jobs on which the company was low bidder, and the dollar spread between the low and second lowest bid. Next, rank the percentage differences from one to ten (one being the smallest and ten being the largest left on the table). An example is shown in Figure 10.2. This example is for a larger general contractor, but the principles apply to any size or type of contractor involved in bidding.

The company's "costs" ($17,170,000) are derived from the company's low bids ($18,887,000) assuming a 10% profit ($1,717,000). The "second bid" is the next lowest bid. The "difference" is the dollar amount between the low bid and the second bid (money "left on the table"). The differences are then ranked based on the percentage of job "costs" left on the table for each. Figure the median difference by averaging the two middle percentages–that is, the fifth and sixth ranked numbers.

$$\text{Median \% Difference} = \frac{5.73 + 6.44}{2} = 6.09\%$$

Job No.	"Cost"	Low Bid	Second Bid	Difference	% Diff.	% Rank	Profit (Assumed at 10%)
1	$ 918,000	$ 1,009,800	$1,095,000	$ 85,200	9.28	10	$ 91,800
2	1,955,000	2,150,500	2,238,000	87,500	4.48	3	195,500
3	2,141,000	2,355,100	2,493,000	137,900	6.44	6	214,100
4	1,005,000	1,105,500	1,118,000	12,500	1.24	1	100,500
5	2,391,000	2,630,100	2,805,000	174,900	7.31	8	239,100
6	2,782,000	3,060,200	3,188,000	127,800	4.59	4	278,200
7	1,093,000	1,202,300	1,282,000	79,700	7.29	7	109,300
8	832,000	915,200	926,000	10,800	1.30	2	83,200
9	2,372,000	2,609,200	2,745,000	135,800	5.73	5	237,200
10	1,681,000	1,849,100	2,005,000	155,900	9.27	9	168,100
	$17,170,000	$18,887,000		$1,008,000			$1,717,000 = 10% of Cost

Figure 10.2

From Figure 10.2, the median percentage left on the table is 6.09%. To maximize the potential returns on a series of competitive bids, a useful formula is needed for pricing profit. The following formula has proven effective.

$$\text{Normal Profit \%} + \frac{\text{Median \% Difference}}{2} = \text{Adjusted Profit \%}$$

$$10.00 + \frac{6.09}{2} = 13.05\%$$

Now apply this adjusted profit percentage to the same list of ten jobs as shown in Figure 10.3. Note that the job "costs" remain the same, but that the low bids have been revised. Compare the bottom line results of Figure 10.2 to those of Figure 10.3 based on the two profit margins, 10% and 13.05%, respectively.

Total volume *drops* from $18,887,000 to $17,333,900. Net profits *rise* from $1,717,000 to $2,000,900.

Profits rise while volume drops! If the original volume is maintained or even increased, profits rise even further. Note how this occurs. By determining a reasonable increase in profit margin, the company has, in effect, raised all bids. By doing so, the company loses two jobs to the second bidder (Jobs 4 and 8 in Figure 10.3).

A positive effect of this volume loss is reduced exposure to risk. Since the profit margin is higher, the remaining eight jobs collectively produce more profit than the ten jobs based on the original, lower profit margin. From where did this money come? The money "left on the table" has been reduced from $1,008,000 to $517,100. The whole purpose is to systematically lessen the

Job No.	Company's "Cost"	Revised Low Bid	Second Bid	Adj. Diff.	Profit [10% + 3.05%]	Total
1	$ 918,000	$ 1,037,800	$1,095,000	$ 57,200	$ 91,800 + $28,000	$119,800
2	1,955,000	2,210,100	2,238,000	27,900	195,500 + 59,600	255,100
3	2,141,000	2,420,400	2,493,000	72,600	214,100 + 65,300	279,400
4	(1,005,000)	(1,136,100)	1,118,000(L)	—	100,500 + 30,600	0
5	2,391,000	2,703,000	2,805,000	102,000	239,100 + 72,900	312,000
6	2,782,000	3,145,100	3,188,000	42,900	278,200 + 84,900	363,100
7	1,093,000	1,235,600	1,282,000	46,400	109,300 + 33,300	142,600
8	(832,000)	(940,600)	926,000(L)	—	83,200 + 25,400	0
9	2,372,000	2,681,500	2,745,000	63,500	237,200 + 72,300	309,500
10	1,681,000	1,900,400	2,005,000	104,600	168,100 + 51,300	219,400
	$15,333,000	$17,333,900		$517,100		$2,000,900

Figure 10.3

dollar amount difference between the low bids and the second low bids. Caution: This is a hypothetical approach based upon the following assumptions.

- Profit must be assumed to be the bid or estimated profit, not the actual profit when the job is over.
- Bidding must be done within the same market in which data for the analysis was gathered.
- Economic conditions should be stable from the time the data is gathered until the analysis is used in bidding. If conditions change, use of such an analysis should be reviewed.
- Each contractor must make roughly the same number of bidding mistakes. For higher numbers of jobs in the sample, this requirement becomes more probable.
- The company must bid additional jobs if total annual volume is to be maintained or increased. Likewise, if net total profit margin is to remain constant, even fewer jobs need be won.
- Finally, the basic cost numbers and sources must remain constant. If a new estimator is hired or a better cost source is found, this technique cannot work effectively until a new track record has been established.

The accuracy of this strategy depends upon the criteria listed above. Nevertheless, it is a valid concept that can be applied, with appropriate and reasonable judgement, to many bidding situations.

Chapter 11

COST CONTROL AND ANALYSIS

An internal accounting system should be used by contractors and construction managers to logically gather and track the costs of a construction project. With this information, a cost analysis can be made about each activity–both during the installation process and at its conclusion. This information or "feedback" becomes the basis for management decisions through the duration of the project. This cost data is also helpful for future bid proposals.

The categories for a mechanical project are major items of construction (e.g., HVAC) which can be subdivided into component activities (e.g., ductwork piping, controls, etc.). These activities should coincide with the system and methods of the quantity takeoff. Uniformity is important in terms of the units of measure and in the grouping of components into cost centers. For instance, cost centers for a mechanical project might include categories such as HVAC, insulation, and plumbing. Activities within one of these categories–plumbing, for example, might include such items as fixtures, water supply, waste, and vent piping.

The major purposes of cost control and analysis are as follows:

- To provide management with a system to monitor costs and progress
- To provide cost feedback to the estimator(s)
- To determine the costs of change orders
- To be used as a basis for progress payment requisitions to the owner, the general contractor, or his representative
- To manage cash flow
- To identify areas of potential cost overruns to management for corrective action

It is important to establish a cost control system that is uniform–both throughout the company and from job to job. Such a system might begin with a uniform Chart of Accounts, a listing of code numbers for work activities. The Chart of Accounts is used to assign time and cost against work activities for the purpose of creating cost reports. A Chart of Accounts should have enough scope and detail so that it can be used for any of the projects that the company may win. Naturally, an effective Chart of Accounts will also be flexible enough to incorporate new activities as the company takes on new or different kinds of projects. Using a cost control system, the various costs can be consistently allocated. The following information should be recorded for each cost component:

- Labor charges in dollars and man-days are summarized from weekly time cards and distributed by code.

- Quantities completed to date must also be recorded in order to determine unit costs.
- Equipment rental costs are derived from purchase orders or from weekly charges issued by an equipment company.
- Material charges are determined from purchase orders.
- Appropriate subcontractor charges are allocated.
- Job overhead items may be listed separately or by component.

Each component of costs–labor, materials, and equipment–is now calculated on a unit basis by dividing the quantity installed to date into the cost to date. This procedure establishes the actual installed unit cost to date.

At this point, it is also useful to calculate the percent complete to date. This is done in two steps, or levels. First, the percent complete for each activity is calculated based on the actual quantity installed to date divided by the total quantity estimated for the activity. Second, the project total percent complete is calculated. To arrive at this number, multiply the percent complete of each activity times the total estimated cost for that activity. Then add these results to a total (sometimes called " earned" dollars) and divide this sum by the estimated total. The result is the project's overall percent complete.

The quantities that remain to be installed for each activity should be estimated based on the actual unit cost to date. This is the projected cost to complete each activity. Due to inefficiencies and mobilization costs as the work begins, it is not practical to use the actual units for projecting costs until an activity is 20% complete. Below 20%, the costs as originally estimated should be used. The actual costs to date are added to the projected costs to obtain the anticipated costs at the end of the project.

Typical forms that may be used to develop a cost control system are shown in Figures 11.1 to 11.6.

The analysis of categories serves as a useful management tool, providing information on a constant, up-to-date basis. Immediate attention is attracted to any center that is projecting a loss. Management can concentrate on this item in an attempt to make it profitable or to minimize the expected loss.

The estimating department can use the unit costs developed in the field as background information for future bidding purposes. Particularly useful are unit labor costs and unit man-hours (productivity) for the separate activities. This information should be integrated into the accumulated historical data. A particular advantage of unit man-hour records is that they tend to be constant over time. Current unit labor costs can be figured simply by multiplying these man-hour standards times the current labor rate per hour.

Frequently, items are added to or deleted from the contract either via field change orders or through contract amendments. Accurate cost records are an excellent basis for determining the cost changes that will result.

As discussed above, the determination of completed quantities is necessary in order to calculate unit costs. These quantities are used to figure the percent complete in each category. These percentages can, in turn, be used to calculate the billing for progress payment requisitions (invoices).

A cost system is only as good as the people responsible for coding and recording the required information. Simplicity is the key word. Do not try to break down the code into very small items unless there is a specific need. Continuous updating of reports is important so that operations which are not in control can be immediately brought to the attention of management.

Means Forms

**DAILY
TIME SHEET**

PROJECT			DESCRIPTION OF WORK								DATE					
FOREMAN											SHEET NO.					
WEATHER CONDITIONS																
TEMPERATURE											TOTALS		RATES		OUTPUT	
NO.	NAME										REG-ULAR	OVER-TIME	REG-ULAR	OVER-TIME		
		HOURS														
		UNITS														
		HOURS														
		UNITS														
		HOURS														
		UNITS														
		HOURS														
		UNITS														
		HOURS														
		UNITS														
		HOURS														
		UNITS														
		HOURS														
		UNITS														
		HOURS														
		UNITS														
		HOURS														
		UNITS														
		HOURS														
		UNITS														
		HOURS														
		UNITS														
		HOURS														
		UNITS														
		HOURS														
		UNITS														
		HOURS														
		UNITS														
		HOURS														
		UNITS														
		HOURS														
		UNITS														
	TOTALS	HOURS														
	EQUIPMENT	UNITS														

Figure 11.1

Also, be sure to draft clear directions and instructions for each phase of the process. Adequate time must be spent to ensure that all who are involved (especially the foremen and supervisors) clearly understand the program.

Productivity and Efficiency

When using a cost control system such as the one described above, the unit costs should reflect standard practices. Productivity should be based on a five day, eight-hour-per-day (during daylight hours) workweek. Exceptions can be made if a company's requirements are particularly and most often unique. Installation costs should be derived using normal minimum crew sizes, under normal weather conditions, during the normal construction season.

All unusual costs incurred or expected should be recorded separately for each category of work. For example, an overtime situation might occur on every job and in the same proportion. In this case, it would make sense to carry the unit price adjusted for the added cost of premium time. Likewise, unusual weather delays, strike activity, owner/architect delays, or contractor interference

Means Forms

WEEKLY TIME SHEET

EMPLOYEE | EMPLOYEE NUMBER | WEEK ENDING

FOREMAN | HOURLY RATE | PIECE WORK RATE

PROJECT	NO.	DESCRIPTION OF WORK	DAY								TOTAL	HOURS	RATE	TOTAL COST	UNIT COST
			HOURS									REG.			
			UNITS									O.T.			
			HOURS									REG.			
			UNITS									O.T.			
			HOURS									REG.			
			UNITS									O.T.			
			HOURS									REG.			
			UNITS									O.T.			
			HOURS									REG.			
			UNITS									O.T.			
			HOURS									REG.			
			UNITS									O.T.			
			HOURS									REG.			
			UNITS									O.T.			
		TOTAL	HOURS									REG.			
			UNITS									O.T.			

PAYROLL DEDUCTIONS	F.I.C.A.	INCOME TAXES FEDERAL	STATE	CITY/COUNTY	HOSPITALIZATION			TOTAL DEDUCTIONS		LESS DEDUCTIONS
	$	$	$	$	$	$	$	$		NET PAY

Figure 11.2

⚓ Means Forms

**LABOR
COST RECORD**

SHEET NO.

DATE FROM:

PROJECT   DATE TO:

LOCATION   BY:

DATE	CHARGE NO.	DESCRIPTION	HOURS	RATE	AMOUNT	HOURS	RATE	AMOUNT	HOURS	RATE	AMOUNT

Figure 11.3

Means Forms

**MATERIAL
COST RECORD**

SHEET NO.

DATE FROM

PROJECT   DATE TO

LOCATION   BY

DATE	NUMBER	VENDOR/DESCRIPTION	QTY.	UNIT PRICE				QTY.	UNIT PRICE				QTY.	UNIT PRICE		

Figure 11.4

74

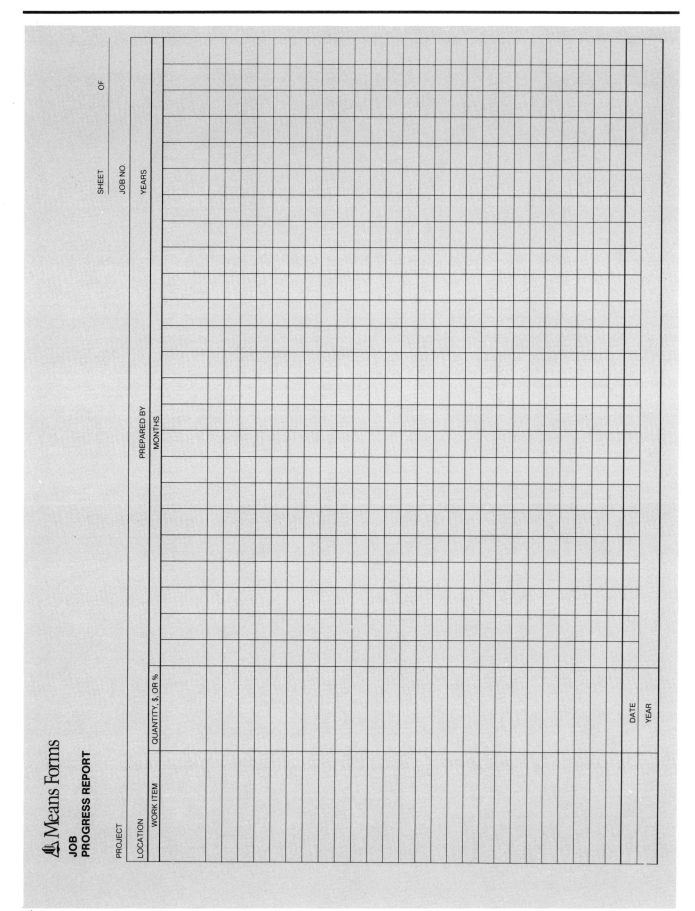

Figure 11.5

⚏ Means Forms

PERCENTAGE
COMPLETE ANALYSIS

PAGE

PROJECT

DATE

ARCHITECT

BY

FROM

TO

NO	DESCRIPTION	ACTUAL OR ESTIMATED	TOTAL PROJECT	THIS PERIOD		PERCENT TOTAL TO DATE										
				QUANTITY	%	QUANTITY	10	20	30	40	50	60	70	80	90	100
		ACTUAL														
		ESTIMATED														
		ACTUAL														
		ESTIMATED														
		ACTUAL														
		ESTIMATED														
		ACTUAL														
		ESTIMATED														
		ACTUAL														
		ESTIMATED														
		ACTUAL														
		ESTIMATED														
		ACTUAL														
		ESTIMATED														
		ACTUAL														
		ESTIMATED														
		ACTUAL														
		ESTIMATED														
		ACTUAL														
		ESTIMATED														
		ACTUAL														
		ESTIMATED														
		ACTUAL														
		ESTIMATED														
		ACTUAL														
		ESTIMATED														
		ACTUAL														
		ESTIMATED														
		ACTUAL														
		ESTIMATED														
		ACTUAL														
		ESTIMATED														
		ACTUAL														
		ESTIMATED														
		ACTUAL														
		ESTIMATED														
		ACTUAL														
		ESTIMATED														
		ACTUAL														
		ESTIMATED														

Figure 11.6

should have separate, identifiable cost contributions; these are applied as isolated costs to the activities affected by the delays. This procedure serves two purposes:

- To identify and separate the cost contribution of the delay so that future job estimates will not automatically include an allowance for these "nontypical" delays, and
- To serve as a basis for an extra compensation claim and/or as justification for reasonable extension of the job.

Overtime Impact

The use of long-term overtime is counter-productive on almost any construction job; that is, the longer the period of overtime, the lower the actual production rate. There have been numerous studies conducted which come up with slightly different numbers, but all reach the same conclusion. Figure 11.7 tabulates the effects of overtime work on efficiency.

As illustrated in Figure 11.8, there can be a difference between the *actual* payroll cost per hour and the *effective* cost per hour for overtime work. This is due to the reduced production efficiency with the increase in weekly hours beyond 40. This difference between actual and effective cost results from overtime work over a prolonged period. Short-term overtime work does not result in as great a reduction in efficiency, and in such cases, effective cost may not vary significantly from the actual payroll cost. As the total hours per

Days per Week	Hours per Day	Production Efficiency					Payroll Cost Factors	
		1 Week	2 Weeks	3 Weeks	4 Weeks	Average 4 Weeks	@ 1½ Times	@ 2 Times
5	8	100%	100%	100%	100%	100%	100%	100%
	9	100	100	95	90	96.25	105.6	111.1
	10	100	95	90	85	91.25	110.0	120.0
	11	95	90	75	65	81.25	113.6	127.3
	12	90	85	70	60	76.25	116.7	133.3
6	8	100	100	95	90	96.25	108.3	116.7
	9	100	95	90	85	92.50	113.0	125.9
	10	95	90	85	80	87.50	116.7	133.3
	11	95	85	70	65	78.75	119.7	139.4
	12	90	80	65	60	73.75	122.2	144.4
7	8	100	95	85	75	88.75	114.3	128.6
	9	95	90	80	70	83.75	118.3	136.5
	10	90	85	75	65	78.75	121.4	142.9
	11	85	80	65	60	72.50	124.0	148.1
	12	85	75	60	55	68.75	126.2	152.4

Figure 11.7

week are increased on a regular basis, more time is lost because of fatigue, lowered morale, and an increased accident rate.

As an example, assume a project where workers are working 6 days a week, 10 hours per day. From Figure 11.8 (based on productivity studies), the actual productive hours are 51.1 hours. This represents a theoretical production efficiency of 51.1/60 or 85.2%.

Depending upon the locale and day of week, overtime hours may be paid at time and a half or double time. For time and a half, the overall (average) *actual* payroll cost (including regular and overtime hours) is determined as follows:

For time and a half:

$$\frac{40 \text{ reg. hrs.} + (20 \text{ overtime hrs. x } 1.5)}{60 \text{ hrs.}} = 1.167$$

Based on 60 hours, the payroll cost per hour will be (on average) 116.7% of the normal rate at 40 hours per week. However, because the actual production (efficiency) for 60 hours is reduced to the equivalent of 51.1 hours, the *effective* cost of overtime is calculated as shown on the next page:

Efficiency and Cost Effects of Prolonged Overtime Work

Days per Week	Hours per Day	Total Hours Worked	Actual Productive Hours	Production Efficiency	Payroll Cost per Hour Overtime after 40 hrs. @ 1-1/2 times	Payroll Cost per Hour Overtime after 40 hrs. @ 2 times	Effective Cost per Hour Overtime after 40 hrs. @ 1-1/2 times	Effective Cost per Hour Overtime after 40 hrs. @ 2 times
	8	40	40.0	100.0%	100.0%	100.0%	100.0%	100.0%
	9	45	43.4	96.5	105.6	111.1	109.4	115.2
5	10	50	46.5	93.0	110.0	120.0	118.3	129.0
	11	55	49.2	89.5	113.6	127.3	127.0	142.3
	12	60	51.6	86.0	116.7	133.3	135.7	155.0
	8	48	46.1	96.0	108.3	116.7	112.8	121.5
	9	54	48.9	90.6	113.0	125.9	124.7	139.1
6	10	60	51.1	85.2	116.7	133.3	137.0	156.6
	11	66	52.7	79.8	119.7	139.4	149.9	174.6
	12	72	53.6	74.4	122.2	144.4	164.2	194.0
	8	56	48.8	87.1	114.3	128.6	131.1	147.5
	9	63	52.2	82.8	118.3	136.5	142.7	164.8
7	10	70	55.0	78.5	121.4	142.9	154.5	181.8
	11	77	57.1	74.2	124.0	148.1	167.3	199.6
	12	84	58.7	69.9	126.2	152.4	180.6	218.1

Figure 11.8

For time and a half:

$$\frac{40 \text{ reg. hrs.} + (20 \text{ overtime hrs.} \times 1.5)}{51.1 \text{ hrs.}} = 1.37$$

Installed cost will be 137% of the normal rate (for labor).

Thus, when figuring overtime, the actual cost per unit of work will be higher than the apparent overtime payroll dollar increase, due to the reduced productivity of the longer workweek. These calculations are true only for those cost factors determined by hours worked. Costs that are applied weekly or monthly, such as equipment rentals, will not be similarly affected.

Retainage and Cash Flow

The majority of construction projects have some percentage of retainage held back by the owner until the job is complete and accepted. This retainage can range from 5% to as high as 15% or 20% in unusual cases. The most typical retainage is 10%. Since the profit on a given job may be less than the amount of withheld retainage, the contractor must wait longer before a positive cash flow is achieved than if there were no retainage.

Figures 11.9 and 11.10 are graphic and tabular representations of the projected cash flow for a small project. With this kind of projection, the contractor is able to anticipate cash needs throughout the course of the job. Note that on the eleventh of May, before the second payment is received, the contractor has paid out about $25,000 more than has been received. This is the maximum amount of cash (on hand or financed) that is required for the whole project. At an early stage of planning, the contractor can determine if there will be adequate cash available or if a loan is needed. In the latter case, the expense of interest could be anticipated and included in the estimate. On larger projects, the projection of cash flow becomes crucial, because unexpected interest expense can quickly erode profits.

A note on the subject of financing may be helpful here. In the above example, assume that the contractor has adequate cash resources to finance the project. Does this mean that the project need not bear any interest charges? No. At the very least, money withdrawn from savings or investments will result in unrealized interest (effective losses). These "losses" should be included in the project records for the purpose of determining actual job profit. For this reason, many companies assign interest charges to their jobs for negative cash flow. Likewise, the project is credited if a positive cash flow is realized. in this way, an owner or manager can readily assess the contribution (or liability) of each job to the company's financial health. For example, if a project has a 5% profit at completion but has incurred 10% in "finance charges," it certainly has not helped the company to stay in business.

The General Conditions section of the specifications usually explains the responsibilities of both the owner and the contractor with regard to billing and payments. Even the best planning and projections are contingent upon the general contractor paying requisitions as anticipated. There is an almost unavoidable adversary relationship between the subcontractor and general contractor regarding payment during the construction process. However, it is in the best interest of the general contractor that the subcontractor be solvent so that delays, complications, and financial difficulties can be avoided prior to final completion of the project. The interest of both parties is best served if information is shared and communication is open. Both are working toward the same goal: the timely and successful completion of the project.

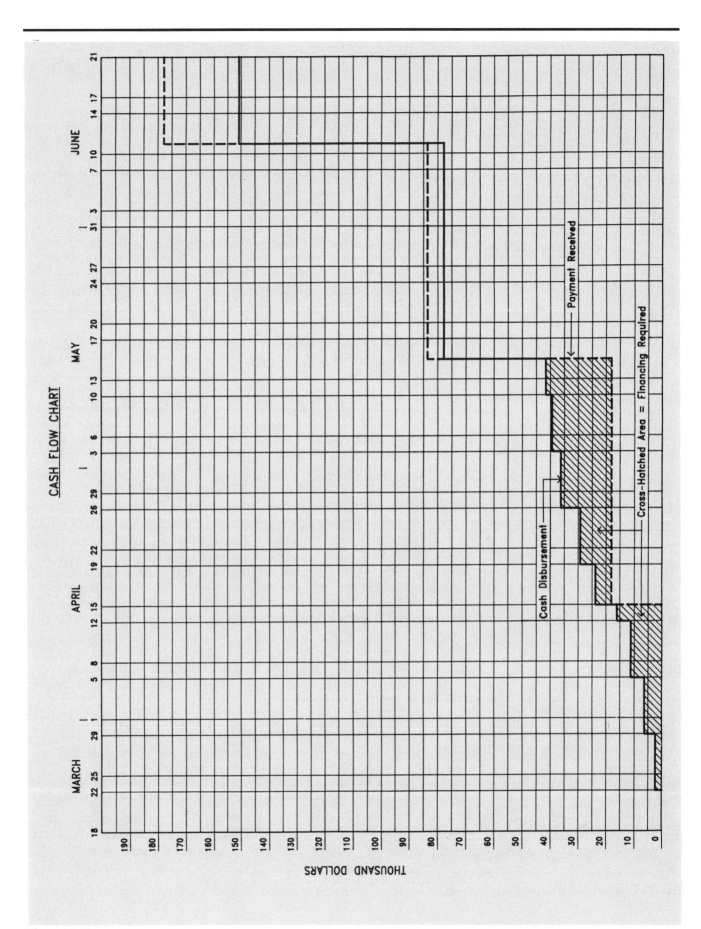

Figure 11.9

Figure 11.10

Date	Payroll Incl. Taxes	Workers Comp.	Monthly Billing + Payment	Retainage	Subs Billing + Payment	Retainage	M+E Incl. Taxes		Accumulated Costs $			
3-22 Payroll	2198	293							— 2198			
3-29 Payroll	3791	521							— 5989			
3-29 Subs Billing					4054	451						
3-29 Monthly Billing			16949	1883					— 10933			
4-5 Payroll	4944	646							— 15404			
4-12 Payroll	4471	578							— 1545			
4-15 Payment			(16949)						— 2509			
Pay Sub					(4054)		(5047)		— 7583			
Pay m+E									— 12054			
4-19 Payroll	4471	578							— 19236			
4-26 Payroll	7182	1331										
4-30 Subs Billing					2283	254			— 22296			
5-3 Monthly Billing			67251	7472					— 25345			
5-3 Payroll	3060	354										
5-10 Payroll	3049	497										
5-15 Payment			(67251)						+ 41906			
Pay Sub					(2283)		(33420)		+ 39623			
Pay M+E									+ 6203			
5-31 Subs Billing			89651	9894	63628	7070			+ 95254			
5-31 Final Billing									+ 31626			
6-15 Payment			(93517)	19249	(60196)	7775	(12967)		+ 254401			
Pay Sub					34932	7775	514434					
Pay M+E	33166	4748	(44166)									

General Conditions
Permit
Supervision Carpenter Foreman 35 days @ $178   475   25401
Temp. Power + Water   6230   44650
Temp. Office + Storage 6 wks @ 32   246 / 192   39852
Clean Up Laborer 2 days @ 128   256   32077
   7399   24678

Cash Retainage
Pay Workers Comp
Pay Subs Retainage
General Conditions

Life Cycle Costs

Life cycle costing is a valuable method of evaluating the total costs of an economic unit during its entire life. Regardless of whether the unit is a piece of excavating machinery or a manufacturing building, life cycle costing gives the owner an opportunity to look at the economic consequences of a decision. Today, the initial cost of a unit is often not the most important cost factor; the operation and maintenance costs of some building types far exceed the initial outlay. Hospitals, for example, may have operating costs within a three year period that exceed the original construction costs.

Estimators are in the business of initial costs. But what about the other costs that affect any owner: taxes, fuel, inflation, borrowing, estimated salvage at the end of the facility's lifespan, and expected repair or remodeling costs? These costs that may occur at a later date need to be evaluated in advance. The thread that ties all of these costs together is the time-value of money. The value of money today is quite different from what it will be tomorrow. One thousand dollars placed in a savings bank at 5% interest will, in four years, have increased to $1,215.50. Conversely, if $1,000 is needed four years from now, then $822.70 should be placed in an account today. Another way of saying the same thing is that at 5% interest, the value of $1,000 four years from now is only worth $822.70 today.

Using interest and time, future costs are equated to the present by means of a present worth formula. Standard texts in engineering economics have outlined different methods for handling interest and time. A present worth evaluation could be used, or all costs might be converted into an equivalent, uniform annual cost method.

The present worth of a piece of equipment (such as a trenching machine) can be figured, based on a given lifespan, anticipated maintenance cost, operating costs, and money borrowed at the current rate. Having these figures can help in determining whether to rent or purchase. The costs between two different machines can also be analyzed in this manner to determine which is a better investment.

Chapter 12

COMPUTERIZED ESTIMATING SYSTEMS

Recently more and more estimators are harnessing the power of the computer as an aid in preparing estimates. This chapter will survey some of the systems currently available. No attempt is made to teach the use of such systems; that would be beyond the scope of this book.

Two important points should be emphasized. First, although computers make estimating easier and faster, their use by an inexperienced estimator can lead to inaccurate results. Remember the guiding principal of all computer applications: *Garbage in, garbage out*. If the information put into the computer is flawed, no amount of computer manipulation will correct it. A thorough knowledge in, and understanding of, estimating is needed to properly use these systems.

Second, use of a computer can be faster in many cases; however, that may not be the case for small estimates. The learning curve is another factor: it may take a while to learn the intricacies of a new computer program. Different programs will take varying amounts of time to learn.

Most mechanical offices today using computerized estimating systems utilize small desktop computers. While there are programs available for larger and more powerful computers, the principles involved are the same and desktop computers have quite enough power to handle large and complex estimates. The computers usually consist of a box housing the electronic "brain," a video display device, and a keyboard to enter information into the system. More sophisticated systems have other devices to input information into the computer such as *mouses* (gadgets that fit into the palm of the user's hand and roll around on a desktop), *stylus pens* (sensitive pens attached by a cable to the computer), and *digitizers* (pads of sensitive material laid out on a desk upon which drawings may be placed and traced into the computer with a special stylus). Some people have a preference for one device over another, and all allow faster input than the keyboard.

The four classes of programs surveyed on the following pages range from the ultra-simple adaptation of a spreadsheet program to systems that automate almost all aspects of the process.

Spreadsheet Applications

A simple, yet powerful, tool for computerizing the estimating process is the computer spreadsheet program (some well-known programs are Lotus 1-2-3™, Quattro™, and Excel™). A computer spreadsheet is a variation on an accountant's ledger sheet consisting of a matrix of rows and columns of blank spaces called cells (shown as rows 1 through 12 and columns A through J in Figure 12.1). Each cell may be either a label or may be related to another cell mathematically. For example, in Figure 12.1, cell A4 (column A, row 4) contains a label with a description: 4″ SWCI hubless pipe. The next cell horizontally, B4, contains a quantity: 10. Continuing horizontally, one encounters a label: LF and a number: $2.97. The next cell, E4, contains the material total cost which is a formula: Quantity x Unit Price, or in spreadsheet language: B4 x D4 (the formula is not displayed, only the result is shown). In this manner an estimate spreadsheet can be built up containing labels, numbers, and formulas. Cells J9, J10, and J12 contain formulas which respectively add the prices for each line item, add markups, and total the estimate.

The computer spreadsheet will only be useful after quantity takeoff and pricing. These will still have to be done manually. Once entered, however, the power of the computer can be utilized not only in computing extensions and totals for large estimates, but in making adjustments to individual line items. Once entered, a formula such as that in cell E4 stays as the mathematical formula B4 x D4. If the value of B4 or D4 changes, the value displayed in E4 changes accordingly. "What-if" analyses can be performed on a spreadsheet by varying some numbers and viewing the results. Variations may include manipulating unit prices to estimate future costs or quantities to perform value engineering or least-cost studies.

To use a spreadsheet effectively, one must familiarize oneself with the "ins" and "outs" of the particular computer program being used. Since most offices

	A	B	C	D	E	F	G	H	I	J
1					MATERIAL			LABOR		TOTAL COST
2	DESCRIPTION	QTY	UNIT	UNIT COST	TOTAL COST	UNIT MH	TOTAL MH	UNIT RATE	TOTAL	MAT/LAB
3	==
4	4″ SWCI HUBLESS PIPE	10	LF	$2.97	$29.70	0.14	1.40	$31.35	$43.89	$73.59
5	4″ SWCI HUBLESS "Y"	1	EA	$6.98	$6.98	1.00	1.00	$31.35	$31.35	$38.33
6	4″ SWCI HUBLESS COUPLINGS	4	EA	$2.58	$10.32					$10.32
7	4″ HANGAR ASSEMBLY	3	EA	$7.11	$21.33	0.36	1.08	$31.35	$33.86	$55.19
8	....
9									SUBTOTAL:	$177.43
10								OVERHEAD & PROFIT @ 21%:		$37.26
11									....
12									TOTAL:	$214.69

Figure 12.1

have spreadsheets for other purposes such as accounting, the cost involved in setup for estimating is solely the time required for personnel to learn the program.

Library-Based Applications

The above example involved utilizing a commercial package that has numerous other uses and tailoring it for estimating. There are programs which are vastly more specialized and exclusively oriented towards estimating, such as Means *Data Source*®, Estimation Inc.'s *Sitemate 45*™, Simple Solution's *The Plumber's Helper*™, Wendes Mechanical Consulting's *Upgrade Super-Duct*™ and *Super-Pipe*™, Exccomate's assembly and subassembly programs, The Tallysheet Corp.'s *Gregg Way* HVAC-PC™ and *Pipe-PC*™, Construction Data Control's *The Bid Team*™, and others. Some of these programs require an estimator to manually take off quantities, others will aid in takeoff. These types of systems provide libraries of cost information from which cost items needed for a particular estimate may be selected.

Means *Data Source*® follows the layout of the Means cost books and can be updated annually. An estimator enters various pieces of information about the cost item selected such as quantity, whether or not the item is being subcontracted, locale, and work markups. The program then computes the cost extensions and totals, and prints out numerous reports ranging from brief summaries to detailed breakdowns (see Figure 12.2).

These library-based applications free an estimator from the burden of researching prices by allowing easy use of a database of national costs adjusted to local areas. Modifications can be made to allow for price or man-hour adjustment such as in-house information maintained by a contractor or for high priced or unusually complicated specialty work. The programs are generally inexpensive, but since they are specialized, they must be purchased specifically for estimating. They are easy to learn but require some basic familiarity with computers.

Sitemate 45™ allows contractors to perform takeoffs and compile bids at the job site, plan room, or at home. *The Plumber's Helper*™ is a master list of most-used parts, arranged in categories such as rough plumbing, finish plumbing, warm air systems, labor hours, etc. Each category includes supplier discount information and user's markup and runs on any IBM compatible PC or Apple Macintosh. *Super-Duct*™ and *Super-Pipe*™ estimating systems produce complete sheet metal and piping estimates. Programs include complete labor and price tables already loaded and ready-to-use, automatic duplications, etc.

Exccomate's enhanced mechanical computer estimating program packages now allow piping/DWV estimates to create families of subassemblies (creating similar assemblies by changing sizes and material types without re-entering material list). Both piping and sheet metal system subtotals may be transmitted to spreadsheets, word processors, and data base management programs automatically. *The Bid Team*™ helps the user design the best, most effective method of performing a takeoff, creating step-by-step procedures based on the user's own methods and terminology.

Another interesting package is Trade Service Corp.'s network software. This system enables several workstations to share a database consisting of prices on more than 85,000 plumbing/HVAC products, updated weekly on diskette. Other features include "custom pricing" columns, vendor comparisons, indexing modes, and standard interface with other programs.

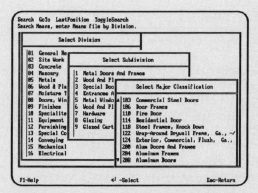

Divisional Breakdown of Database Information

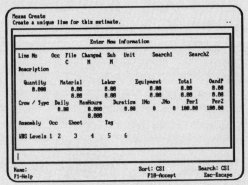

Line Item Selection from Database

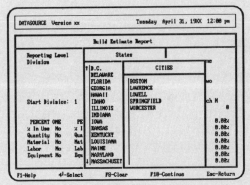

Information Available for Each Line Item

Cost Index for Specific Localities

Figure 12.2

86

Another Means system that is more flexible than the *Data Source®* software, meant for small to medium-sized firms, is *Astro* II. This software uses *Data Source®* but provides more features, more data, and more flexibility.

Input Device Applications

There are several programs currently on the market which take the library-based applications one step further by allowing direct input from a drawing to the computer. They usually use a digitizer (described above) to trace drawings directly into the computer. If a building perimeter is traced, wall length, building area, and volume are computed automatically. The system will look up costs, man-hours, and crew size information from a library database and produce numerous reports.

This type of system eliminates the double work of taking off quantities manually and then typing them into the computer. The takeoff is faster, but care should be exercised since a mistake can throw off the estimate just as easily as in the manual method and might be harder to find. These systems are highly specialized, and the program itself might cost several times more than the computer hardware it runs on. They also require that time and effort be spent in learning all the features and usually take much longer than the previous two types of applications.

Other features commonly found on these more sophisticated programs are the ability to convert the data generated by the takeoff into a format that allows "what-if" analyses as do spreadsheets. Programs exist which allow the cost and man-hour data to be used to prepare realistic cost-loaded construction schedules, budget-tracking, and payment loading throughout a project.

Quick Pen International's *Pegasus™* is a lightweight estimating system that can be carried to the job site, home, or office. It can be set up in minutes to prepare final estimates. This system uses a stylus device for fast, convenient takeoffs of fixtures, equipment, and assemblies. *Super-Duct™* and *Super-Pipe™* (mentioned above) can use digitizer boards and pens, or a compatible keyboard for fast, easy takeoff. GTCO Corp.'s *Super L Series Digitizer™* combined with a computer provides accurate length, area, volume, and count measurements. The user positions plans on the digitizing tablet with the aid of active area marks, then performs takeoffs by touching the corners or tracing items to be measure with an electronic stylus.

While digitizer systems are available in many sizes, the decision to purchase a digitizer board should be carefully evaluated. The input can be less accurate than desired, and the boards are costly. This type of system is not a cost-effective option for the small mechanical firm.

Direct Input from CAD

No discussion of computer-aided estimating would be complete without mention of programs which allow direct input from computer-aided design packages (CAD). As long as drawings are still generated by manual draftsmen, the information must be manually taken off or entered into the computer with a digitizer. When the drawing is done on a computer screen, however, the takeoff can proceed automatically and simultaneously with the design. Once quantified, the program proceeds in a similar manner to the above described applications (using database cost information, generating reports, producing schedules, etc.).

These applications are highly specialized, are the most expensive to purchase and setup, and require the user to be somewhat familiar with CAD systems as well as estimating. They are not yet a common sight in construction offices but are definitely the wave of the future.

Summary

The first step in determining the correct hardware and software choices is to define how a firm's estimates fit into the overall project management and business management systems:

- Will you use the estimate to control the project?
- Will your job accounting system be based on your estimates as well?
- How can you interrelate these functions?

There are integrated software packages available for either general or specific applications. Software and hardware in today's market are highly efficient and useful–if you are willing to adapt your current system. In most cases, this is worthwhile but a sufficient training period and learning curve must be anticipated.

Of course, a well organized accounting, estimating, and management system should be in place. All too often, a computer and software are brought in to cure problems that demand a more basic solution. To make the most appropriate computer decisions for any firm, it is best to consult an independent agent (not a salesman) through consulting, seminars, or publications on choosing computer systems. While the cost of such agents adds to the budget, it can save thousands in the long run. With rapid technological advances developing almost daily, it is nearly impossible to make a decision without an informed guide. Deciding which of these options to pursue will depend largely upon a firm's specific needs and goals. The choice should involve a good deal of research.

One final consideration is maintaining state-of-the-art equipment: New technological advances render existing systems obsolete almost as soon as they are purchased. Some firms deal with this issue by budgeting for purchase of new or upgraded software (and even hardware) every two to three years. All software and hardware options should be reviewed for expandability. In any case most mechanical firms can expect to spend a certain portion of revenue annually on computer products in the years to come.

Part II
COMPONENTS OF MECHANICAL SYSTEMS

Chapter 13
PIPING

Piping is the common thread that ties plumbing, fire protection, heating, and cooling together into the category known as "the mechanical trades." This chapter contains descriptions of different types of pipe, valves, fittings, and other appurtenances which make up the piping systems assembled by the mechanical trades. The appropriate units of measure, step-by-step takeoff procedures for recording materials, appropriate labor units, and any pertinent cost modifications are indicated for each component of the piping system.

The fluid conveyed, location used, and the pressure contained usually determine what piping materials and joining methods are required. Steel pipe, for example, is often used for handling steam, chilled water, hot water, compressed air, gas, fuel oil, storm drains, and sprinkler systems. This chapter describes the cost estimating procedure for each type of pipe, (e.g., steel pipe) rather than repeat the pricing procedure for every function of that particular pipe.

Fittings are used in piping systems to change direction (elbows, bends, tees, and crosses), to divert or merge (wyes, tees, and laterals), to connect (couplings, unions, flanges, hubs, and sleeves), to terminate (plugs, caps, and blind flanges), to change from one joining method or material to another (adapters and transitions) or to change size (reducers, increasers, and bushings).

Nipples are short pieces of threaded pipe with a maximum length of 12", usually included in the overall pipe totals as increased footage, rather than being taken off and priced separately. An estimator may prefer to take them off individually; however, this can be a time-consuming procedure.

Valves in piping systems control the flow of the conveyed fluid or regulate its pressure. Valves are produced in a wide variety of materials, end connections, and configurations, to be compatible with the rest of the piping system. Valves are manufactured for both manual and automatic (motorized) operation.

Check valves prevent reversal of flow. *Gate, globe, ball,* and *butterfly valves* and *plug cocks* stop or start the flow. A gate valve should be either fully open or fully closed. Globe, ball, and butterfly valves and plug cocks are also designed to throttle the flow. Specially constructed *reducing* and *relief valves* are available to regulate the pressure within a piping system.

Supports, anchors, and *guides* support and direct the expansion and contraction of pipe lines subject to the temperature changes of the conveyed fluid. The type of material and hardware used to support a piping system can significantly affect the installed cost. As a general rule, supports (or hangers) are made of the same material as the piping system.

Steel supports are used with ferrous piping; copper or copper-clad supports are used with nonferrous piping. Plastics are an exception because of the many forms of plastic available and the lack of a complete range of plastic hangers. Another exception is glass pipe for acid waste where metal or plastic supports are used.

Anchors and guides are usually indicated on the plans. Pipe hangers are only described in the specifications. Spacing distance is determined by the piping material and size as well as the fluid conveyed, and in the case of plastic pipe, the ambient temperature.

Steel Pipe

Steel pipe is often referred to as "black steel," "black iron," "carbon steel," "full weight," "merchant pipe," or "standard pipe" in the construction industry. Although such terms are regional in origin and usage, they refer to the same product.

Steel pipe usually comes from the mill with a black external coating for protection during shipment and storage. Galvanized steel pipe is shipped with a zinc coating on both the interior and exterior surfaces for protection against corrosion. Galvanized pipe is used for critical duties, such as drainage, waste, vent, water service, and cooling towers (condenser water), where the pipe will be exposed to the atmosphere.

There are two forms of steel pipe, welded and seamless, according to American Society of Testing and Materials (ASTM) specifications. Three ASTM designations will be discussed here; A106, A53, and A120. Seamless pipe is primarily made to A106 and A53 specifications; welded pipe is made to both the A53 and AI20 specifications. The ASTM designation and manufacturer's data will be stenciled along the exterior wall of the pipe, or in smaller diameters, on the bundle tag.

A120 pipe is traditionally for noncritical* use, when the pipe is not intended to be bent or coiled. Pipe made to the A53 specification is also used in the above-mentioned situations, but this type of pipe can be bent or coiled. In recent years, dual stenciled or dual certified pipe has been made available at the same low cost of the A120, but with all the requisites for A53 up through 4″ diameter. Seamless A106 pipe is used for higher pressures and temperatures, with the advantage of being flexible. A106 pipe is available in plain end only (must be field threaded).

A120 pipe is now being withdrawn from the marketplace. All applications for which it was specified now require A53. Specification ASTM 120 is no longer being published.

*Note: Noncritical use of pipe referred to in this book means low pressure gas, steam, or water systems normally found in plumbing, heating, cooling, and fire protection use.

American Standards Association (ASA) schedule numbers classify wall thicknesses of steel pipe up through 24″ O.D. (outside diameter). In diameters 1/8″ through 10″, schedule 40 is synonymous with standard weight, and schedule 80 with extra heavy (XH), or extra strong (XS) through 8″. For larger sizes, the wall thickness varies depending on the anticipated pressures, temperatures, or stress conditions.

Lighter wall thickness pipe is used in noncritical plumbing, heating, cooling, or fire protection applications. With the advent of roll grooving, and to compete with plastic and thin wall copper piping, light wall steel pipe (e.g., schedule 10) has come into accepted use, particularly in fire protection systems.

Steel pipe, regardless of wall thickness, retains its designated outside diameter. For example, 3″ pipe, regardless of wall thickness or schedule, is always 3-1/2″ in diameter, as shown in Figure 13.1. This is necessary for threading, socket welding, and other fitting and mating purposes. The inside diameter decreases or increases depending on the wall thickness specified.

Steel pipe is produced in three lengths: a *uniform* length of 21'; *random* length, which varies from 16' to 22'; or cut specifically to the customer's required length. Double lengths are available by special order only.

Steel piping is shipped in bundles for sizes 1-1/2" and smaller. Figure 13.2 is a bundling schedule for 21' uniform lengths. Piping is also available from local suppliers in individual lengths.

The pipe end finishes are determined by the proposed joining or coupling method. End finishes can be plain (or bald) end, bevelled end for welding, threaded and coupled (T&C), cut grooved, or roll grooved. Cut or roll grooved ends are used with grip type bolted or hinged fittings. Piping larger than 12" is threaded on special order only.

Units of Measure: Steel piping is measured in linear feet (L.F.). Associated fittings, supports, and joints, are recorded as each.

Labor Units: The following procedures are generally included in the per foot labor cost of piping:

- Unloading and relocating from receiving or storage area to the point of installation
- Setup of scaffolding, staging, ladders, or other work platforms
- Layout of piping route

These additional procedures are also recorded and included in the piping estimate:

- Joining of straight lengths every 20'
- Installation of fittings, valves, and appurtenances
- Installation of supports, anchors, and guides
- Testing and cleaning individual systems

Takeoff Procedure: A takeoff sheet should be set up for each piping system to record the various footages required for each diameter of pipe (shown in Figure 13.9). To figure the footages, the piping runs should be scaled or otherwise measured, beginning at the source of each system. By starting at the source of each system, the larger, more costly pipe will be taken off first. This does not hold true, however, when taking off condensate return, drain, or waste piping, where the larger pipe is installed at the termination point.

Each system should be marked with a colored pencil as it is taken off. A color code should be developed, to distinguish between piping materials or piping systems, which can be standardized for use throughout the company or throughout the estimating department. Horizontal piping runs should be recorded and measured first, then vertical risers or drops. The total should be rounded off to coincide with the average lengths available from local suppliers (e.g., 20') for steel pipe.

Height Modifications: Additional labor costs are incurred when piping is installed over 10' above the work area. For anticipated work over 10', add the percentages shown in Figure 13.3 to compensate for the difference between actual cost and usual labor cost figures.

Screwed Fittings and Flanges for Steel Pipe

Fittings for use with steel pipe come in a variety of shapes and materials to accommodate the end finish of the piping and the fluid conveyed. Fittings are available for any end finish found on steel pipe, including threaded, welded, flanged, grooved, or plain end. Figures 13.4 and 13.5 show different fittings used with steel pipe.

Weights and Dimensions of Carbon Steel Pipe—Seamless and Welded

A.S.A. Pipe Schedule

Pipe Size	O.D. in Inches	5	10	20	30	40	Std.	60	80	XS	100	120	140	160	X.X.S.
1/8	.405	.035 .1383	.049 .1863			.068 .2447	.068 .2447		.095 .3145	.095 .3145					
1/4	.540	.049 .2570	.065 .3297			.088 .4248	.088 .4248		.119 .5351	.119 .5351					
3/8	.675	.049 .3276	.065 .4235			.091 .5676	.091 .5676		.126 .7388	.126 .7388					
1/2	.840	.065 .5383	.083 .6710			.109 .8510	.109 .8510		.147 1.088	.147 1.088				.187 1.304	.294 1.714
3/4	1.050	.065 .6838	.083 .8572			.113 1.131	.113 1.131		.154 1.474	.154 1.474				.218 1.937	.308 2.441
1	1.315	.065 .8678	.109 1.404			.133 1.679	.133 1.679		.179 2.172	.179 2.172				.250 2.844	.358 3.659
1¼	1.660	.065 1.107	.109 1.806			.140 2.273	.140 2.273		.191 2.997	.191 2.997				.250 3.765	.382 5.214
1½	1.900	.065 1.274	.109 2.085			.145 2.718	.145 2.718		.200 3.631	.200 3.631				.281 4.859	.400 6.408
2	2.375	.065 1.604	.109 2.638			.154 3.653	.154 3.653		.218 5.022	.218 5.022				.343 7.444	.436 9.029
2½	2.875	.083 2.475	.120 3.531			.203 5.793	.203 5.793		.276 7.661	.276 7.661				.375 10.01	.552 13.70
3	3.5	.083 3.029	.120 4.332			.216 7.576	.216 7.576		.300 10.25	.300 10.25				.437 14.32	.600 18.58
3½	4.0	.083 3.472	.120 4.973			.226 9.109	.226 9.109		.318 12.51	.318 12.51					.636 22.85
4	4.5	.083 3.915	.120 5.613			.237 10.79	.237 10.79	.281 12.66	.337 14.98	.337 14.98		.437 19.01		.531 22.51	.674 27.54
5	5.563	.109 6.349	.134 7.770			.258 14.62	.258 14.62		.375 20.78	.375 20.78		.500 27.04		.625 32.96	.750 38.55
6	6.625	.109 7.585	.134 9.289			.280 18.97	.280 18.97		.432 28.57	.432 28.57		.562 36.39		.718 45.30	.864 53.16
8	8.625	.109 9.914	.148 13.40	.250 22.36	.277 24.70	.322 28.55	.322 28.55	.406 35.64	.500 43.39	.500 43.39	.593 50.87	.718 60.93	.812 67.76	.906 74.69	.875 72.42
10	10.75	.134 15.19	.165 18.70	.250 28.04	.307 34.24	.365 40.48	.365 40.48	.500 54.74	.593 64.33	.500 54.74	.718 76.93	.843 89.20	1.000 104.1	1.125 115.7	
12	12.75	.165 22.18	.180 24.20	.250 33.38	.330 43.77	.406 53.53	.375 49.56	.562 73.16	.688 88.51	.500 65.42	.844 107.2	1.000 125.5	1.125 139.7	1.312 160.3	
14	14.0		.250 36.71	.312 45.68	.375 54.57	.437 63.37	.375 54.57	.593 84.91	.750 106.1	.500 72.09	.937 130.7	1.093 150.7	1.250 170.2	1.406 189.1	
16	16.0		.250 42.05	.312 52.36	.375 62.58	.500 82.77	.375 62.58	.656 107.5	.843 136.5	.500 82.77	1.031 164.8	1.218 192.3	1.437 223.5	1.593 245.1	
18	18.0		.250 47.39	.312 59.03	.437 82.06	.562 104.8	.375 70.59	.750 138.2	.937 170.8	.500 93.45	1.156 208.0	1.375 244.1	1.562 274.2	1.781 308.5	
20	20.0		.250 52.73	.375 78.60	.500 104.1	.593 122.9	.375 78.60	.812 166.4	1.031 208.9	.500 104.1	1.280 256.1	1.500 296.4	1.750 341.1	1.968 379.0	
24	24.0		.250 63.41	.375 94.62	.562 140.8	.687 171.2	.375 94.62	.968 238.1	1.218 296.4	.500 125.5	1.531 367.4	1.812 429.4	2.062 483.1	2.343 541.9	

Top number is wall thickness. Bottom number is pounds per foot.

Figure 13.1

Cast-iron screwed fittings and flanges are produced in standard black (125 lb.) or extra heavy (250 lb.) pressure classes and in galvanized finish for corrosive atmospheres. Cast iron (also known as "gray iron") is nonductile, therefore it is not used in systems subject to excessive shock or hammering. (Malleable fittings are used for these more critical conditions.) The nonductability of cast iron, however, becomes an asset when dismantling or replacing a fitting that is "frozen" or "cooked" in place by prolonged heat. In this case, the fitting will shatter after several hammer blows if the joint does not submit to a pipe wrench.

Special drainage or pitched fittings are designed to give an unobstructed flow by having an inside diameter (I.D.) the same as the I.D. of the pipe. These fittings are used for plumbing drains, vents, and vacuum cleaning systems. Tees and elbows have pitched threads to assure a 1/4" per foot pitch of the drain line.

Bundling Schedule			
Number of Pieces — Average Feet and Weight of Steel Pipe Per Bundle			
Size (Inches)	Pieces per Bundle	Average per Bundle	
		Feet	Weight (Lbs.)
Standard Weight Pipe			
1/8	30	630	151
1/4	24	504	212
3/8	18	378	215
1/2	12	252	214
3/4	7	147	166
1	5	105	176
1¼	3	63	144
1½	3	63	172
Extra Strong Pipe			
1/8	30	630	195
1/4	24	504	272
3/8	18	378	280
1/2	12	252	275
3/4	7	147	216
1	5	105	228
1¼	3	63	189
1½	3	63	229

Figure 13.2

These fittings do not have a pressure rating and are normally shipped with a coating of black enamel. Plain or galvanized coating is also available by special order.

Some local plumbing code requirements call for drainage fittings with combinations of bell and spigot (soil pipe fittings), and threaded ports. Combination fittings are not stock items, and are available by special order only.

Malleable iron screwed fittings are used in gas piping, smaller sizes of sprinkler piping, fuel oil piping, gasoline, and any system subject to shock or physical abuse. Malleable iron fittings are produced in standard (150 lb.) and extra heavy (300 lb.) pressure classes, except for unions, which come in classes of 150, 200, 250, and 300 lb. These fittings are available in most of the same sizes and configurations as cast-iron fittings, except the unions, which are exclusively malleable iron.

Malleable iron screwed fittings are similar in appearance to cast-iron fittings, except for the bead on the fitting ends. Cast-iron fittings have a heavy bead and are of a closed pattern. Malleable fittings have a narrow bead and an open pattern. The 300 lb. malleable iron fitting is the exception, bearing the pattern usually found on cast-iron fittings (a heavy bead and a closed pattern). Malleable iron fittings are available with a black or galvanized finish.

Flanged cast-iron fittings, *threaded cast-iron flanges*, and *flanged unions* are used with threaded steel pipe. Flanged fittings are most often used in large diameter piping systems, although they are made in smaller sizes, from 1-1/2" diameter and larger. These fittings are used to mate with flanged valves, pumps, and other mechanical equipment provided with flanged ports. The flanges and fittings have 125 lb. or 250 lb. ratings. A 250 lb. flange has a larger O.D. than a 125 lb. flange, and also has a larger bolt circle, requiring more bolts to make up the flanged joint.

Height Modifications	
10-15 feet	add 10 percent
15-20 feet	add 20 percent
20-25 feet	add 25 percent
25-30 feet	add 35 percent
30-35 feet	add 40 percent
35-40 feet	add 50 percent
over 40	add 55 percent

Figure 13.3

Threaded and Coupled Steel Pipe
(showing merchants coupling)

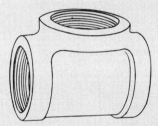

Union–Malleable Iron

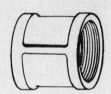

Coupling–Malleable Iron

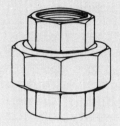

Tee-Cast Iron
(Note: closed pattern and heavy bead)

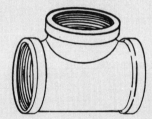

Tee-Malleable Iron
(Note: open pattern and narrow bead)

45° Elbow - Malleable Iron

Cast Iron and Malleable Fittings for T & C Steel Pipe

Figure 13.4

Flanging is frequently used for offsite prefabricated piping (e.g., fire protection sprinkler systems). Flanged elbows are available in standard or long radius configurations. The advantage of flanged fittings is the comparative ease with which a flanged joint may be assembled or disassembled.

Ductile iron fittings are not usually encountered in building construction. Both 300 lb. screwed and 150 lb. flanged are available in limited sizes for use in more critical systems.

Units of Measure: All fittings are taken off and recorded as each. The number of joints will be determined by this fitting count.

Material Units: In addition to the cost of the fitting, material is needed to prepare and make up each joint. Threaded joints require cutting oil, cotton waste or wipes, pipe dope (pipe joint compound), or a suitable tape or paste lubricant or sealant. Joints in screwed piping usually require a certain quantity of pipe nipples. These are not shown on the drawings, but must be included in the material and labor costs.

Flanged joints require bolts, nuts, and gaskets. Threaded steel flanges can be substituted for cast-iron flanges, but the cost of a steel flange is considerably

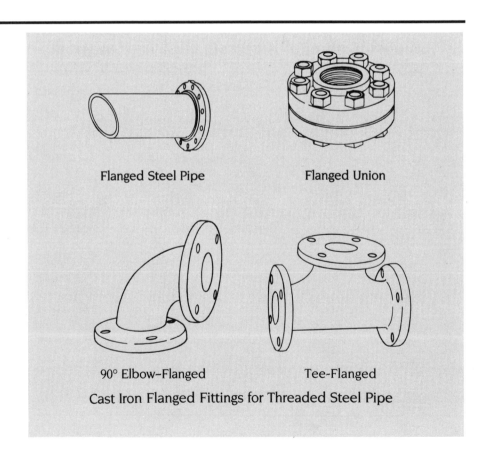

Flanged Steel Pipe   **Flanged Union**

90° Elbow–Flanged   **Tee-Flanged**

Cast Iron Flanged Fittings for Threaded Steel Pipe

Figure 13.5

higher than cast iron. In some flanged installations, it is possible to estimate the number of companion flanges or sets of bolts, nuts, and gaskets (B.N.&G.) by doubling the count of flanged valves or strainers.

Each assembly should be checked carefully to determine the exact number of flanges or joints required. An example is shown in Figure 13.6. The sketch indicates two butterfly valves, one check valve, two flexible connections, and two flanged reducers. If the flanged items are doubled, it would seem that 14 flanges and 14 sets of B.N.&G. are required. A closer look, however, reveals that eight companion flanges and 13 sets of B.N.&G. would be sufficient to button up this assembly. If the discharge side butterfly valve and the check valve were buttoned together, two more flanges, two joints, and one set of BN&G could be eliminated.

Labor Units: The labor units for fittings usually involve making up a tight joint.

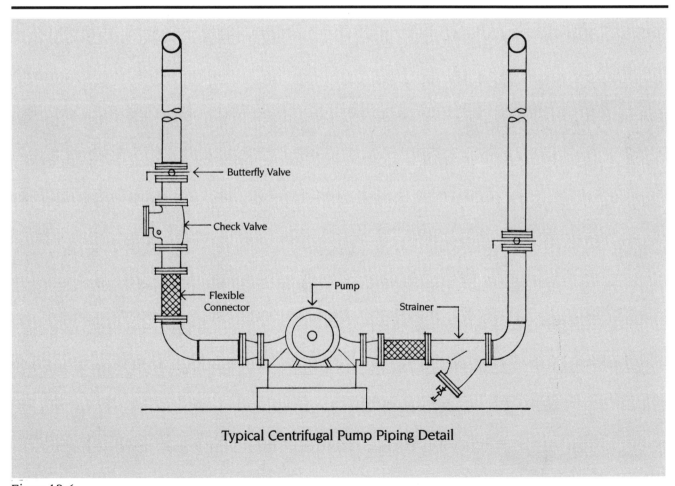

Typical Centrifugal Pump Piping Detail

Figure 13.6

For larger fittings, weight becomes an added factor in handling, installing, and supporting. The field or shop cutting and threading must also be recorded.

When estimating by the joint method, each fitting, valve, or other piping accessory is counted and recorded by pipe size, material, and the total number of joints per fitting; each elbow having two joints, tees having three, crosses having four, etc.

Rather than take off all of the screwed fittings and nipples, some estimators have an allowance built into the piping labor factor based on historical labor data for smaller sizes of T&C piping (2" and less). In this case, an allowance based on a certain percentage of the T&C pipe material cost is added for the screwed fittings and nipple material costs. Until an estimator's database includes information on past labor and fitting costs, the time-consuming individual takeoff and pricing procedure must be followed.

Takeoff Procedure: The fittings may be recorded below the accumulated footage on the takeoff sheet for steel piping systems. If the job is of significant size, separate takeoff sheets for fittings are necessary, and should be attached to the pipe takeoff sheets for each system. The same procedure as that used for pipe should be followed, treating larger sizes first, using the color code to indicate that each fitting and intermediate joint has been recorded. Intermediate joints are couplings connecting straight sections of pipe. In this case, threaded and coupled (T&C) pipe is supplied with a steel coupling attached to each length. Therefore, only the labor to make up both joints of the coupling need be recorded. For gas piping, the wrought steel or merchants coupling is removed and replaced with a malleable iron coupling. In a flanged piping system, a flanged union or a pair of companion flanges couple the straight lengths of pipe and still allow the capability for future dismantling.

Steel Weld Fittings and Flanges

Steel welding fittings (shown in Figure 13.7) are produced in standard weight and extra heavy weight for normal piping systems encountered in the building construction industry. Welding fittings are manufactured in higher ratings for industrial or process use. A lightweight fitting is also available by special order that is compatible with Schedule 10 pipe wall thickness. This type of fitting is most often used for gas distribution piping.

The standard weight line is used for general purposes. Most elbows are of the long radius pattern, where the center-to-face dimension is 1-1/2 times the nominal pipe diameter.

Steel welding flanges (shown in Figure 13.7) are produced in standard weight (150 lb.) and extra heavy (300 lb.), as well as the higher ratings for power and process service. The 150 lb. weld neck flanges are most commonly used. Despite the differences in pressure ratings, the 150 lb. series weld flanges mate with 125 lb. cast-iron or steel threaded flanges and the 300 lb. series are compatible with the 250 lb. flanges, because they have the same bore, O.D., and bolt circle.

Units of Measure: All fittings are taken off and recorded as each. The number of joints and the inches of welding will be determined by this fitting count plus the number of intermediate welds required.

Material Units: Additional material is necessary to prepare and make up each joint. This cost must be estimated in addition to the actual purchase cost of a fitting or flange. Welded joints require welding rod, oxygen, and acetylene for gas welding. For arc welding, electrodes (coated rod) and a power source are needed. These are the two most commonly used welding procedures for mild carbon steel pipe and fittings in the field. Other materials to include in the welding estimate are backing rings (chill rings) if specified. These are counted per joint. Chill rings allow for uniform spacing of the two pipe ends to be welded, resulting in a smooth interior surface and eliminating the necessity of tack welding. Backing rings are not normally specified for noncritical piping used in mechanical systems.

An experienced estimator has learned that there are some welding fittings never used unless space conditions mandate: reducing weld elbows and welding tees which reduce on the run.

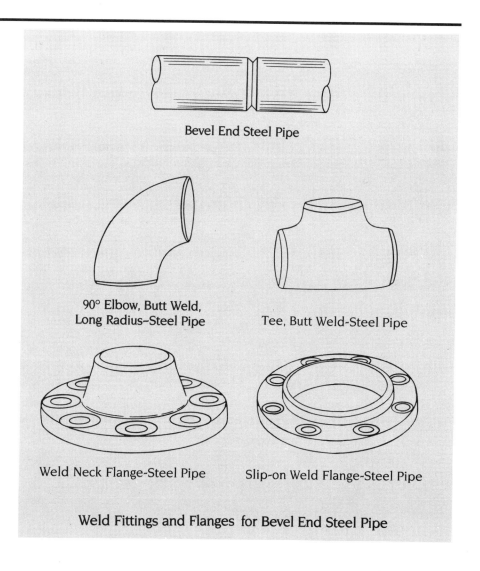

Bevel End Steel Pipe

90° Elbow, Butt Weld,
Long Radius–Steel Pipe

Tee, Butt Weld-Steel Pipe

Weld Neck Flange-Steel Pipe

Slip-on Weld Flange-Steel Pipe

Weld Fittings and Flanges for Bevel End Steel Pipe

Figure 13.7

Reducing weld elbows are not usually stock items because they are prohibitively priced. It is cheaper, both material-wise and labor-wise, to use a straight size elbow and a concentric weld reducer if space permits.

Welding tees reducing on the run are also prohibitively priced and should be used with straight size on the run plus a reducer, either concentric or eccentric, depending on the fluid being piped and the position of the fitting. When a reducer is used in vertical piping, it should be concentric, which is less expensive than eccentric. For horizontal piping, an eccentric reducer is necessary. For steam piping, the eccentric reducer is installed with the flat on the bottom to maintain a flat surface for the condensate flow. For forced hot and cold water and condensate return, the flat is installed at the top to prevent air pockets from forming.

Welding tees reducing more than two sizes on the outlet are extremely expensive and sometimes not available at all. For example, a 10" main requiring a 3/4" outlet would require a 10" x 4" tee plus a 4" x 1-1/2" reducer plus a 1-1/2" by 3/4" reducer. This would result in an outlet 6" long and 26-1/4 diameter inches of welding. A preferable alternative would be to use outlet fittings which are made to be welded directly into an opening cut in the pipe itself. Some trade names for those fittings are weldolets, threadolets, and sockolets, depending on the type of pipe connection required at the outlet. A weldolet for the above example would require two diameter inches of welding. Field expedients for these outlets are often nipples cut in half for a male outlet, or any of the surplus wrought steel merchant pipe couplings that accumulate on a job or in a shop for a female outlet. These field expedients should be used only for noncritical piping systems. The contract specifications should be checked, as they may prohibit the use of these field expedients.

Welding material costs are estimated by daily consumption; the pounds of wire or rod per day and the number of cubic feet of oxygen and acetylene, or gallons of fuel for engine driven generators. The material costs for gas or arc welding should usually be the same or close enough for estimating purposes.

Labor Units: The labor for weld fittings entails more than just the actual time required to gas or arc weld the two pieces together. Joint preparation, cutting, clamping, bevelling, chipping, cleaning, and measuring are all entailed. Daily output of a welder is generally accepted as the best way to estimate the welding labor for a piping system. The welder's output is determined by the diameter inches of welding accomplished per day. The term "diameter inches" refers to the nominal diameter of the pipe and considers the number of passes per joint as well as the circumference. These two factors determine the labor per joint. For example, a 4" diameter pipe weld is four diameter inches. For branch welds, the diameter inches are recorded and doubled to allow for the cutting or burning of the hole. For example, a 1-1/2" diameter branch would be recorded as three diameter inches of weld.

The labor for weld flanges would include aligning, welding, and bolting up each joint. Bolting requirements are shown in Figure 13.8. Slip-on flanges require two welds, an internal and an external. In low pressure heating and cooling applications it is not uncommon to tack weld the hub and complete the normal interior weld. This practice, although not recommended, may save a considerable amount of time.

The daily output of a welder varies from mechanic to mechanic. Thirty diameter inches per day for a welder and a layout man working together is considered a good overall average for an eight-hour day (16 man-hours). Do not anticipate that a welder will spend eight hours under the hood. Preparation and positioning of the assembly to be welded is time-consuming. A piping fit, prefabricated on the bench or floor and hoisted into place, increases daily output.

Takeoff Procedure: The fittings, flanges, nozzles, branch welds, and butt welds needed for intermediate joints should be recorded by size on the takeoff sheet for steel piping (shown in Figure 13.9). From this fitting list, the diameter inches are determined for the total welding labor and material cost.

150 Lb. Steel Flanges				
Pipe Size	Diam. of Bolt Circle	Diam. of of Bolts	No. of Bolts	Bolt Length
1/2	2⅜	1/2	4	1¾
3/4	2¾	1/2	4	2
1	3⅛	1/2	4	2
1¼	3½	1/2	4	2¼
1½	3⅞	1/2	4	2¼
2	4¾	5/8	4	2¾
2½	5½	5/8	4	3
3	6	5/8	4	3
3½	7	5/8	8	3
4	7½	5/8	8	3
5	8½	3/4	8	3¼
6	9½	3/4	8	3¼
8	11¾	3/4	8	3½
10	14¼	7/8	12	3¾
12	17	7/8	12	4
14	18¾	1	12	4¼
16	21¼	1	16	4½
18	22¾	1⅛	16	4¾
20	25	1⅛	20	5¼
22	27¼	1¼	20	5½
24	29½	1¼	20	5¾
26	31¾	1¼	24	6
30	36	1¼	28	6¼
34	40½	1½	32	7
36	42¾	1½	32	7
42	49½	1½	36	7½

Steel, cast iron, bronze, stainless, etc. Flanges have identical bolting requirements per pipe size and pressure rating.

Figure 13.8

Grip Type Mechanical Joint Fittings

The grip type of fitting is used in the fire protection field, building service systems (i.e., storm and sanitary drains), potable, heating, cooling, and condenser water systems. (Not available for steam or gas piping.)

Grip type fittings (shown in Figure 13.10) are designed for use with plain end, grooved, roll grooved, or bevel end steel pipe, and are available with plain or galvanized finish. This kind of fitting is produced for use with lighter schedules of piping, as well as with standard weight (Schedule 40). The wide variety of sizes and configurations of grip fittings allows a complete system to be assembled using only these fittings.

Special tools are required for field or shop cutting, rolling of grooves, and drilling outlets in the pipe wall where "hook type" or other mechanically joined outlets are substituted for tees or welded nozzles. In 1990 a new development in the grip type fitting was introduced for copper tubing to eliminate the need for solder or brazed joints.

Units of Measure: All fittings are taken off and recorded as each. The number of joints, grooves, or holes to be cut is determined by this fitting count.

Material Units: Little else is required, other than the special tools previously mentioned, to prepare the grip type joint. The only exception is adapter flanges (available in standard or extra heavy); the installer should furnish the normal quantity of nuts and bolts, but the gasket is supplied with the adapter flange.

Labor Units: The labor units for grip type fittings include pipe end preparation. Even plain end pipe requires cleaning and deburring. The fittings are then mechanically joined to the pipe using factory supplied bolts, pins, or quick disconnect lever handles. A system based on diameter inches, similar to welding, is often used with a claimed productivity of 90" to 100" per day.

Takeoff Procedure: Grip joint systems are recorded by size for each fitting or outlet on the takeoff sheet for steel piping. Straight lengths of pipe are joined by one coupling, but each pipe end must be prepared per joint. Intermediate joints for long runs of straight pipe are recorded separately because the grooving labor is not required (assuming that the pipe is purchased already grooved). Therefore, only the labor to bolt on the coupling is included in the estimate.

Cast-Iron Pipe

Cast-iron pipe, also known as gray iron pipe, is used in soil pipe systems for sanitary waste, vent systems, and storm drain piping systems. Cast-iron soil pipe is manufactured in both service weight and extra heavy, or full weight. Extra heavy soil pipe is generally used in underground installations. A corrosion-resistant high silicon content soil pipe is available from certain manufacturers for use in acid type waste systems and corrosive atmospheres.

Cast-iron pipe is available with bell and spigot joints, which are made up or poured with hot head, or by push-on type gasketed compression joints. Plain end, or no-hub cast-iron soil pipe is joined by applying bolted clamp type couplings and neoprene gaskets. The clamp type couplings are produced in rigid cast-iron or flexible stainless steel bands, depending on use and local acceptance.

Soil pipe is produced in both 5' and 10' lengths, in diameters of 2", 3", 4", 5", 6", 8", 10", 12", and 15". The 5' lengths are manufactured in double or single hub (hub being another term for the bell shaped end).

Units of Measure: Cast-iron pipe is measured in linear feet (L.F.). Associated fittings, supports, joints, and other appurtenances are recorded as each.

Means® Forms

PIPING SCHEDULE

PAGE 1 of 1

JOB Office Building HVAC

DATE 3/17/91

SYSTEM Dual Temp. - Hot/Chill Water - 2½" + up welded 2" + down T+C

BY JJM

	PIPE DIAMETER IN INCHES																	
			12	10	8	6	5	4	3 1/2	3	2 1/2	2	1 1/2	1 1/4	1	3/4	1/2	
Black Steel								26 50 23 74		14 14 (28)	14 14 (28)	28 180 (208)			20 (20)			Air Vents 卌 III
								33 29 (235)										
Ells								卌 卌 卌 卌 卌 I (26)										Thermometers 卌 卌 III
Tees								卌 卌 卌 II (17)		II (2)	II (2)							
Reducers								IIII (4)		II (2)	II (2)							Gauges 卌 III
Gate Valves								卌 III (8)					卌 卌 II (12)					
Check Valves								IIII (4)										Drain Valves 卌 卌 卌 卌 I
Balancing Cocks								IIII (4)										
Strainers								IIII (4)										
Flexible Connectors								卌 III (8)										C.I. Flanged Elbows 4 - 4"
Butt Welds								107 10	卌 I	卌 卌	II							
Weld Flanges								10										
Branch Welds																卌 卌		
Bolts, nuts Gaskets								40										

Figure 13.9

Labor Units: The following procedures are generally included in the per foot labor cost:

- Unloading and relocating from receiving or storage area to point of installation
- Setup of scaffolding, staging, ladders, or other work platform
- Layout of piping route

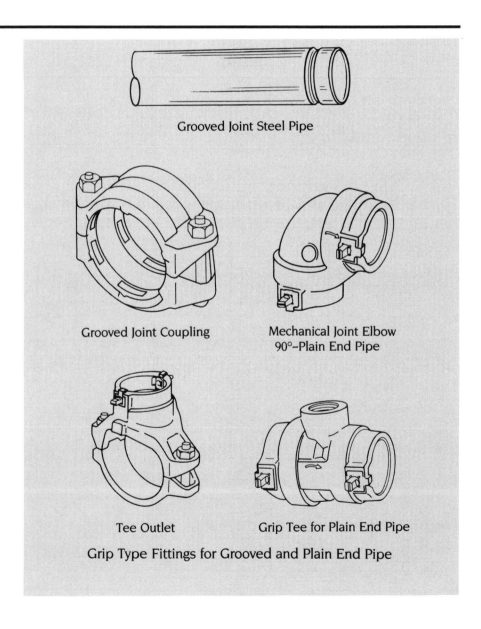

Grooved Joint Steel Pipe

Grooved Joint Coupling

Mechanical Joint Elbow
90°–Plain End Pipe

Tee Outlet

Grip Tee for Plain End Pipe

Grip Type Fittings for Grooved and Plain End Pipe

Figure 13.10

These additional procedures should also be recorded and included in the piping estimate:

- Joining of straight lengths every 5'-10'
- Installation of fittings, valves, supports, and anchors
- Leak and flow testing of the assembled system

Takeoff Procedure: A takeoff sheet should be set up for each piping system and footages recorded for each diameter of pipe, beginning with the waste system at its termination point. This will be the largest diameter pipe found in the system. If a portion of the system is to be tarred or coated, it should be taken off separately. Underground pipe should also be recorded separately, as its joining, supporting, and laying will be estimated differently.

Each system should be marked on the plans in a different color pencil. The pipe totals should be rounded off to coincide with 5' or 10' lengths available for each category, type, and size of pipe.

Cast-Iron Soil Pipe Fittings

Fittings for cast-iron soil pipe, because of their designated use in drainage and venting, have many more configurations (shown in Figure 13.11) than do other types of fittings. A manufacturer's catalog showing all the available fittings and patterns is a must for an estimator's reference library. Soil pipe fittings used for a change in direction are called bends rather than elbows. Other piping systems have 90°and 45° elbows whereas drain waste and vent systems have quarter, eighth, fifth, sixteenth, and, for special use, half and sixth bends. Quarter bends are also available in long turn and short turn patterns. Tees and wyes are used in these systems, plus variations peculiar to DWV use, such as sanitary tees for vertical use. The traditional pouring of lead and oakum joints is being replaced, where allowed by local code, with push-on gasketed joints or no-hub coupled joints.

Combination fittings which conserve space and joint makeup labor can often be used in cast-iron soil pipe systems. Examples are combination wye and eighth bends, single and double vent branches, offsets, running traps, and many more (available on special order). These fittings are all available with special cleanout or vent ports, and in straight or reducing sizes. Soil pipe fittings reduce on the spigot end only.

Cast-iron soil pipe fittings are manufactured in bell and spigot type (in service or extra heavy weight) and no-hub (in service weight).

Units of Measure: Soil pipe fittings and combination fittings are taken off and recorded as each.

Material Units: The number of joints or no-hub couplings will be determined from the fitting count, as well as any special supports, cleanout access, or extension covers.

Labor Units: Unloading, sorting, and bringing the fittings to the point of installation should be included in the installation cost. Joint makeup, testing, and inspection by local municipal authorities, should also be considered.

Takeoff Procedure: Each fitting and combination should be recorded on the soil pipe estimate sheet (or one attached to it) by size and type. Line drawings usually do not show every fitting or possible combination of fittings. This is where knowledge of local codes and available fittings can be useful to the mechanical estimator. The plumbing fixture count will also aid in determining many fitting requirements.

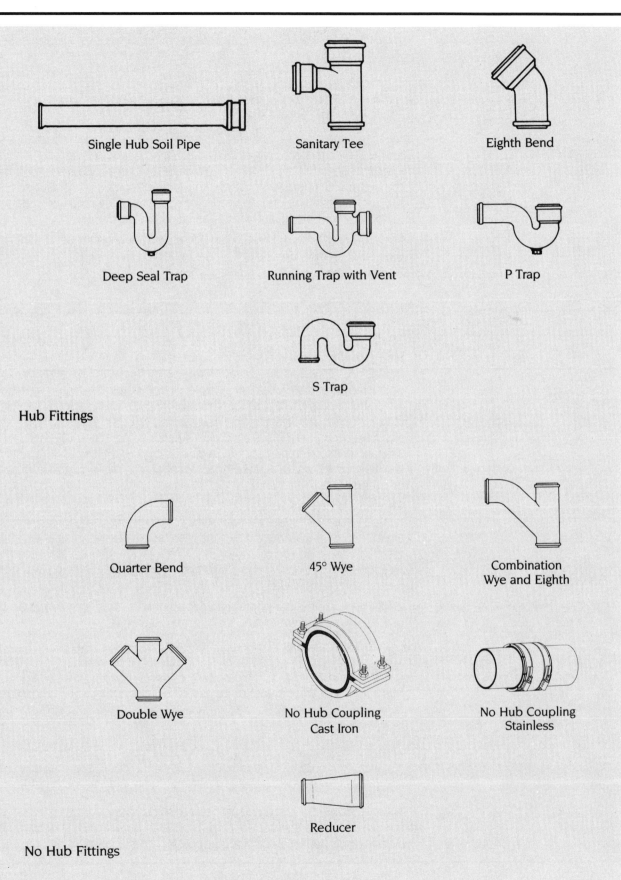

Single Hub Soil Pipe

Sanitary Tee

Eighth Bend

Deep Seal Trap

Running Trap with Vent

P Trap

S Trap

Hub Fittings

Quarter Bend

45° Wye

Combination
Wye and Eighth

Double Wye

No Hub Coupling
Cast Iron

No Hub Coupling
Stainless

Reducer

No Hub Fittings

Typical Cast Iron Soil Pipe Fittings for Hub or No Hub Use

Figure 13.11

Copper Pipe and Tubing

Copper pipe is not as commonly used in the construction industry today as it was in the past. Copper is now relegated to very special systems and trim around boiler controls and the like. Copper, yellow brass, and red brass (85% copper) pipe have, for the most part, been replaced by copper tubing. These materials are still available, should the need for them arise, with threaded brass fittings 125 lb. MIP (malleable iron pattern), and 250 lb. CIP (cast-iron pattern).

Copper pipe, having the same O.D. as steel pipe, will accept threaded valves and fittings. Labor units for threaded copper pipe are similar to those for threaded steel.

Because copper tubing has a thinner wall and smaller O.D. than pipe, it cannot be threaded and must instead be assembled using soldered, flared or brazed fittings and valves. The tubing comes in several classes and types as well as tempers.

The first class is *water tubing*, which is produced in types K, L, M, and DWV. Type K heavy wall tubing is used in interior and underground systems, available in soft or hard temper. Type L medium wall tubing is used in interior systems, available in hard or soft temper. Type M light wall tubing is used in low pressure systems, and cannot be used in refrigeration systems. Type M is available in hard temper only. Type DWV light wall tubing is used in plumbing drainage, waste, and vent systems, and is available in hard temper only. The Copper Development Association and the BOCA National Plumbing Code allow types L and M in underground systems, types M and DWV in chilled water systems and forced hot water systems excluding high rise and other pressure restrictions.

The second class of copper tubing is type ACR, produced for use in *air-conditioning and refrigeration systems*. Type ACR tubing has the same wall thickness as type L, and is shipped cleaned, capped, and charged with nitrogen, available in hard or soft drawn.

Water tubing is sized nominally. Types K and L are produced in 1/8" through 12" diameters, type M in 1/4" through 12", DWV in 1-1/4" through 8" diameters. ACR tubing is sized by outside diameter (O.D.), produced in 1/8" through 4-1/8" O.D. Hard drawn tubing is manufactured in 20' lengths; soft, or annealed, is manufactured in straight lengths or coils. Figure 13.12 shows commercially available lengths of copper tubing.

Units of Measure: Copper pipe and tubing is measured in linear feet (L.F.). Associated fittings, supports, and joints are recorded as each.

Labor Units: The following procedures are generally included in the per foot labor cost:

- Unloading and relocating from receiving or storage area to point of installation
- Setup of scaffolding, staging, ladders, or other work platforms
- Layout of piping route

These additional procedures must be recorded and included in the piping estimate:

- Joining straight lengths every 20' (or 10' in larger sizes)
- Installation of fittings, valves, appurtenances, supports, anchors, and guides
- Testing and cleaning of systems (special emphasis on cleaning and chlorinating potable systems)

Takeoff Procedure: A takeoff sheet should be set up for each piping system and record the various footages for each diameter and type of pipe. Hot, cold,

Commercially Available Lengths		
Tube	Hard Drawn	Annealed

Type K	Straight lengths up to:
	    8-inch diameter 20 ft.
	    10-inch diameter 18 ft.
	    12-inch diameter 12 ft.
		Straight lengths up to:
		    8-inch diameter 20 ft.
		    10-inch diameter 18 ft.
		    12-inch diameter 12 ft
		Coils up to:
		    1-inch diameter 60 ft.
		                             and 100 ft.
		    1¼ and 1½ inch
		        diameter 60 ft.
		    2-inch diameter 40 ft.
		                             45 ft.

Type L	Straight lengths up to:
	    10-inch diameter 20 ft.
	    12-inch diameter 18 ft.
		Straight lengths up to:
		    10-inch diameter 20 ft.
		    12-inch diameter 18 ft.
		Coils up to:
		    1-inch diameter 60 ft.
		                             and 100 ft.
		    1¼ and 1½
		        diameter 60 ft.
		    2-inch diameter 40 ft.
		                          or 45 ft

Type M	Straight lengths
	    All diameters 20 ft.
		Straight lengths up to:
		    12-inch diameter 20 ft.
		Coils up to:
		    1-inch diameter 60 ft.
		                             and 100 ft
		    1¼ and 1½-inch
		        diameter 60 ft.
		    2-inch diameter 40 ft.
		                             45 ft.

DWV	Straight lengths: All diameters 20 ft.	Not available —
ACR	Straight lengths: 20 ft.	Coils: 50 ft.

Figure 13.12

drain, and vent systems should be listed separately. Underground pipe should also be listed separately due to differences in support and installation procedures. Each system should be marked on the plans as it is taken off, beginning with the larger diameter pipe and proceeding in the direction of the flow (except for drain piping, where the larger pipe is at the termination point). Round off individual totals to coincide with 20' lengths, (60' or 100' coils) available from local suppliers. The height modifications given for steel pipe apply to copper pipe and tubing as well (refer back to Figure 13.3).

Copper Fittings and Flanges

Threaded copper and brass pipe both require the same fitting types; cast bronze threaded fittings, available in the 125 lb. rating in the open pattern (with a light bead). These fittings are often referred to as "malleable pattern bronze fittings." Extra heavy cast bronze fittings (250 lb.) are also available for use with copper pipe, in the cast-iron pattern (closed pattern with a heavy bead). Drainage and flanged fittings are also made which are similar to their cast-iron counterparts. For special requirements, threadless fittings are produced for brazing to pipe.

Fittings and flanges for copper tubing are soldered. Solder end (sweat) fittings are produced for use in both pressure and drainage systems. In soldered pressure systems only one standard, or line of fittings, is produced for types K, L, and M tubing. More critical pressures and temperatures are managed by using solder with higher melting points or by brazing. For most plumbing, heating, and cooling systems, soft solder is more than adequate for these noncritical applications.

At this writing, the use of lead in solder for potable water systems has been banned in most areas of the country. Instead, solder comprised of a mixture of tin and antimony or other substitutes should be used.

In general, solder type fittings are produced in wrought copper and cast bronze. Both are priced equally and have similar installation costs. In the larger sizes, some manufacturers do not make fittings in wrought copper, and cast bronze must be used. Cast bronze, however, requires more heat to make the joint but does not cool down quickly (which can be important when soldering large diameter pipe). Drainage fittings are also produced in wrought copper and cast bronze. Since drainage fittings do not have a pressure factor, soft solder can be used in making up the joints.

Flared fittings are used with annealed copper tubing (soft drawn) in water, refrigeration, and gas systems. These fittings (shown in Figure 13.13) are used when soldering or brazing is not practical, and ease of disassembly is a requirement. Special tools are needed, in this case, to flare the tubing end, which becomes an integral part of the joint. These fittings are limited in size from 3/8" to 2" nominal.

Compression type fittings are also used with soft drawn tubing for some oil burner work, domestic refrigeration, and instrumentation.

Units of Measure: All fittings are taken off and recorded as each. The number of joints is determined by this fitting count. The number of fittings and joints may be greatly reduced by the judicious use of bends, end swaging (expanding), and drilled and formed branch outlets. These items should be recorded in the same manner as fittings (as each) to anticipate the labor units necessary to perform the work.

Material Units: In addition to the actual cost of the fittings are the materials necessary to prepare the joints. For threaded bronze fittings, material unit costs

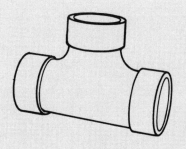

Tee–Solder Joint

45° Elbow–Solder Joint

90° Elbow–Solder Joint

Solder Type Fittings for Copper Tube

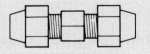

Coupling–Flare Type

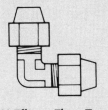

90° Elbow–Flare Type

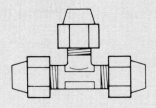

Tee–Flare Type

Fittings for Flared Copper Tube

Figure 13.13

include nipples, cutting oil, and joint compound, similar to threaded steel joints. For solder type fittings, sandcloth, flux, solder, and a heating source (gas) are needed. A chart of solder and flux consumption for these joints is shown in Figure 13.14. *Note*: In most areas lead free solder is required by law in portable water systems.

Labor Units: The labor units for threaded bronze fittings include:

- Tightening and sealing each threaded joint
- Cutting and threading, when required for intermediate joints in the shop or field

These considerations are identical to cast-iron and malleable iron fittings for steel pipe, both threaded and flanged. The total number of joints is determined from the fitting and valve counts, plus any intermediate joints in long straight runs of pipe.

Estimated Quantity of Soft Solder Required to Make 100 Joints	
Size	Quantity in Pounds
3/8"	.5
1/2"	.75
3/4"	1.0
2"	1.5
1¼"	1.75
1½"	2.0
2"	2.5
2½"	3.4
3"	4.0
3½"	4.8
4"	6.8
5"	8.0
6"	15.0
8"	32.0
10"	42.0

1. The quantity of hard solder used is dependent on the skill of the operator, but for estimating purposes, 75% of the above figures may be used.

2. Two oz. of Solder Flux will be required for each pound of solder.

3. Includes an allowance for waste.

4. Drainage fittings consume 20% less.

Figure 13.14

Solder joint fittings require a unique method of joint preparation: a square cut tube end, deburred, cleaned with a sandcloth or commercial (acid) cleaner; application of flux; insertion into the similarly cleaned hub of the fitting or valve; application of heat; and filling the space between the hub and tube with solder. Allowances should also be considered for cooling or quenching a soldered assembly if it is to be handled immediately.

Flared fitting joint assembly consists of a square cut of the tube end, deburring, and slipping the flare nut onto the tubing before flaring the end of the tube. After the flare is completed, the nut is mechanically tightened to the male thread of the fitting. Flared joints require no other materials besides proper cutting and flaring tools and wrenches.

Takeoff Procedure: Fittings, valves, appurtenances, bends, swaged joints, field drawn tees, and test nipples should be recorded on the takeoff sheet for copper pipe or tubing, by size and system. From this count, the number of joints and the labor for each system can be determined. By recording the systems separately, the same takeoff sheets can be used for purchasing and scheduling during construction.

Plastic Pipe and Tubing

Plastic and thermoplastic (material which softens when subjected to heat, rehardening when cooled) pipe and tubing have revolutionized mechanical piping systems over the past few decades, and will continue to do so in the years to come as the plastics industry grows. Because of its relative "newness" to the market, a wide variety of materials, types, and joining procedures are available, all seeking their share of this expanding market. For example, the most familiar plastic piping material found in mechanical systems is polyvinyl chloride (PVC) pipe, used in drain, waste, and vent (DWV) and cold water systems. PVC competes with acrylonitrile butadiene styrene (ABS) piping, especially in DWV applications. Polypropylene is another competitor in DWV applications. Chlorinated PVC (CPVC) pipe has been developed for use in hot water systems; polyethylene is used in cold water systems; and polybutylene is used in hot and cold water systems. CPVC and polybutylene are recently approved for use in wet type fire protection systems, while polyethylene is commonly used in gas and water services.

As with other types of piping systems, there are both advantages and disadvantages to the use of plastic pipe. The light weight of plastic pipe is a major installation laborsaving advantage over more traditional heavier materials. The flexibility of some types of plastic is another laborsaving advantage. This flexibility eliminates some bends and elbows, and reduces the need for expansion control or water hammer eliminators. The disadvantages of some plastics are vulnerability to sunlight, ambient temperatures, abrasion, and piercing by sharp objects.

Plastic pipe comes in rigid 10' and 20' lengths, or in coils of 1,500', 1,000', 600', 500', 400', 300', 200', and 100' lengths, depending on diameter. Plastic pipe and tubing are also produced in piping Schedules 40 and 80, plus in the O.D. sizes comparable to those of copper, and in a wall thickness rating method beyond scheduled pipe classifications, known as SDR (standard dimension ratio). The SDR method was developed to standardize the inside and outside dimensions of plastic piping which does not conform to the pipe schedule method. It is determined by dividing the average O.D. by the minimum wall thickness.

Pipe diameters are generally comparable to those found in steel, copper, and cast-iron systems. Some manufacturers, however, do not produce a full line, skipping some intermediate sizes due to closeness in diameter or infrequency of use.

Units of Measure: Plastic pipe and tubing are taken off and recorded per linear foot. Associated fittings, joints, supports, valves, and appurtenances are counted as each. In some instances, constant or continuous support may be required, which should also be recorded by linear foot.

Labor Units: The following procedures are generally included in the per foot installation labor cost of plastic pipe and tubing:

- Unloading, protecting, relocating from receiving or storage area to point of installation
- Setup of scaffolding, staging, ladders, or other work platforms
- Layout of piping route

These additional procedures should also be recorded and included in the piping estimate:

- Joining or coupling intermediate lengths
- Curing of solvent cement or otherwise preparing the joints for fusion welding, threading, etc.
- Installation of fittings, valves, appurtenances, supports, anchors, and guides
- Testing and flushing of systems (especially portions concealed below grade in the building construction)

Takeoff Procedure: A takeoff sheet for each piping system should be set up. If more than one type of plastic pipe or joining method is to be used in the same system, each type should be listed separately. The various footages required should be recorded for each pipe diameter. Color coding the systems as they are taken off helps track the progress of the takeoff, as well as serving to define each system so it can be quickly identified on the plans later on. The largest size pipe should be taken off first, and the totals rounded off to coincide with pipe lengths available. Height modifications (shown in Figure 13.3) may need to be considered. Underground or structurally supported lengths should be listed separately because they have unique support considerations.

Fittings for Plastic Pipe and Tubing

There are at least three times as many joining methods for plastic pipe as there are different types of plastic. There are fittings available to accommodate any end finish or joining procedure required for the various types and sizes of plastic pipe or tubing. This may include threaded, flanged, compression, socket (solvent) weld, fusion weld, butt weld, brass insert, plastic insert, clamp type, grip type, flared, gasketed fittings, and many combinations thereof. Some of the common fitting and joining methods for plastic pipe are shown in Figure 13.15.

Units of Measure: Fittings and flanges are taken off and recorded as each. The number of joints, cuts threads, and welds will be determined by this fitting count.

Material Units: Unique tools for cutting, deburring, and preparing are required for plastic fittings. Solvent welds also require primers, solvent cements, and applicators. Fusion welds require special torches and tips. Electrical resistance fusion kits consisting of power units and fusion insert coils are needed for

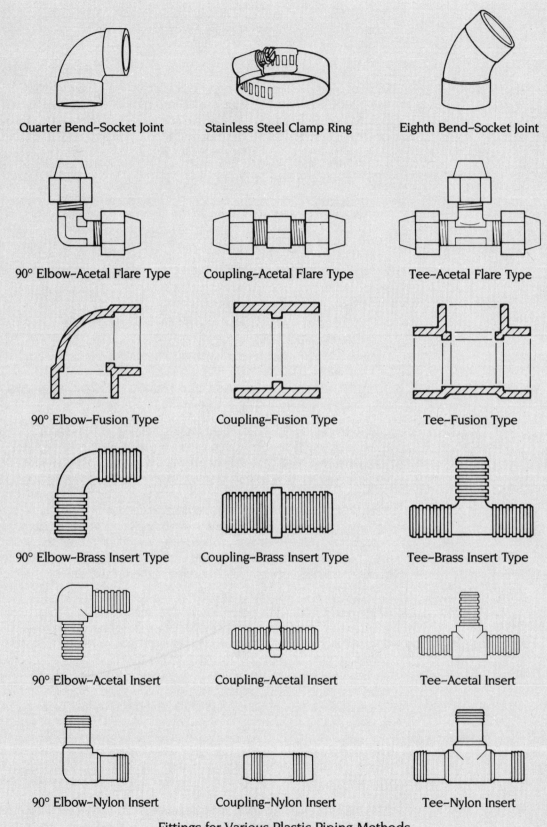

Quarter Bend–Socket Joint

Stainless Steel Clamp Ring

Eighth Bend–Socket Joint

90° Elbow–Acetal Flare Type

Coupling–Acetal Flare Type

Tee–Acetal Flare Type

90° Elbow–Fusion Type

Coupling–Fusion Type

Tee–Fusion Type

90° Elbow–Brass Insert Type

Coupling–Brass Insert Type

Tee–Brass Insert Type

90° Elbow–Acetal Insert

Coupling–Acetal Insert

Tee–Acetal Insert

90° Elbow–Nylon Insert

Coupling–Nylon Insert

Tee–Nylon Insert

Fittings for Various Plastic Piping Methods

Note: Threaded PVC, CPVC etc. fittings resemble their metal counterparts.

Figure 13.15

some polypropylene drain waste and vent systems for use with acids and other corrosives. Threaded joints require pipe joint compound or tape; flanged joints require bolts, nuts, and gaskets; and some compression or clamp type joints require gaskets.

Labor Units: Labor units for plastic fittings are similar to those of metal pipe for flanged or threaded joints. For solvent weld fittings, the simple but essential procedure of priming and applying the solvent must be adhered to. Fusion welding procedures vary widely with each manufacturer. It is important to follow the manufacturer's instructions and to verify that the fittings are compatible with the pipe or tubing being installed. Local codes should be consulted before selecting the piping material and joining method. It is also critical in plastic pipe joint makeup that dirt or any foreign matter be removed. A bad plastic joint usually cannot be redone; it must be removed and replaced.

Takeoff Procedure: Plastic fittings, valves, appurtenances, bends, and intermediate joints should be recorded by size, system, and joining method on the takeoff sheet for plastic pipe or tubing. From this count, the number of joints or diameter inches can be determined, and the labor units can be estimated. It is advantageous to record the various systems on separate takeoff sheets for use during construction planning and purchasing.

Preinsulated Pipe or Conduit

Conduit, or preinsulated pipe, is available in various combinations of pipe and insulation materials. It consists of a prefabricated core of pipe (one or more) coated with insulation. The core is then protected with an outer jacket of heavy gauge sheet metal or a rigid plastic casing (shown in Figure 13.16). Conduits have been developed for underground piping systems, overhead lines between buildings, along waterfront docks and piers, spanning rivers, railroad tracks, etc., by using a bridge for support. Conduit systems may contain one or more pipe (either nested or individually insulated), and prefabricated elbows, tees, expansion loops, anchors, and building entry leakplates.

Units of Measure: All conduit is taken off by individual sizes per linear foot. Prefabricated fittings, manholes, and the like should be recorded as each.

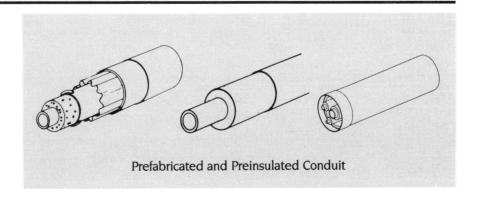

Prefabricated and Preinsulated Conduit

Figure 13.16

Material Units: The conduit supplier should furnish all material required to make a complete section (including slings for handling) and any unique tool for connecting. All the material for field joint assembly, insulation, and outer closure should also be part of the supplier's job quotation.

Labor Units: The labor units required here are to physically place the sections of conduit in the trench or on the proper supports. Diameter and weight will determine whether the installation is to be by manual labor or the use of a crane or if powered lifting equipment is required. The pipe joints are prepared and made up in accordance with manufacturers' instructions and are pressure tested prior to insulation.

After testing and insulation are completed the outer closure or casing is installed. Whether steel or plastic, the linear inches of welding this closure must be taken off and recorded for material and labor costs. The exterior closure itself is then tested with a pressure test (using compressed air). In addition to the manufacturer's drawings and instructions, a field service engineer is normally provided to direct the installer and certify that the installation conforms to the manufacturer's recommendations.

Takeoff Procedure: The conduit is normally taken off by the manufacturer's representative, but the estimator should also do a takeoff as both a double check and to determine labor units, incidental supports needed, etc. This takeoff should be listed on its own takeoff sheet and the labor should be listed separately for future observation.

Valves

A valve is a mechanical device which controls the flow, pressure, level, or temperature of the fluid being conveyed or contained in a piping system. Valves can be operated manually or automatically. An automatic valve may be operated by a motor, or by the temperature, pressure, liquid level, or direction of the liquid flow. A manual valve may have a wheel, lever, wrench, or gear operator.

Valves are used in virtually all piping systems with the exception of drain, waste, and vent; backwater valves which are sometimes used in drain or waste piping systems being the exception.

There are valves for each pipe and tubing size and for each method of connection, whether it be flanged, screwed, weld, solder, compression, grip type, or mechanical joint installation. In mechanical systems valves are usually iron body, bronze, copper, or plastic. Valve sizes are rated as standard (in the 100 lb. to 175 lb. range), or extra heavy (in the 200 to 300 lb. range). Valves described in the following paragraphs are shown in Figure 13.17.

Gate valves provide full flow, minute pressure drop, minimum turbulence and minimum fluid trapped in the line. The flow is controlled by raising or lowering a gate via a screw mechanism. Gate valves should always be fully opened or fully closed. They are normally used when operation is infrequent.

Globe valves are designed primarily for throttling flow. A plug is raised or lowered via a screw mechanism into a seat. The configuration of the globe valve causes restricted flow which results in increased pressure drop. Besides throttlings, globe valves can also shut off the flow.

The fundamental difference between the *angle valve* and the globe valve is the fluid flow through the angle valve. When a 90° turn is required, an angle

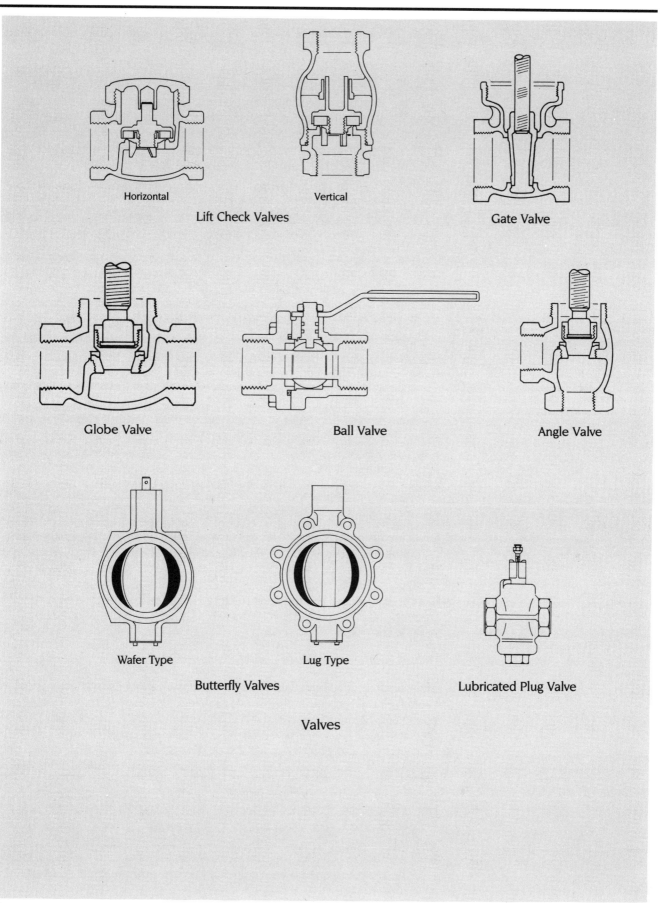

Horizontal   Vertical

Lift Check Valves   **Gate Valve**

Globe Valve   **Ball Valve**   **Angle Valve**

Wafer Type   Lug Type

Butterfly Valves   **Lubricated Plug Valve**

Valves

Figure 13.17

valve offers less resistance to flow than a globe valve and elbow combination. Use of an angle valve also reduces the number of joints and installation time.

Swing check valves are designed to prevent backflow by automatically seating when the direction of the fluid is reversed. Swing check valves are usually installed with gate valves, as they provide comparable full flow. Usually recommended for lines where flow velocities are low, they should not be used on lines with pulsating flow. Swing check valves are recommended for horizontal installation, or in vertical lines where flow is upward.

Lift check valves have diaphragm seating arrangements similar to globe valves and are recommended for preventing backflow of steam, air, gas, and water, and on vapor lines with high flow velocities. For horizontal lines, horizontal lift checks should be used. Vertical lift checks should be used for vertical lines.

Ball valves are light and easily installed, and because of modern elastomeric seats, provide tight closure. Flow is controlled by rotating up to 90° a drilled ball which fits tightly against resilient seals. This ball seats with the flow in either direction, and the valve handle indicates the degree of opening. Recommended for frequent operation, readily adaptable to automation, it is ideal for installation where space is limited.

Butterfly valves provide bubble tight closure with excellent throttling characteristics. They can be used for full-open, closed, and throttling applications. The butterfly valve consists of a disc within the valve body which is controlled by a shaft. In its closed position, the valve disc seals against a resilient seat. The disc position throughout the 90o rotation is visually indicated by the position of the operator. A butterfly valve is only a fraction of the weight of a gate valve and, in most cases, requires no gaskets between flanges. Recommended for frequent operation, it is adaptable to automation. Wafer and lug type bodies, when installed between two pipe flanges, can be easily removed from the line. The pressure of the bolted flanges holds the valves in place. Locating lugs makes installation easier.

Lubricated *plug valves* are similar to ball and butterfly valves in operation and configuration. Because of the wide range of services to which they are adapted, they may be classified as all purpose valves. Plug valves can be safely used at all pressures and vacuums, and at all temperatures up to the limits of available lubricants. They are the most satisfactory valves for the handling of gritty suspensions and many other destructive, erosive, corrosive, and chemical solutions.

Units of Measure: Valves are taken off and recorded as each.

Material Units: The material cost of valves usually involves only the cost of the valve itself, except in electric or pneumatic operated systems. Electric or pneumatic operation requires the addition of a motor or actuator, installed after the valve body has been piped. Remote chain wheel operators and gear operators are also purchased and installed separately.

Labor Units: The labor involved in installing a valve is usually the makeup and handling of one, two, or three piping connections. Valves which have only one piping connection include safety or relief valves, drain cocks, sill cocks, hose bibbs, vacuum breakers, air vents, and ball cocks. Valves having two piping connections are the type most often encountered. Two connection valves are gates, globes, checks, plug cocks, butterfly, ball, stop and waste, balancing, radiator control, flushometer, fire, fusible oil, backflow preventer, pressure regulating, angle stop, etc. Three-way or mixing valves include shower valves, motor operated temperature control valves, and blending valves for fuel.

As with fittings and piping, weight is always a contributing factor to the labor cost of installing valves, as well as height above the floor or work platform (for height modifications, see Figure 13.3).

Takeoff Procedure: The valves and accessories for each system should be recorded by size and type on the takeoff sheet used for pipe and fittings (shown in Figure 13.9). An estimator may have to include valves not shown on the plans, but which may be necessary to isolate portions of a piping system for testing and draining.

Pipe Hangers and Supports

Piping for mechanical installations is supported by a wide variety of hangers and anchoring devices. The location of piping supports is an important consideration, and many different building components may be used for anchoring pipe hangers. Some of the commonly used locations for anchoring devices (shown in Figure 13.18) include the roof slab or floor slab, structural members, side walls, another pipe line, machinery, or building equipment. The pipe hanger material is usually black or galvanized steel. For appearance or in corrosive atmospheres, chromé or copper plated steel, cast iron, or a variety of plastics may also be used.

Another consideration is the method of anchoring the hanger assembly to the building structure. The method selected for anchoring the hanger depends on the type of material to which it is being secured, usually concrete, steel, or wood. If the roof or floor slab is constructed of concrete, formed, and placed at the site, then concrete inserts may be nailed in the forms at the required locations prior to the placing of concrete. These inserts (shown in Figure 13.18) may be manufactured from steel or malleable iron, and they are either tapped to receive the hanger rod machine thread or contain a slot to receive an insert nut. Because the slotted type of insert requires separate insertable nuts, for the various rod diameters, only one size of insert need be warehoused. For multiple side by side runs, long, slotted insert channels of up to 10′ increments are available with several types of adjustable insert nuts.

When precast slabs are used, the inserts may be installed on site in the joints between slabs, drilled, or shot into the slab itself. Electric or pneumatic drills and hammers are available to drill holes for anchors, shields, or expansion bolts. Another method of installing anchors on site utilizes a gunpowder-actuated stud driver, which partially embeds a threaded stud into the concrete.

If the piping is to be supported from the sidewalls, similar methods of drilling, driving, or anchoring to those mentioned above are used for concrete walls. Where hollow-core masonry walls are to be fitted, holes may be drilled for toggle or expansion-type bolts or anchors.

Attaching the pipe supports to steel or wood requires different methods from those used for concrete. When the piping is to be supported from the building's structural steel members, a wide variety of beam clamps, fish plates (rectangular steel washers), and welded attachments can be employed. If the piping is being run in areas where the structural steel is not located directly overhead, then intermediate steel is used to bridge the gap. This intermediate steel is usually erected at the piping contractor's expense. If the building is constructed of wood, then lag screws, drive screws, or nails are used to secure the support assembly.

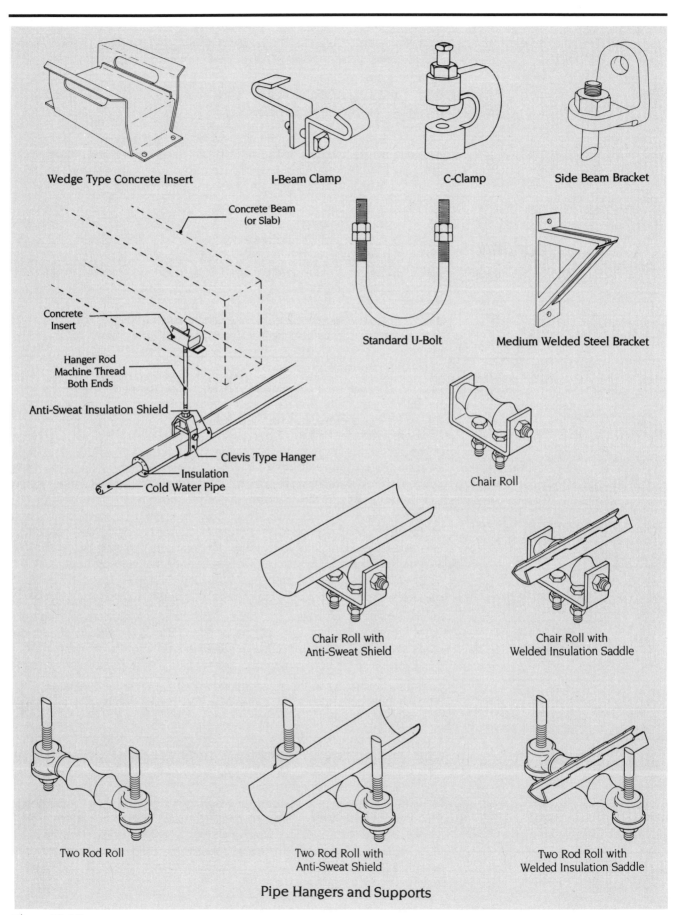

Wedge Type Concrete Insert

I-Beam Clamp

C-Clamp

Side Beam Bracket

Concrete Beam
(or Slab)

Concrete
Insert

Hanger Rod
Machine Thread
Both Ends

Anti-Sweat Insulation Shield

Clevis Type Hanger

Insulation
Cold Water Pipe

Standard U-Bolt

Medium Welded Steel Bracket

Chair Roll

**Chair Roll with
Anti-Sweat Shield**

**Chair Roll with
Welded Insulation Saddle**

Two Rod Roll

**Two Rod Roll with
Anti-Sweat Shield**

**Two Rod Roll with
Welded Insulation Saddle**

Pipe Hangers and Supports

Figure 13.18

122

From the anchoring device, a steel hanger rod, threaded on both ends, extends to receive the pipe hanger. One end of this rod is threaded into the anchoring device, and the other is fastened to the hanger itself by a washer and a nut. For cost effectiveness and convenience, continuous thread rod may be used. The pipe hanger itself (shown in Figures 13.18 and 13.19) may be a ring, band, roll, or clamp, depending on the function and size of the piping being supported. Spring-type hangers are also used when it is necessary to cushion or isolate vibration.

Piping systems subject to thermal expansion must often absorb this expansion by use of piping loops, bends, or manufactured expansion joints. The piping must be anchored to force the expansion movement back to the joint. In order to prevent distortion of the piping joint itself, alignment guides are installed. Figure 13.20 shows alignment guides used in these types of installations.

Specifically designed pipe hangers are used for fire protection piping with underwriters and factory mutual approvals.

In wood frame residential construction, holes drilled through joists or studs have often been the sole support for certain pipes. These pipes may occasionally be reinforced with wedges cut from a 2" x 4". A piece of wood cut and nailed between two studs is often used to support the plumbing stubouts to a fixture. Prepunched metal brackets are available to span studs for both 16" and 24" centers. These brackets can be used with both plastic or copper pipe and maintain supply stub spacing 4", 6", or 8" on centers, giving perfect alignment and a secure timesaving support. Plastic support and alignment systems are also available for any possible plumbing fixture rough-in. Some typical pipe supports for wood construction are shown in Figure 13.21.

Units of Measure: Pipe hangers and supports are taken off, recorded, and priced as each. As a general rule, the number of pipe hanger assemblies can be approximated from the piping totals and the required hanger spacing (e.g., one hanger every 10'). The fixture count will give an accurate measure for bracket supports or alignment systems.

Material Units: Hanger assemblies should be prepriced for each different size and type of piping. The assembly price should include the insert or beam clamp, vertical rod, nuts, hanger, and, if necessary, an insulation shield. The estimator will decide, from experience, which hanger assemblies should be prepriced, depending on the type of work or system a firm might specialize in. This method can save an estimator many hours of repetitively pricing each nut, bolt, and washer for each and every size of pipe and for each estimate prepared.

Labor Units: The following procedures must be considered and included in the labor unit per hanger assembly, depending on the job conditions:

- Sleeving and inserting for each concrete slab when it is being formed, otherwise, drilling or driving expansion shields or studs
- Cutting the hanger rods (cutting and threading if labor conditions do not permit the use of threaded rod)
- Installing and adjusting the height of each assembly to maintain the desired pitch

The type of hanger selected can significantly influence the time and cost of piping installation.

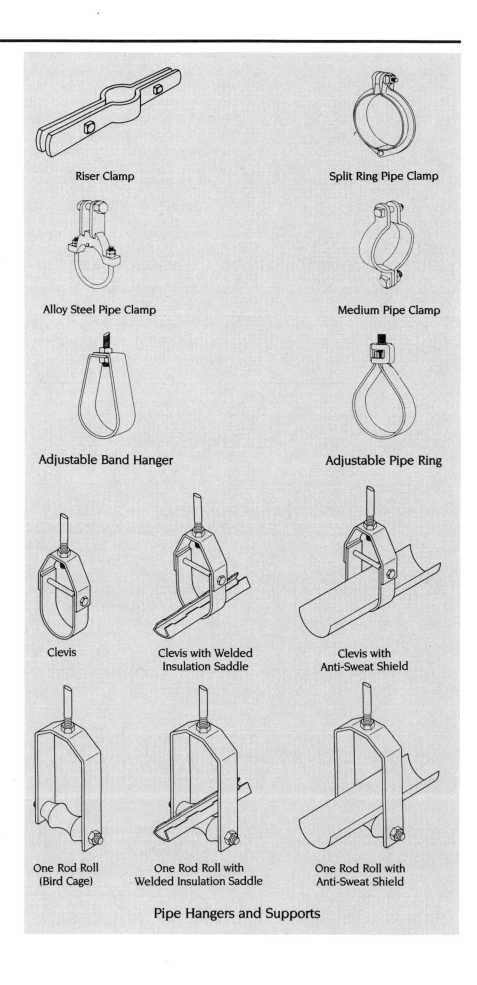

Pipe Hangers and Supports

Figure 13.19

124

Residential or light commercial construction (wood frame) permits the installation of simple supports or hangers after the piping is in place. Underground piping may not require supports depending on the type of pipe or tubing used and the condition of the pipe bed. Masonry blocks or bricks are commonly used if the pipe bed is not satisfactory. Steel pipe (insulation protection) saddles will require the added labor of tack welding the saddles to the pipe.

Saddles for cold water piping may be installed by the pipe coverer outside the insulation (installation instructions should first be read).

Piping support may be inconsequential or it may be critical to an estimate; do not overlook it.

Takeoff Procedure: From the piping takeoff totals, the number of hanger assemblies can be determined, by size and system, as shown in Figure 13.22.

Cast-iron soil pipe should be supported every 5'. Thermoplastic piping may have to be supported as much as every 3', depending on ambient temperatures. In many cases, it may be more cost effective to support plastic pipe using continuous angle, channel, or other rigid members between widely spaced hangers.

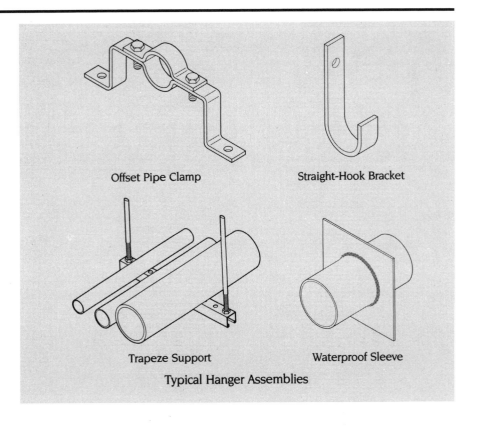

Offset Pipe Clamp

Straight-Hook Bracket

Trapeze Support

Waterproof Sleeve

Typical Hanger Assemblies

Figure 13.20

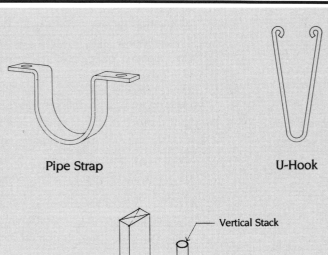

Pipe Strap

U-Hook

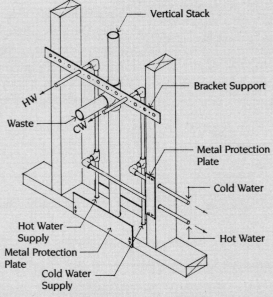

Vertical Stack

Bracket Support

HW

Waste

CW

Metal Protection
Plate

Cold Water

Hot Water Supply

Metal Protection
Plate

Cold Water
Supply

Hot Water

Stubout Bracket Support

Plastic Drive Hooks

Figure 13.21

In addition to the table shown in Figure 13.22, additional support must be added for runouts, change of direction, large valves, headers, etc. A short-cut estimating method for new construction is to use a 10' average span for all sizes and types of piping from the takeoff totals.

Alternatives: In place of individual hangers, trapeze or gang hangers, consisting of two rods and one horizontal member, may be used to support several lines at a labor and material savings. Field fabrication is typically required to make up these assemblies (shown in Figure 13.20).

Support Spacing for Metal Pipe or Tubing			
Nominal Pipe Pipe Size Inches	Span Water Feet	Steam, Gas, Air Feet	Rod Size
1	7	9	↑
1½	9	12	⅜"
2	10	13	↓
2½	11	14	↑
3	12	15	½"
3½	13	16	↓
4	14	17	↑
5	16	19	⅝" ↓
6	17	21	¾"
8	19	24	↑
10	20	26	⅞"
12	23	30	↓
14	25	32	↑
16	27	35	1"
18	28	37	
20	30	39	↓
24	32	42	↑
30	33	44	1¼" ↓

Figure 13.22

Chapter 14

PUMPS

A pump is a mechanical device used to convey, raise, or variate the pressure of fluids, such as water, oil, or sewage. In the building construction industry, pumps are employed in dewatering the site of excavation, increasing water pressure for potable or fire protection systems, circulating potable, cooling, or heating water, transferring fuel oil, draining wet basements, or ejecting sewage from sub-basement pits. Representative pumps are shown in Figure 14.1.

The most commonly used pumps are centrifugal, driven by an electric motor. In some instances fire pumps may be driven by diesel engine as a precaution against electrical failure. Most pump bodies are constructed of cast iron. For potable water pumps, bronze is usually substituted for cast iron at an increase in cost.

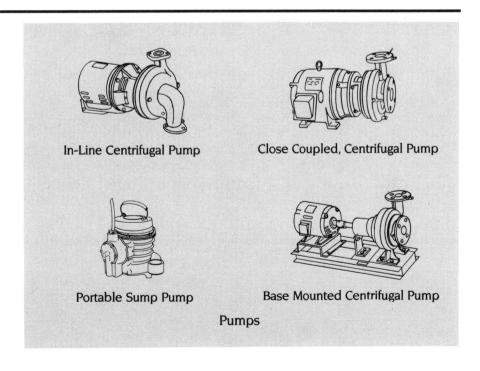

In-Line Centrifugal Pump

Close Coupled, Centrifugal Pump

Portable Sump Pump

Base Mounted Centrifugal Pump

Pumps

Figure 14.1

To insure proper operation, a check valve should be installed on the discharge side of the pump for backflow protection. If a throttling or flow regulating valve is required, it should also be on the discharge side to prevent pump cavitation.

Units of Measure: Pumps or pumping units are taken off, recorded, and priced as each.

Material Units: As a general rule, the pump and its motor are obtained as a complete unit with no other exterior trim or fittings. Some exceptions to this rule are vacuum heating pumps, condensate return, or boiler feed systems. These pumps may be purchased mounted on a steel or cast-iron tank or receiver, complete with interconnecting piping, valves, gauges, thermometers, controls, etc.

Fire pumps, sewage pumps or ejectors, and factory-assembled or prepackaged water pressure booster systems are also available. These prepackaged assemblies usually include more than one pump, as well as the aforementioned piping and trim.

An estimator's library should include manufacturers' catalogs of pumps and pump packages. These will show pipe sizes, weights, horsepower, and trim, and aid in visualizing the installation of the specified equipment or its equivalent. Many pump installations also require in-line vibration isolating flexible connections and inertia bases.

Labor Units: Installing in-line pumps is comparable to a valve installation, depending on pipe size and weight. Base-mounted pumps normally require the preinstallation of a masonry foundation and anchor bolts. For larger base-mounted pumps, extraordinary vibration elimination bases may be specified and included in both the material and labor cost.

The size and weight of a pump or prepackaged assembly will dictate the crew size and the mechanical equipment or tools required to unload, store, move to installation area, and set in place on the base or foundation. When weight and physical size indicate a larger crew than would normally be available, or if the item is impossible to handle by a normal crew, a rigging service should be considered.

The piping connection labor is taken from the actual joint connections at the suction and discharge terminals.

Takeoff Procedure: Pumps, like other mechanical equipment or fixtures, should be taken off and colored in prior to the piping system takeoff. In some instances, an equipment schedule is listed on the plans. This schedule should be checked carefully against the drawings. A separate pricing list should be made for equipment. Taking off equipment first has some advantages, listed below:

- The estimator becomes familiar with both the plans and the proposed building geography.
- The pieces of equipment become terminal points for the piping system takeoff.
- Vendors can be contacted to prepare quotations prior to bid day, while the estimator is busy with the time-consuming piping takeoff.

Chapter 15
PLUMBING SYSTEMS

The plumbing portion of a mechanical estimate includes the piping and fixtures for the potable water, domestic hot water, storm, sanitary waste, and interior gas piping systems. In buildings such as hospitals, schools, or laboratories, medical gas and acid waste piping might be added to this list. Piping systems have already been discussed in detail in Chapter 13, "Piping." Plumbing fixtures, appliances, drains, carriers, and other plumbing accessories will be discussed in this chapter.

Knowledge of applicable local plumbing codes is a must. The estimator should never assume that the designer has complied with these codes. Fire standpipes, which are also the work of the plumber, will be covered in Chapter 16, "Fire Protection."

Fixtures and Appliances

Plumbing fixtures and appliances are the receptacles connected to and served by the water distribution, drainage, waste, and vent systems. Fixtures normally used in plumbing systems of commercial or residential buildings include bidets, bubblers or drinking fountains, emergency eyewashes, interceptors, lavatories, receptors, shower stalls, sinks, tubs, urinals, wash centers, and water closets. Appliances for these buildings may include electric water coolers, dishwashers, dryers, washing machines, water heaters, or water storage tanks.

Plumbing fixtures requiring cold water, waste, and vent connections are shown in Figure 15.1. The plumbing fixtures shown in Figure 15.2 require both hot and cold water supplies, plus waste and vent connections. Figure 15.3 shows plumbing appliances requiring only water piping connections, and no waste or vent. Both gas and oil water heaters require fuel piping and smoke pipe or flue connections. The water and relief piping connectors are all necessary, regardless of the type of fuel or electricity being used. The plumbing appliances shown in Figure 15.4 require cold water supply plus waste and vent connections. Figure 15.5 shows plumbing appliances requiring hot and cold water supplies, mixing valve, waste, and vent connections.

Electric water coolers may have remotely located chiller sections to serve more than one dispensing unit. In this case, additional insulated piping must be included in the estimate.

The plumbing contractor is usually responsible for the connection of piping appliances and systems furnished by others. These may include vending machines, kitchen and cafeteria items, heating and cooling systems, window washing apparatus, etc. The additional piping for these systems is not usually shown on the plumbing drawings; however, the estimator should read through all drawings for these plumbing connections.

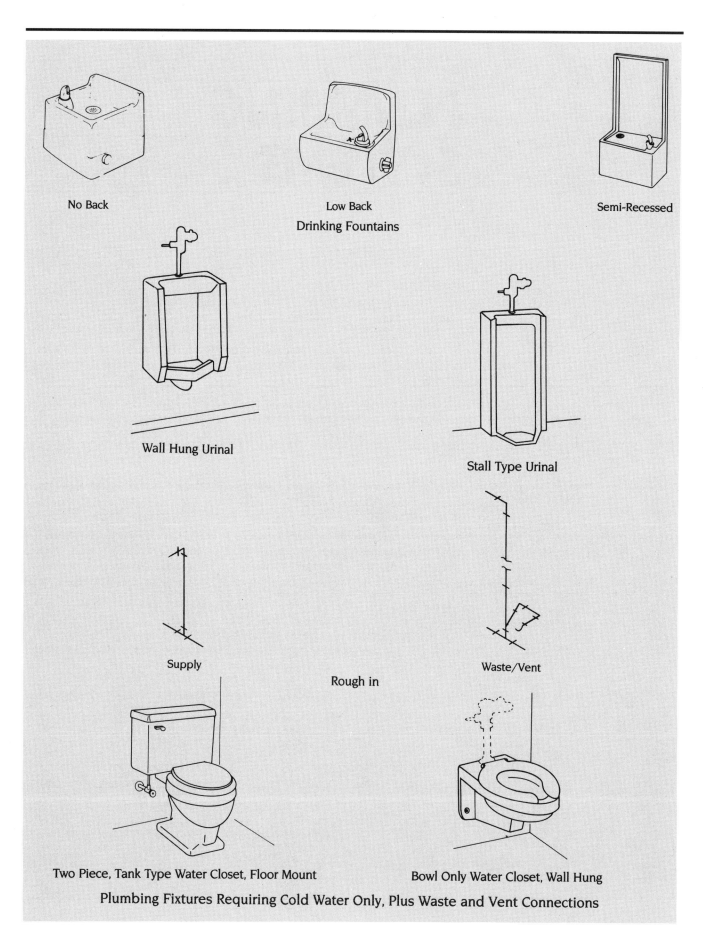

No Back

Low Back

Drinking Fountains

Semi-Recessed

Wall Hung Urinal

Stall Type Urinal

Supply

Rough in

Waste/Vent

Two Piece, Tank Type Water Closet, Floor Mount

Bowl Only Water Closet, Wall Hung

Plumbing Fixtures Requiring Cold Water Only, Plus Waste and Vent Connections

Figure 15.1

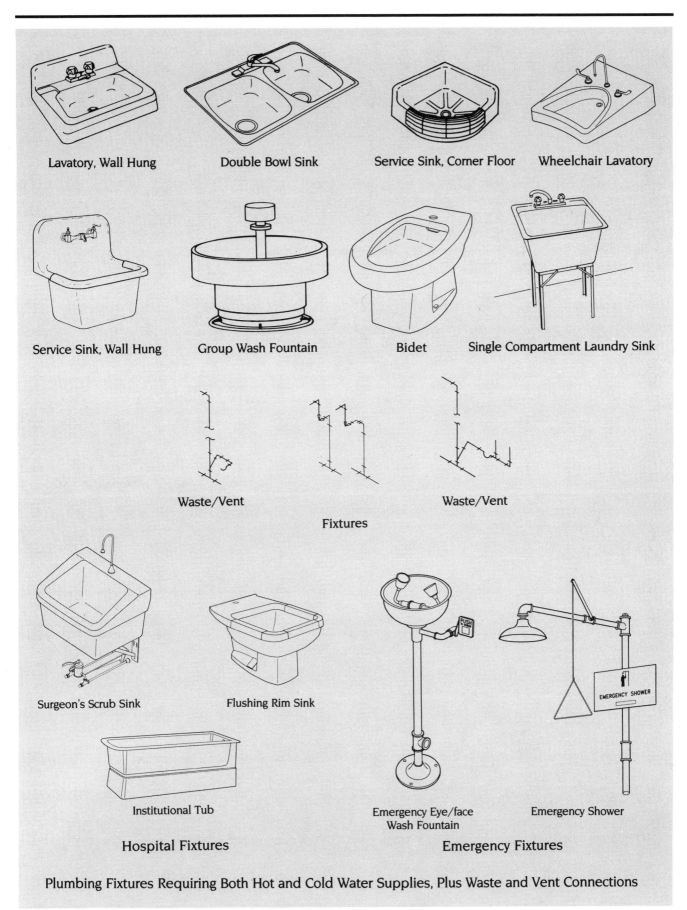

Lavatory, Wall Hung

Double Bowl Sink

Service Sink, Corner Floor

Wheelchair Lavatory

Service Sink, Wall Hung

Group Wash Fountain

Bidet

Single Compartment Laundry Sink

Waste/Vent

Waste/Vent

Fixtures

Surgeon's Scrub Sink

Flushing Rim Sink

Emergency Eye/face
Wash Fountain

Emergency Shower

Institutional Tub

Hospital Fixtures

Emergency Fixtures

Plumbing Fixtures Requiring Both Hot and Cold Water Supplies, Plus Waste and Vent Connections

Figure 15.2

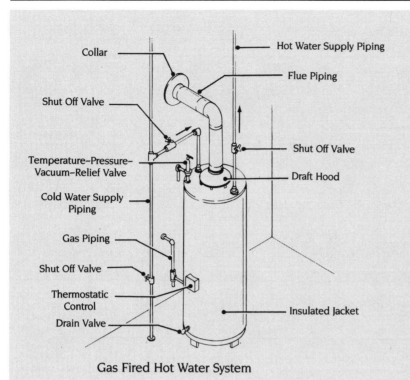

Gas Fired Hot Water System

Collar

Hot Water Supply Piping

Flue Piping

Shut Off Valve

Temperature-Pressure-Vacuum-Relief Valve

Shut Off Valve

Draft Hood

Cold Water Supply Piping

Gas Piping

Shut Off Valve

Thermostatic Control

Drain Valve

Insulated Jacket

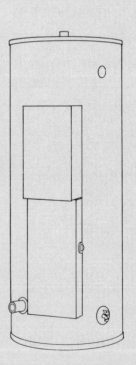

Gas Fired Water Heater–Commercial

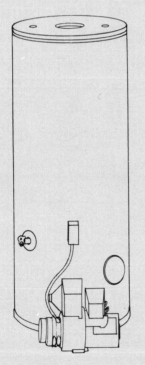

Oil Fired Water Heater–Commercial

Electric Water Heater–Commercial

Plumbing Appliances Requiring Only Water Piping Connections, and No Waste or Vent Connections

Figure 15.3

Units of Measure: Plumbing fixtures are taken off and counted as each.

Material Units: Each plumbing fixture manufacturer offers various optional "trim" packages for each type of fixture. A trim package for a lavatory, for example, might include the faucets, pop-up waste, supply stops, flexible supplies, and a trap. Plating or other finishes and styles are optional. These trim packages should be included in the individual fixture prices.

Labor Units: Fixture installation is performed in two phases at two separate stages of construction. Initially, a fixture is *roughed in* prior to installation of the finished wall or floor. Roughing in is the installation of the pipes that are to be concealed in the wall or beneath the floor. After the wall or floor finish is

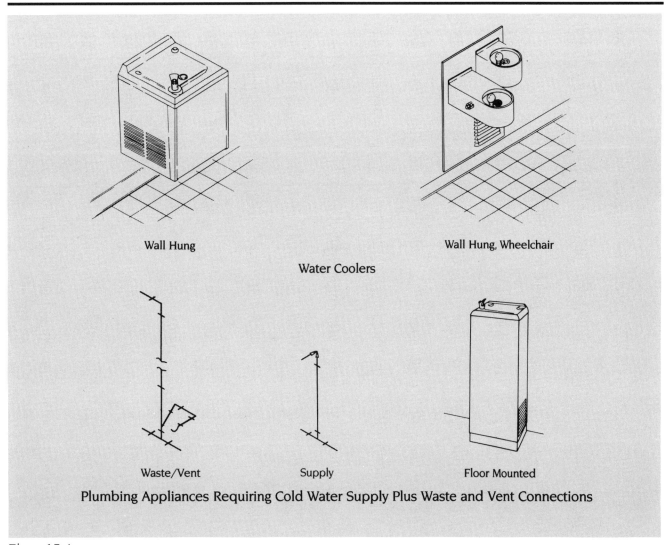

Wall Hung

Wall Hung, Wheelchair

Water Coolers

Waste/Vent   Supply   Floor Mounted

Plumbing Appliances Requiring Cold Water Supply Plus Waste and Vent Connections

Figure 15.4

completed, test nipples, drain connections, carrier arms, or studs will extend through the wall, ready for the actual setting and connecting of the fixture or appliance, the second phase of installation.

Takeoff Procedure: A separate fixture takeoff sheet should be utilized for each floor. After completion of the takeoff, the quantities from each floor should be totaled on a summary sheet. Similar fixtures should be listed separately if they have different trim packages or colors.

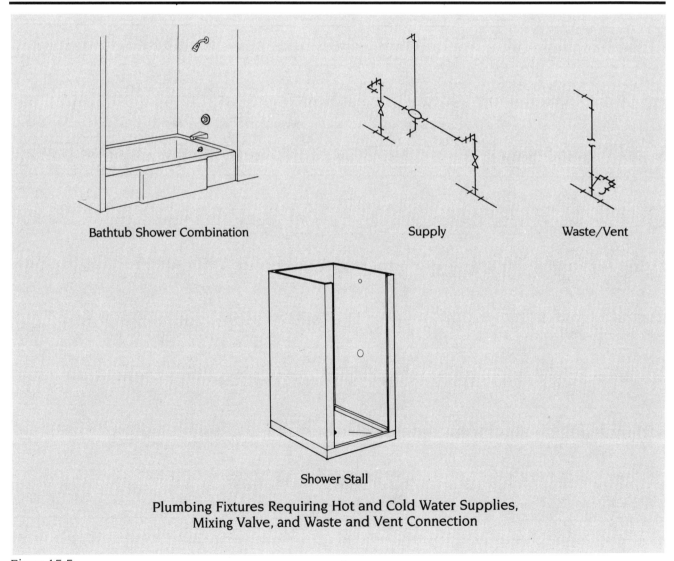

Bathtub Shower Combination   Supply   Waste/Vent

Shower Stall

Plumbing Fixtures Requiring Hot and Cold Water Supplies, Mixing Valve, and Waste and Vent Connection

Figure 15.5

Drains and Carriers

Drains and carriers can have a significant impact on the plumbing estimate, for both material and labor. Drains are used extensively throughout a building, including roof, floor, parapet, area, trench, yard, deck, swimming pool drains, and floor sinks. Also used in buildings are wall and yard hydrants, backwater valves, cleanouts, and interceptors. The remainder of this labor-intensive group includes fixture carriers and over 100 special drain waste and vent fittings to be combined with the water closet carriers. All of the above mentioned items are normally available from a single source of supply.

Storm drainage systems convey rainwater from roofs and the upper portions of structures where the water could cause damage to the building or constitute a hazard to the public. The water is discharged into a storm sewer or other approved location where it will not cause damage. The drainage system is not usually connected to a public sanitary sewage line.

Above ground piping for storm drains may be fabricated using brass, cast iron, copper, galvanized steel, lead, ABS, or PVC. Materials acceptable for underground use include cast iron, heavy wall copper, ABS, PVC, extra-strength vitrified clay pipe, and concrete pipe.

Stormwater is usually conducted away from the drainage area at the same rate that it collects. The discharge capacity is based on the size of the horizontal area to be drained, the design rate of the system, and the average rainfall rate. When the roof area abuts a sidewall, 50% of the wall area should be added to the roof area to determine the area to be drained. Local government departments, plumbing codes, and weather bureaus can usually supply climatological data on rainfall. This flow rate is the basis for sizing the system. The chart in Figure 23.19 is a guide for sizing. The chart is based on a maximum rate of rainfall of 4" per hour. Where maximum rates are more or less than 4" per hour, the figures for drainage should be adjusted. This can be done by multiplying by a factor of 4 and dividing by the local rate in inches per hour. The chart is applicable for round, square, or rectangular rainwater pipe. All of these shapes are considered equivalent if they can enclose a circle equal to the leader diameter.

Traps are not required for stormwater drains connected to storm sewers. Leaders and drains connected to a combined sewer and floor drain, which is then connected to a storm drain, need to be trapped. The size of the trap for an individual conductor should be the same size as the horizontal drain to which it is connected. The illustration in Figure 15.6 is a typical roof storm drainage system.

Floor and area drains provide adequate liquid drainage of floors and other surfaces in and around buildings. The local plumbing code will determine the minimum standards for size, location, and type. The code will also determine when and where interceptors and separators (for grease, oil, sand, hair, etc.) should be used. It also determines the when and where of traps, cleanouts, and backwater valves. Another major consideration in drain selection is the capability of the grate surface to be able to withstand the intended pedestrian or vehicular traffic.

Floor drains are manufactured in various metals, including bronze, alloy and cast iron tops, ductile iron, brass, and alloy bodies. The material used is governed by the drain location, (e.g., shower rooms, toilet rooms, hospitals, zoos, kitchens, etc.). A floor sink differs from a floor drain; a floor sink has rounded edges and an acid-resistant finish for improved sanitation and to allow thorough cleansing. Floor sinks, or receptors, should be used in food preparation areas, hospitals, and laboratories. Flushing rims for floor sinks and drains, can washers, and trap primers, all entail additional piping.

A plumbing estimator's library should contain complete drain and carrier manufacturers' catalogs. The estimator should be familiar with the accessories and various combinations which will affect the installation labor. Figure 15.7 shows common drains and castings. Carriers are shown in Figure 15.8.

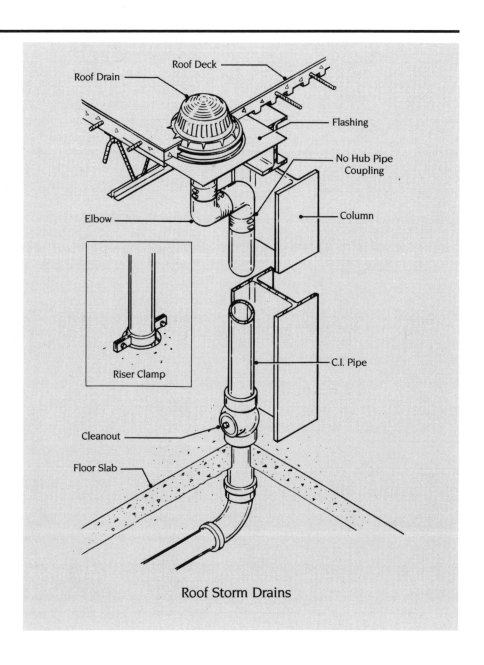

Roof Storm Drains

Figure 15.6

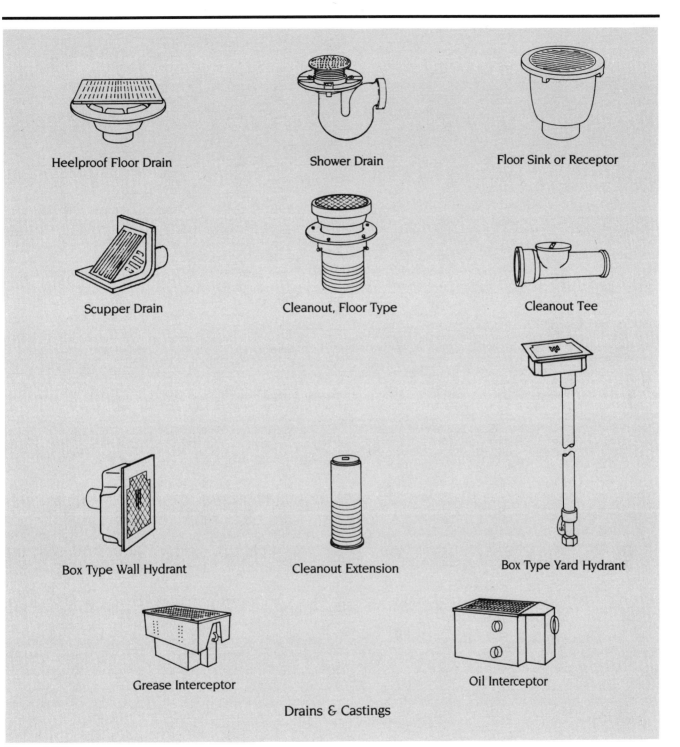

Heelproof Floor Drain

Shower Drain

Floor Sink or Receptor

Scupper Drain

Cleanout, Floor Type

Cleanout Tee

Box Type Wall Hydrant

Cleanout Extension

Box Type Yard Hydrant

Grease Interceptor

Oil Interceptor

Drains & Castings

Figure 15.7

139

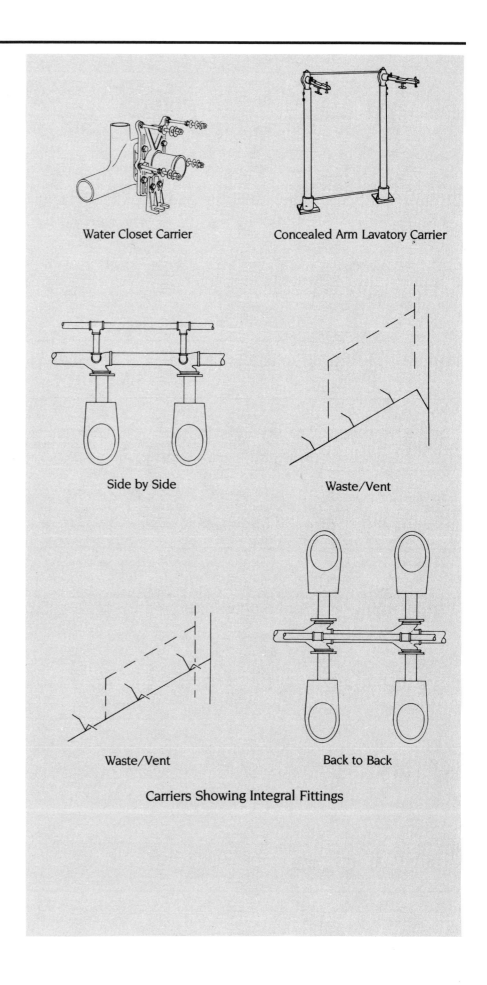

Water Closet Carrier

Concealed Arm Lavatory Carrier

Side by Side

Waste/Vent

Waste/Vent

Back to Back

Carriers Showing Integral Fittings

Figure 15.8

Cleanouts are an integral part of the drainage system, as they provide access for removal of accumulated solids. It is good practice, and often a code requirement, to include cleanouts and extensions at changes of direction greater than 45° and at appropriate distances in straight piping.

Piping connections for drains and cleanouts are available in sizes of 1-1/2", 2", 3", 4", 5", 6", and 8" for threaded, gasketed, caulked, or no hub connections.

Hydrants supply potable water to building exteriors and to hose outlets for remote building areas. For exterior walls in climates subject to freezing temperatures, nonfreeze features are mandatory. Automatic backflow shutoff, vacuum breaker, and pressure relief options are also available. Yard and post hydrants are used outside the building structure, and offer safety features similar to those of wall hydrants.

Wall hydrants have 3/4" IPS female and 1" male connections on the inlet, and 3/4" male hose connections on the outlet. Yard hydrants, both post type and box type, are available with 3/4" through 2" IPS female inlets and 3/4" through 2" male hose connections. Depth of bury is a consideration when taking off yard hydrants.

Carriers have evolved from simple supports for water closets, lavatories, sinks, and perineal baths, to being integrated into the drainage system itself. These devices secure wall-mounted plumbing fixtures above the floor. Special carriers are available for drinking fountains, electric water coolers, wheelchair lavatories, and penal or institutional fixtures. Floor-mounted back outlet water closets may also be integrated in the carrier support system. The use of carriers and wall-hung fixtures not only provides a neater and more sanitary environment, but saves man-hours of roughing and installation time.

Carrier closet fittings have optional waste and vent locations to accommodate any code requirement. They are made in over 100 variations and are available with 2", 3", 4", 5", and 6" pipe connections.

Carriers will accommodate all types of metal and plastic pipe accepted by code, and can accept back to back or side by side multiple closet fixture installations. Figure 15.8 shows water carriers with integral waste fittings for back to back and side by side group installation.

Units of Measure: Drains, cleanouts, hydrants, interceptors, and carriers are all estimated as each.

Material Units: Using the specifications as a guide and a manufacturer's catalog as a reference, the options and/or accessories for the various drains and carriers can be visualized and listed to comply with the designer's intent.

Labor Units: The following procedures must be considered and included in the labor estimate per drain for sleeving or cutting through each floor or roof:

- Setting the drain body and trim
- Returning at a later time to add the strainer, grate, funnel, or dome

Wall hydrants require the setting of the wall unit and making the inlet water connection. The installation of the wall unit must be coordinated with the construction of the wall. Setting may be accomplished by the wall construction contractor or by the plumber.

Interceptors are installed on the floor, partially recessed, flush with the floor, or in a pit or chamber beneath the floor. Size and weight must be considered in locating and setting an interceptor. Piping connections consist of vent, waste, inlet, and skimmer ports. When volatile gasses are to be piped, prevailing codes (be sure to check) may require a multiple venting arrangement. As with drains and wall units, installation should be coordinated with the general contractor.

Carrier labor considerations range from very simple installation of hangers and arms for sinks, lavatories, water fountains, or urinals, to complex integrated systems consisting of wall or floor-mounted carriers anchored to the building floor behind the finished wall, or in the case of substantial wall construction, bolted to or through the wall itself. More complex water closet carriers with integral waste and vent fittings are secured to the floor, piped into the waste and vent rough-in, all within the wall cavity or pipe chase. The fixture stubs and waste extension protrude through the finished wall for final fixture mounting.

Takeoff Procedure: Roof and area drains are taken off from the plumbing roof plans. The floor drains and interior specialty drains are taken off from the various floor plans and detail drawings, just as the backwater valves and interceptors might be.

Yard and wall hydrants are found on the ground floor plan and the site plan. Interior wall hydrants for special use might be found in swimming pool areas, kitchens, etc. These specialties can all be recorded on the same takeoff and summary sheet.

Carrier totals should be taken from the estimator's fixture takeoff sheet and listed with the drains for pricing.

Chapter 16

FIRE PROTECTION

Fire protection systems in the mechanical contracting industry include the use of standpipes, sprinklers, fire and smoke control dampers, fans, controllers, and alarms. This chapter contains descriptions of the materials and labor unique to installing standpipe systems, sprinkler systems, chemical systems, and combinations thereof. Piping, sheet metal, and other components have been or will be addressed in their own respective chapters.

Standpipe Systems

Standpipe systems, properly designed and maintained, are effective and valuable timesaving aids for extinguishing fires. This is especially true in the upper stories of tall buildings, the interior of large commercial or industrial malls, or other areas where construction features or access make the laying of temporary hose lines time-consuming and/or hazardous. Adequate pressure is obtained by city water pressure, a reservoir at the roof or upper story of the building, or by booster pumps. Standpipes are frequently installed with automatic sprinkler systems for maximum fire protection.

Standpipe systems with fire hose valves, hose cabinets, fire department, and pumper connections are installed by plumbers. Sprinkler piping systems with spray heads, special valve arrangements, alarms, air compressors, and controls are installed by sprinkler fitting specialists. The standard for standpipe system design is set by the National Fire Protection Association (NFPA), Volume 14. However, the local authority having jurisdiction must be consulted for special conditions, local requirements, and approval.

There are three general classes of service for standpipe systems, listed below:

- **Class I** systems are used by fire departments and personnel with special training for heavy hose (2-1/2" hose connections).
- **Class II** systems are used by building occupants until the arrival of the fire department (1-1/2" hose connector with hose).
- **Class III** systems are used by either the fire department and trained personnel or by the building occupants (both 2-1/2" and 1-1/2" hose connections or one 2-1/2" hose valve with an easily removable 2-1/2" by 1-1/2" adapter).

Standpipe systems are also classified by the method that water is supplied to the system. The four basic types are listed below:

- **Type 1** denotes a wet standpipe system with its supply valve open and water pressure maintained at all times.
- **Type 2** is a standpipe system arranged through the use of approved devices to admit water to the system automatically when a hose valve is opened.
- **Type 3** is a standpipe system arranged to admit water to the system through manual operation of approved remote control devices located at the hose stations.

• **Type 4** is a dry standpipe having no fixed water supply.

Standpipe systems (shown in Figure 16.1) are usually made of steel pipe. Other forms of ferrous or copper piping may be used where acceptable to local authorities. Joints may be flanged, threaded, welded, soldered, grooved, or any method compatible with the approved piping materials.

Units of Measure: Fire hose standpipe systems are taken off according to their components. The pipe itself is taken off and recorded by the linear foot. Valves, fittings, siamese connections, cabinets, etc., are taken off as each. Hoses and hose racks are priced as each assembly, which includes selections of hose length in 50', 75', and 100' increments.

Material Units: Material units for pipe and fittings are described in Chapter 13, "Piping." Care should be taken when pricing the required valves, adapters, couplings, finishes, type of hose, and type of nozzle. Hose threads in all standpipe systems must comply with the requirements of the local fire department. Local authorities may require that fire hose cabinets contain enough space, or a separate compartment, for a portable fire extinguisher. UL approval or listing may also be required for some of the component items. Materials and components that are UL listed are usually more expensive than nonlisted items.

Labor Units: Labor units for the piping portions of standpipe protection systems are described in Chapter 13, "Piping." Standpipe risers are required by law or code for high-rise buildings. They are installed piecemeal during construction as each floor is finished to ensure that there is always a live hose station within one or two floors of the top. Normally closed valves on this riser should not be left open, as this will inhibit flow to upper floors during a fire. These valves should, instead, be wired closed and tagged.

Glass fronts should not be installed on recessed hose cabinets until construction is in the finishing stages. These items will add man-hours to the overall installation labor. Any mechanical system that is to perform temporary duties during construction will also require additional labor: for installation, final cleaning, and possible replacement or repair before final acceptance. Labor units for standpipe systems should be given special consideration during the estimating process.

Takeoff Procedure: Piping portions of these systems are taken off by the linear foot. All other components and accessories are taken off and recorded as each. Each standpipe system should be listed on separate summary sheets, totaled, and priced. The totals from each summary sheet should then be listed separately on the plumbing estimate sheet.

Sprinkler Systems

The *wet pipe sprinkler system* (shown in Figure 16.2) is the most popular type of automatic fire suppression system in use today. This type of piping system is filled with water under pressure and connected to a municipal or private supply. Water is released immediately through one or more sprinkler heads when opened by fire or heat. A fusible element melts, opening the sprinkler head when a specified temperature is reached. Once opened, the head or heads must be replaced before the system can be reactivated. Quick response and low first cost are the main reasons for the popularity of the wet pipe system.

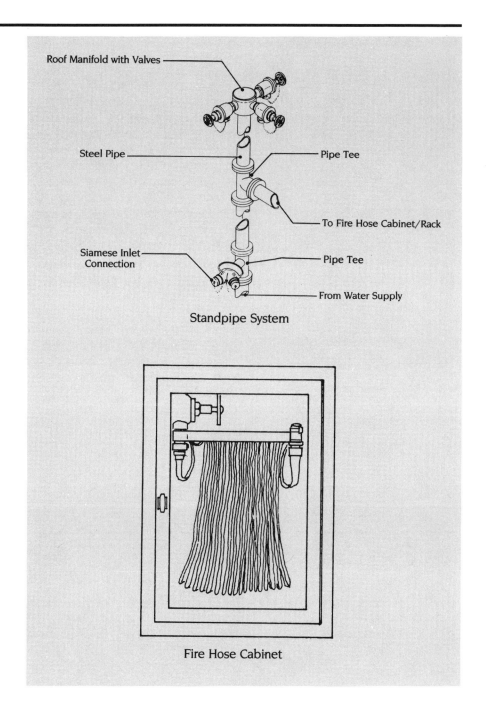

Roof Manifold with Valves

Steel Pipe

Pipe Tee

To Fire Hose Cabinet/Rack

Siamese Inlet
Connection

Pipe Tee

From Water Supply

Standpipe System

Fire Hose Cabinet

Figure 16.1

In areas subject to freezing, *dry pipe sprinkler systems* (shown in Figure 16.3) are used. The dry pipe system is similar to the wet pipe system, except that the dry pipe system is filled with compressed air. The compressed air is maintained under a constant pressure higher than the available water pressure. Water is kept out of the system by means of a valve which contains the opposing pressures. When one or more sprinkler heads open, after sensing fire or heat, the air pressure drops. Water enters the piping system and is released through the sprinkler heads. The dry pipe system does not respond as quickly as the wet pipe system and costs more to install because of the required compressed air and special dry pendent heads. Further, dry pendent heads and the piping must be installed without low points or pockets that cannot be drained. The dry type valve itself must be enclosed in an area not subject to freezing.

For areas where water damage must be kept to a minimum during fire suppression, a *firecycle system* is employed (shown in Figure 16.4). The firecycle system is a dry system with preaction fire detection devices and a control valve. Electrical controls in this system have the capability to close the flow control valve after fire detectors sense that the fire is out. A time-delay feature built into the system allows the water to flow through the open type sprinkler head for a predetermined period before closing the flow valve. Should the fire re-ignite, the valve opens and the cycle begins again. Battery backup is a requirement for this type of system.

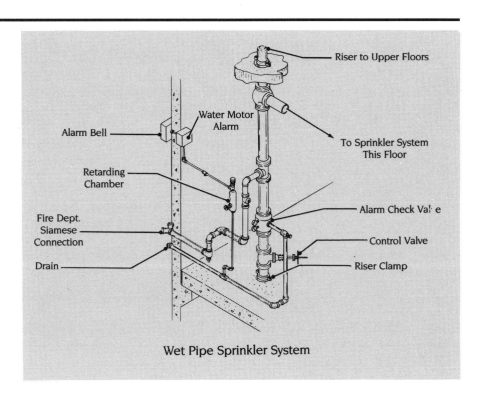

Wet Pipe Sprinkler System

Figure 16.2

The firecycle system is always available for subsequent fires since it does not have to be shut down for head replacement, as do systems employing sprinkler heads with fusible links. This type of system also eliminates the need for manual shut-off valves and position indicators.

Preaction systems (shown in Figure 16.5) are used in areas subject to freezing. They are also used where accidental damage to sprinkler heads or piping, and subsequent water leakage, is unacceptable. Because the preaction system is more sensitive to fire than a sprinkler head, it provides a quicker response than the dry pipe system.

The preaction system is filled with air that does not have to be pressurized. Conventional closed-type sprinkler heads are used; however, the initial detection of heat or fire occurs via heat-activated devices that are more sensitive than the sprinkler heads. These devices detect a rapid temperature rise and open the preaction sprinkler valve. This valve fills the system with water and activates the alarm. With a further rise in temperature, one or more sprinkler heads in the fire area will open and begin the extinguishing process. This type of system has two important advantages: early warning, which allows occupants to evacuate; and quick notification of firefighting personnel. Like the dry pipe

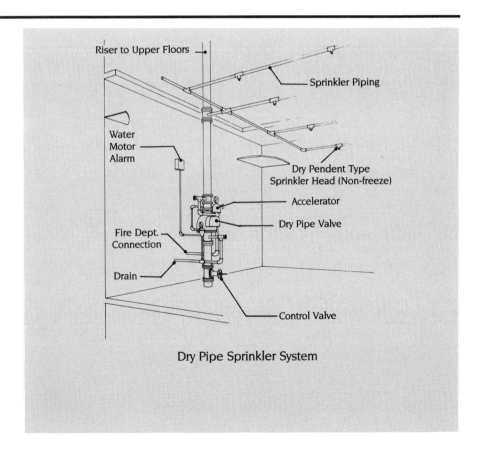

Dry Pipe Sprinkler System

Figure 16.3

system, if the preaction system is installed in an area subject to freezing, pendent heads (if used) must be of the dry type and the other drainage provisions must be adhered to. The preaction valve must be installed in a nonfreeze area.

A *deluge system* (shown in Figure 16.6) is a dry system connected to a water supply. This water supply is contained by a preaction deluge valve which is controlled by heat-activated devices. This type of system employs open-type nozzles rather than fusible heads. When the deluge valve opens, it admits water to the entire system, or valved zone. The valve then discharges water from all the nozzles in the protected area. By using preaction fire detection devices and by wetting down the entire area instantly, the deluge system prevents the spread of the fire. For this reason, the deluge system is particularly suitable for storage or work areas involving flammable liquids.

All of the aforementioned types of sprinkler systems can be combined with fire standpipe systems. Sprinkler systems should be designed and installed in accordance with the current provisions of the National Fire Protection Agency, Standard number 13. Traditional piping methods and materials, black steel threaded and flanged piping, are often replaced with thin wall steel pipe and-grip type fittings. The abundance and frequency of sprinkler heads, each requiring a tee, almost mandates the use of threaded pipe and fittings or

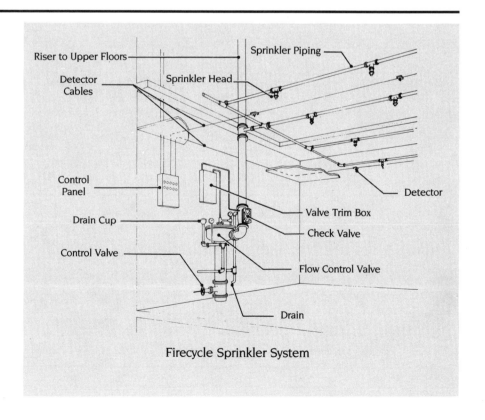

Firecycle Sprinkler System

Figure 16.4

the grip-type system. In some situations, other acceptable materials include copper tubing and several types of plastic tubing. Local building codes govern the acceptability of these materials.

System Classification

Rules for installation of sprinkler systems vary depending on the classification of building occupancy falling into one of the following three categories.

Light Hazard Occupancy: The protection area allotted per sprinkler should not exceed 200 S.F. with the maximum distance between lines and sprinklers on lines being 15'. The sprinklers do not need to be staggered. Branch lines should not exceed eight sprinklers on either side of a cross main. Each large area requiring more than 100 sprinklers and without a subdividing partition should be supplied by feed mains or risers for ordinary hazard occupancy. Building types included in this group are:

- Auditoriums
- Churches
- Clubs
- Museums
- Nursing Homes
- Offices

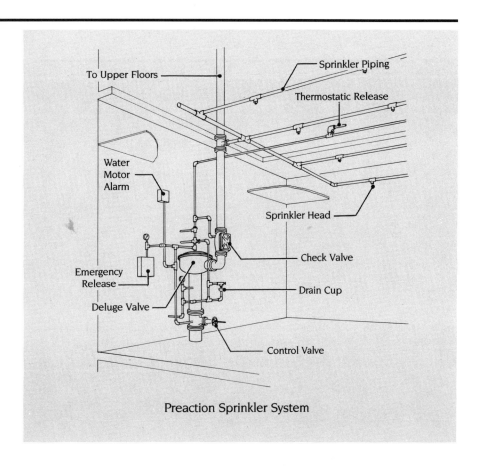

Preaction Sprinkler System

Figure 16.5

- Educational
- Hospital
- Institutional
- Libraries (except large stack rooms)

- Residential
- Restaurants
- Schools
- Theaters

Ordinary Hazard Occupancy: The protection area allotted per sprinkler shall not exceed 130 S.F. of noncombustible ceiling and 120 S.F. of combustible ceiling. The maximum allowable distance between sprinkler lines and sprinklers on line is 15'. Sprinklers shall be staggered if the distance between heads exceeds 12'. Branch lines should not exceed eight sprinklers on either side of a cross main. Building types included in this group are:

- Automotive garages
- Bakeries
- Beverage manufacturing
- Bleacheries
- Boiler houses
- Canneries
- Cement plants
- Clothing factories
- Cold storage warehouses
- Dairy product mfg.

- Electric generating stations
- Feed mills
- Grain elevators
- Ice manufacturing
- Laundries
- Machine shops
- Mercantiles
- Paper mills
- Printing and publishing
- Distilleries

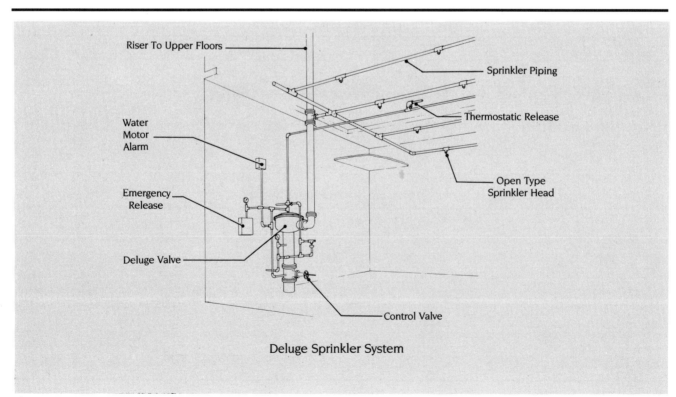

Deluge Sprinkler System

Figure 16.6

- Dry cleaning
- Shoe factories
- Wood product assembly

Extra Hazard Occupancy: The protection area allotted per sprinkler shall not exceed 90 S.F., of noncombustible ceiling and 80 S.F. of combustible ceiling. The maximum allowable distance between lines and between sprinklers on lines is 12'. Sprinklers on alternate lines shall be staggered if the distance between sprinklers on lines exceeds 8'. Branch lines should not exceed six sprinklers on either side of a cross main. Included in this group are:

- Aircraft hangars
- Chemical works
- Explosives mfg.
- Linoleum mfg.
- Linseed oil mills
- Volatile or flammable liquid manufacturing and use
- Paint shops
- Shade cloth mfg.
- Solvent extracting
- Varnish works
- Oil refineries

Units of Measure: For bidding purposes, sprinkler systems are often estimated by the square foot of floor area to be protected. Square foot prices are developed from the contractor's own records for each type of system and material specified. While it is preferable to use prices from completed projects, the estimator may also wish or need to consult cost publications such as *Means Plumbing Cost Data*. A detailed unit price estimate may be required if the square foot cost data is not sufficient or relevant to the project. An item by item takeoff and pricing exercise is required for a unit price estimate. The pipe is taken off by the linear foot, according to the specified material and joining methods. Fittings, valves, and accessories are recorded as each.

Material Units: Material units for steel, copper, and plastic piping have been described in Chapter 13, "Piping." Alarm valves, controllers, and sprinkler heads should be recorded on the takeoff sheets by type and class. From these totals, a list of required accessories (e.g., escutcheons or guards) can be determined. A chrome finish may be specified for heads in various areas. Wax coated heads are utilized in corrosive atmospheres. In climates subject to freezing, special nonfreeze pendent heads must be used for drop ceiling installations. These nonfreeze pendent heads are complex and expensive to purchase and to install because the exact measure of the drop to a finished ceiling must be determined prior to installation. Final adjustment of the telescoping escutcheon allows up to 1/2" play for field adjustment. Not shown on the drawings but usually required are spare heads, cabinets, wrenches, and other miscellaneous devices; check the specifications for these requirements.

Labor Units: Labor units for piping are covered in Chapter 13, "Piping," under the appropriate piping material classification. Fabrication and installation of sprinkler piping, however, are unique. Most sprinkler piping is cut and fabricated at the shop or at a job site pipe shop. Fabricated piping assemblies are bundled together and marked or coded by the location of final installation. The pipe sizes, measures, and coding schedule are listed on a layout plan prepared by the sprinkler contractor from the architectural plans. These plans are coordinated with the electrical and sheet metal contractors to avoid scheduling conflicts. In some instances, shop welding is used in the pipe fabrication, however, welding is not permitted for modification or repair of an existing system. An installation layout plan does not have to be prepared by the contractor until construction is to begin. At this time, a complete installation plan is required. If a detailed unit price estimate is made, some of the installation plans may be required in advance. A square foot estimate normally requires little advance planning.

Since sprinkler heads are precisely located by code, and the piping has to maintain a required pitch, other trades must defer to the sprinkler piping route.

Takeoff Procedure: For a unit price estimate, the sprinkler heads are taken off first and summarized by type for pricing. The main valves, alarms, and controllers should be listed accordingly. The pipe, fittings, control or zone valves are taken off as described in Chapter 13, "Piping." From these totals the necessary amounts of hangers and other appurtenances are estimated.

For a square foot estimate, building area is taken off in square feet. This is used along with the system type and materials to determine cost.

Chemical, Foam, and Gas Fire Suppression

Chemicals, foams, and gasses of various types delivered through portable or fixed spray systems are used to extinguish fires in special applications. Fixed spray systems include foams and gasses, primarily carbon dioxide or halon, which extinguish fires by suffocation. Portable spray systems include foams and gasses, as well as a variety of dry chemicals and pressurized water.

Halon Fire Suppression

Halon fire suppression systems are used because they are fast, effective, and clean (shown in Figure 16.7). These systems are called "clean" because they leave no residue that must be cleaned up or that could contaminate valuable items (i.e., records or electronic equipment). Because halon gas is a nonconductor of electricity and is several times as heavy as air, it permeates the working area and penetrates cabinets or other electric or electronic enclosures where chemical powders cannot.

Depending on its concentration, halon gas ranges from nontoxic to low toxicity. Halon 1301 is normally used in a concentration of 5% to 7%. This concentration has no effect on personnel in the area, for a limited period of time.

Seven to ten percent concentration requires evacuation within one minute of exposure; concentrations above 10% require evacuation prior to discharge. Halon 1301 is also colorless and has minimal visual impedance to hamper evacuation.

Halon may be applied to fires by portable extinguishers, local application of strategically placed cylinders (or prepackaged systems), or by a centrally located cylinder or battery of storage cylinders connected to a piping distribution system and discharge nozzles. The local application or modular system is the most cost effective type of system, because it eliminates the piping installation costs encountered in a centrally located system.

Detection and actuation are critical requirements for fast extinguishing, to eliminate not only fire damage but also the accompanying risks of smoke, heat, carbon monoxide, and oxygen depletion. Fire and smoke detectors are wired to a control panel that activates the alarm systems, verifies or proves the existence of combustion and releases the extinguishing agent, all in a matter of seconds.

The halon fire suppression system is most effective in an enclosed area. The control system may also have the capability of closing doors and shutting off exhaust fans. These systems have been commonly used in the following places:

- Aircraft (both cargo and passenger)
- Libraries and museums
- Bank and security vaults
- Electronic data processing
- Transformer and switchgear rooms
- Tape and data storage vaults (rooms)
- Telephone exchanges
- Laboratories
- Radio and television studios
- Flammable liquid storage areas

Although halon fire suppression systems have been popular in protecting sensitive equipment and areas, it is soon to be replaced. Halon 1301 is a member of a family of halogenated agents that has a detrimental effect on the earth's ozone layer. For this reason its production has been frozen by international agreement. In a few years its use will be prohibited completely. At this time, halon 1301 can be used only if no other agent can provide adequate protection. Manufacturers are currently developing a replacement which will hopefully be adaptable to existing halon system configurations.

Units of Measure: Self-contained or modular fire suppression systems are taken off and priced as each. For larger "pre-engineered" or "engineered" systems (central storage configuration) components should be taken off individually as each. Pipe should be taken off by the linear foot (see Chapter 13 for piping and components).

Material Units: A halon piping system (or other fixed spray system) is similar to a sprinkler piping system in material costs; hangers and supports are

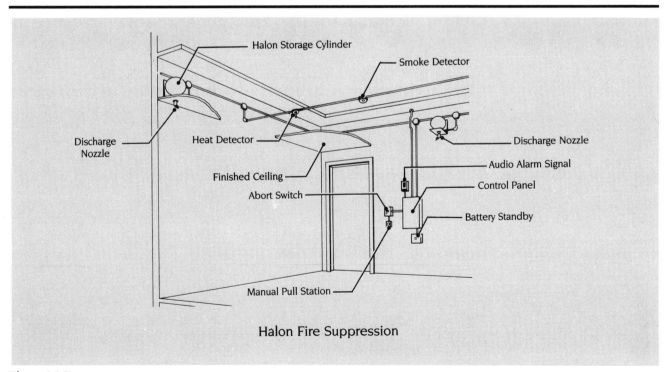

Halon Fire Suppression

Figure 16.7

determined from the piping totals and cylinder locations. Dispersion nozzles (rather than heads) are utilized, although not as frequently as heads are used in sprinkler systems. Other materials to be taken off and priced are release stations, detectors, deflectors, alarms, annunciators, abort switches, cylinders (containers), and controllers.

Labor Units: Before labor units can be estimated, discharge nozzles and detectors should be located and a piping route laid out from the nozzles back to the storage cylinders. The quantities should also be taken off to determine fabrication labor. Steel T&C (threaded and coupled) piping and grip type connectors are most often used in these systems.

Takeoff Procedure: The takeoff should begin with the nozzles and other piping components, and proceed to the pipe and fitting takeoff. Any valves used in these systems will be found at the storage cylinders or manifolds. Piping takeoff should proceed as described in Chapter 13, "Piping."

Other Fire Extinguishing Systems

In addition to the methods of fire protection already covered in this chapter, the plumbing contractor may be required to furnish portable extinguishers, and the sheet metal contractor may be required to install or furnish kitchen hoods with built-in fire dampers, spray assemblies, and fan disconnect apparatus.

Other spray systems similar to halon are used in the form of portable extinguishers or fixed pipe installations. These spray systems include foam used primarily for fuel fires and other flammable liquids, and carbon dioxide. This type of system extinguishes fires by smothering. Because carbon dioxide is slightly toxic it is limited in use to fires in classifications B or C (listed below). It should also be noted that carbon dioxide dissipates and may allow re-ignition.

A third type of extinguisher covers a variety of dry chemicals and powders which are available for the range of fire classifications. Care should be taken to match the proper system with the expected hazard. To facilitate proper use of extinguishers on different types of fires, the NFPA Extinguisher Standard has classified fires into the following four types.

- **Class A** fires involve ordinary combustible materials (such as wood, cloth, paper, rubber, and many plastics) requiring the heat-absorbing (cooling) effects of water, water solutions, or the coating effects of certain dry chemicals which retard combustion.
- **Class B** fires involve flammable or combustible liquids, flammable gasses, greases, and similar materials where extinguishment is most readily secured by excluding air (oxygen), inhibiting the release of combustible vapors, or interrupting the combustion chain reaction.
- **Class C** fires involve live electrical equipment where safety to the operator requires the use of electrically nonconductive extinguishing agents. (Note: When electrical equipment is de-energized, the use of Class A or B extinguishers may be indicated).
- **Class D** fires involve certain combustible metals (such as magnesium, titanium, zirconium, sodium, potassium, etc.) requiring a heat-absorbing extinguishing medium not reactive with the burning metals.

Some portable fire extinguishers are of primary value on only one class of fire; some are suitable on two or three classes of fire; none is suitable for all four classes of fire.

Class A and Class B fires. Color coding is part of the identification system, and the triangle (Class A) is colored green, the square (Class B) red, the circle (Class C) blue, and the five-pointed star (Class D) yellow. These symbols are shown in Figure 16.8.

Units of Measure: Portable fire extinguishers and related accessories are taken off and priced as each. Fixed pipe systems, such as halon systems, are taken off as each or by component, depending on system configuration.

Material Units: Material units for portable fire extinguishers are relatively simple. Individual wall-hung extinguishers, when specified, will be furnished with mounting brackets. These same brackets may be used for mounting within certain types of extinguisher cabinets as well.

Specifications for cabinets may require one or more extinguishers or hose and valve combinations.

The cabinets would have the same material units as previously outlined under standpipe systems. Cabinet doors and trim may be painted steel, aluminum, or stainless steel. The cabinet front may be solid panel, glass, wire glass, plexiglass, or a combination of solid panel and glass. Decals, blankets, spanner wrenches, axes, and alarms are among the safety and operating options available.

Fixed or permanent systems will have piping and heads or nozzles similar to the apparatus in the halon systems. Kitchen exhaust hoods would probably arrive on the job with the piping assembly in place. Unique fuel loading areas might require a piping system and components (covered in Chapter 13, "Piping").

Labor Units: Labor considerations for portable extinguishers include the following:

- Anchoring the wall bracket to the building structure
- Securing the extinguisher in its bracket

Fire Classification Symbols

Figure 16.8

arrive on the job with the piping assembly in place. Unique fuel loading areas might require a piping system and components (covered in Chapter 13, "Piping").

Labor Units: Labor considerations for portable extinguishers include the following:

- Anchoring the wall bracket to the building structure
- Securing the extinguisher in its bracket

Pressurized water type extinguishers are shipped empty and must be charged in the field. Any extinguishers that have been set in place for standby use during construction will have to be removed, cleaned, and inspected or recharged before final acceptance.

Recessed cabinets are built into the walls by the General Contractor. The finished frames and doors are installed by the plumber as the extinguishers are placed. This is similar to the procedures previously outlined for fire hose cabinets.

The labor units for permanent installations are similar to the units for halon systems. Labor units for piping components will be found in Chapter 13, "Piping."

Takeoff Procedure: Portable fire extinguishers and special built-in permanent spray systems are often not shown on the mechanical drawings, however, the specifications will indicate the types and locations required. The architectural floor plans and elevations may indicate each extinguisher station. The plumbing drawings will indicate whether or not extinguishers will be housed with fire hoses (in the same enclosure). Spray systems for kitchen exhaust hoods should be found either on the kitchen or HVAC equipment drawings or both. Piping arrangements for fuel storage and piping systems will be indicated on the drawings.

The estimator, in beginning the takeoff, should separate the extinguishers by type and floor. Permanent spray installations should have a pipe and fittings takeoff as described in Chapter 13, "Piping." The spray heads and any valves, pumps, or storage cylinders should be recorded and priced separately from any other piping system.

The project documents should be carefully read to determine who is responsible for supply and/or installation of these special items.

Chapter 17

HEATING, VENTILATING AND AIR-CONDITIONING

The word air-conditioning is often used to mean air cooling. Actually, the field includes much more then just cooling; heating, cleaning, humidifying, dehumidifying, and replacement of the air within a structure or contained space are other functions performed by mechanical air-conditioning systems. The heating or cooling medium might be air, water, or steam, and the energy source, coal, electricity, gas, or oil. The conveyance of these liquids and gases (piping and ductwork) is covered under other sections of this book. This chapter contains descriptions of equipment (both unitary and packaged), and hydronic and air handling systems.

Heating Boilers

Heating boilers are designed to produce steam or hot water. The water in the boilers is heated by coal, oil, gas, wood, electricity, or a combination of these fuels. Boiler materials include cast iron, steel, and copper. Several types of boilers are available to meet the hot water and heating needs of both residential (see Figure 17. 1) and commercial buildings (see Figures 17.2 and 17.3).

Cast-iron sectional boilers (see Figures 17.1, 17.2, and 17.3) may be assembled in place or shipped to the site as a completely assembled combustion package. These boilers can be made larger on site by adding intermediate sections.

Steel boilers (see Figure 17.2 and 17.3) are usually shipped to the site completely assembled. Large steel boilers may be shipped in segments for field assembly and testing. The components of a steel boiler consist of tubes within a shell, plus a combustion chamber. If the water being heated is inside the tubes, the unit is called a *water tube boiler*. If the water is contained in the shell and the products of combustion pass through tubes surrounded by this water, the unit is called a *fire tube boiler*. Water tube boilers are manufactured with copper tubes or coils. Electric boilers have elements immersed in the water and do not fall into either category of tubular boilers.

Heating boilers are rated by their hourly output expressed in "British Thermal Units" (Btu). The output available at the boiler supply nozzle is referred to as the gross output. The gross output in Btu per hour divided by 33,475 indicates the boiler horsepower rating. The net rating of a boiler is the gross output less allowances for the piping tax and the pickup load. The net load should match the building heat load.

Due to the high cost of fuels, efficiency of operation is a prime consideration when selecting boiler units, and because of this, recent innovations in the manufacturing field have provided the opportunity to install efficient and compact boiler systems.

The Department of Energy has established test procedures to compare the "Annual Fuel Utilization Efficiency" (AFUE) of comparably sized boilers. Better insulation, heat extractors, intermittent ignition, induced draft, and automatic draft dampers contribute to the near 90% efficiencies claimed by manufacturers today.

In the search for higher efficiency, a new concept in gas-fired water boilers has recently been introduced. This innovation is the pulse condensing type boiler, which relies on a sealed combustion system rather than on a conventional burner. The AFUE ratings for pulse-type boilers are in the low to mid 90% range. Pulse-type boilers cost more initially than conventional types, but savings

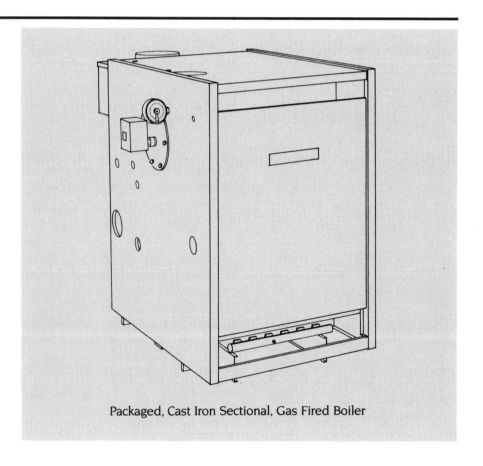

Packaged, Cast Iron Sectional, Gas Fired Boiler

Figure 17.1

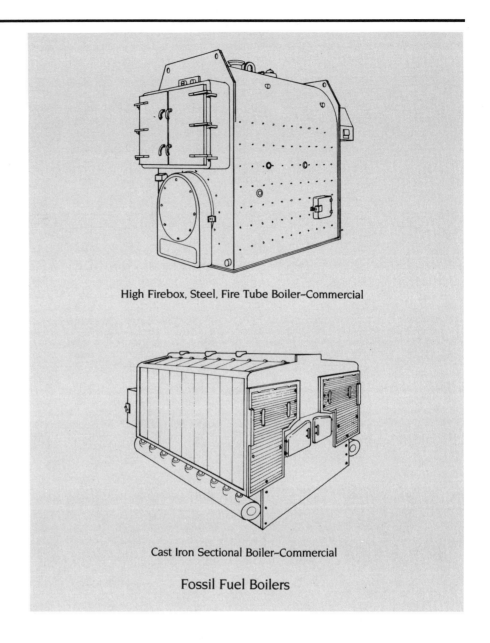

High Firebox, Steel, Fire Tube Boiler–Commercial

Cast Iron Sectional Boiler–Commercial

Fossil Fuel Boilers

Figure 17.2

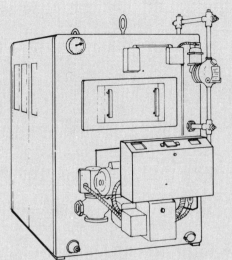

Packaged, Cast Iron Sectional,
Gas/Oil Fired Boiler–Commercial

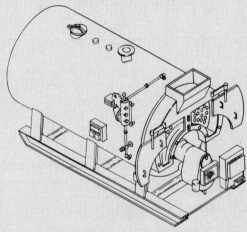

Packaged, Oil Fired,
Modified Scotch Marine Boiler–Commercial

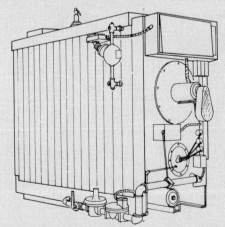

Packaged, Gas Fired, Steel, Watertube Boiler–Commercial
(can be rolled or skidded through a standard
doorway by removing gas train and exterior trim)

Fossil Fuel Boilers

Figure 17.3

in other areas help to offset this added cost. Because these units vent through a plastic pipe to a side wall, no chimney is required. The pulse-type boiler also takes up less floor space, and its high efficiency saves on fuel costs.

Another innovation in the boiler field is the introduction from Europe of wall-hung, residential-size boilers. These gas-fired, compact, efficient (up to 80% AFUE) boilers may be directly vented through a wall or to a conventional flue. Combustion makeup air is directed to a sealed combustion chamber similar to that of the pulse-type boiler. Storage capacity is not needed in these boilers because the water is heated instantaneously as it flows from the boiler to the heating system. The boiler material consists mostly of steel. Heat exchanger water tubes are made of copper or stainless steel, with some manufacturers using a cast-iron heat exchanger. Governing conditions in boiler selection include the following:

- **Accessibility**: Both for installation and for future replacement, cast-iron boiler sections can be delivered through standard door or window openings. Some steel water-tube boiler sections can be delivered through standard door or window openings. Some steel water-tube boilers are made long and narrow to fit through standard door openings.
- **Economy of installation**: Packaged boilers that have been factory fired require minimal piping, flue, and electrical field connections.
- **Economy of operation**: In addition to the AFUE ratings, boiler output should be matched as closely as possible to the building heat loss. Installation of two or more boilers (modular), piped and controlled so as to step-fire to match varying load conditions, should be carefully evaluated. Only on maximum design load would all boilers be firing at once. This method of installation not only increases boiler life, but also provides for continued heating capacity in the event that one boiler should fail.

The two boilers shown in Figure 17.2 may be trimmed out to fire any fossil fuel and generate either steam or hot water. The cast-iron boiler may be shipped completely assembled, or the nine "pork chop" sections (18 castings) of the boiler may be assembled in place by two men.

Truly packaged boilers are factory fired and tested. They arrive on the job ready to be rigged or manhandled into place. Connections required are minimal; fuel, water, electricity, supply and return piping, plus a flue connection to a stack or breeching. Figure 17.3 shows three different packaged burner units.

If steam or hot water service is available to a building from a remote source of supply, the need for a boiler is eliminated. If the proposed system is to be forced hot water and the remote source of supply is to be steam or high temperature hot water, a shell and tube type heat exchanger (converter) must be provided. A heat exchanger would also be required if the building is served by a low temperature chilled water service or brine for building cooling.

Units of Measure: Boilers and heat exchanges are taken off and recorded as each.

Material Units: Manufacturers' catalogs should be consulted to determine options available for the specified boiler or its equal. Steel boilers are usually less expensive than cast-iron boilers, however, it is not up to the estimator to substitute on so grand a scale. A cast-iron boiler shipped "knocked down" in sections may appear to be less expensive than a packaged unit, but labor and material costs to assemble the unit should be considered before a choice between the two options is made.

Quotations should be carefully examined to ensure that the proper steam or water trim, insulated jacket, operating controls, burner package, and delivery to the site or a staging area (i.e., the contractor's or rigger's yard) are included in the price. In some instances the burner gas, oil, or combination must be purchased from a separate source. Allowance must be made for a proper mounting front plate, refractory, and possible electrical work. An oil-fired unit will require an oil storage tank, and a pumping and piping system to the boiler. Grates and stokers should be added to the boiler cost for coal-fired burners. The job specifications should be thoroughly read to ensure that all requirements are met.

Labor Units: Residential boilers are normally placed by two men while still in the shipping crate. Larger boilers, whether packaged or knocked down, require mechanical aids for unloading and positioning. These tools may range from rollers assisted by a come-along, rope or chain fall, or a rigging crew, to a crane or even a helicopter, depending on location. There is often a piece of excavating or hoisting equipment on the job which can be used to set boilers, oil tanks, chillers, cooling towers, air handling units, etc., at a considerable cost savings. This advantage, however, cannot be reliably preplanned, therefore, the estimator should not consider it in the estimating procedure.

Takeoff Procedure: Boilers or boiler systems can be directly entered on the estimate or summary sheet as they will usually be taken off from the same plan. If the estimate is for tract housing or multi-family units with many boilers, a separate takeoff sheet is necessary to list all the boilers. Fuel oil storage and piping systems should also be handled on a separate sheet.

Hydronic Terminal Units

Hot water or steam systems transfer heat to desired locations via radiators, convecters, coils, heat exchangers, non-ducted fan coil units, humidifiers, water heaters, condensate meters, and fuel oil heaters, as well as to hospital and kitchen equipment. Figure 17.4 illustrates typical layouts for hydronic systems.

Radiation is broken down into two classes, direct and indirect. Direct radiation includes freestanding or wall hung cast-iron sectional tubular radiators, cast-iron baseboard, fin and tube bare elements or with expanded metal grilles. Also available (manufactured in Europe) are stamped steel panels containing channels or flattened tubes for water or steam circulation, classed as direct radiation.

Indirect radiation, which heats by air circulating through its enclosure by convection, includes cast iron-or copper heating elements enclosed in sheet metal casings called convecters. Convecters may be freestanding, wall hung, semi- or fully-recessed. Residential copper tube aluminum fin baseboard radiation encased in a metal enclosure as well as commercial fin tube (both copper and steel) are methods of indirect radiation. Two types of hydronic heating terminal units are shown in Figure 17.5.

Duct coils are suspended in the supply ductwork. Suspended, floor, or wall mounted unit heaters are heating coils and a fan within the same enclosure.

Units of Measure: Freestanding and wall hung radiators are taken off as each and listed according to type, size, or output. Unit heaters and duct coils are recorded in the same manner. Baseboard radiation, fin tube, and panel must be taken off as each, but also recorded by linear foot of enclosure. In the case of wall to wall, radiation is recorded by the linear foot of heating element (and the number of rows). All other terminal units should be recorded as each and

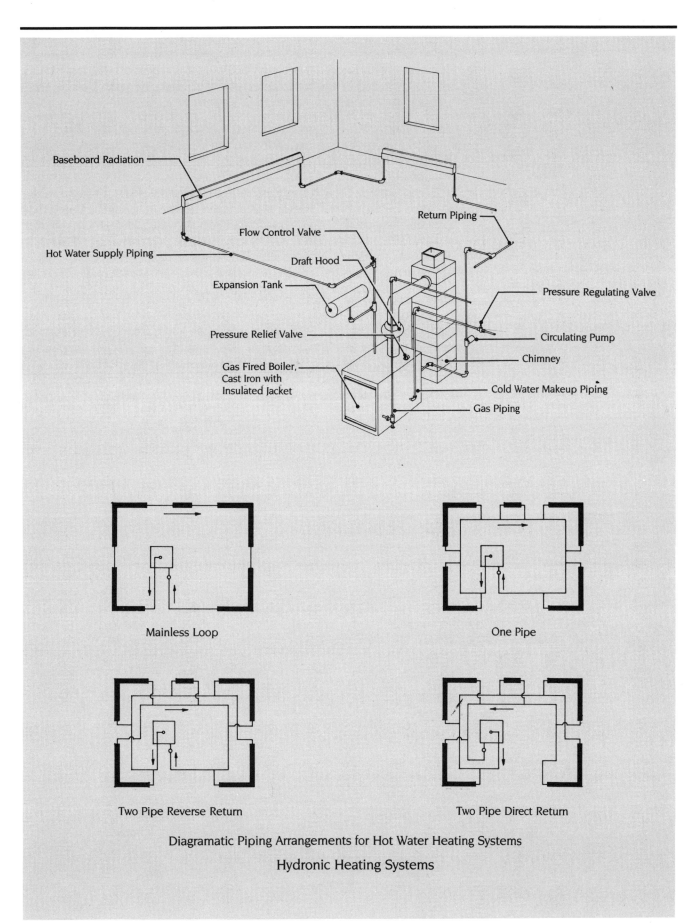

Baseboard Radiation

Hot Water Supply Piping

Flow Control Valve

Draft Hood

Expansion Tank

Pressure Relief Valve

Gas Fired Boiler,
Cast Iron with
Insulated Jacket

Return Piping

Pressure Regulating Valve

Circulating Pump

Chimney

Cold Water Makeup Piping

Gas Piping

Mainless Loop

One Pipe

Two Pipe Reverse Return

Two Pipe Direct Return

Diagramatic Piping Arrangements for Hot Water Heating Systems

Hydronic Heating Systems

Figure 17.4

kept with their specialized system (e.g., hospital equipment, fuel oil heaters, etc.). Cast-iron radiators are priced by output in square feet of radiation. Fin tube radiation is priced by the linear foot plus the various trim items. Unit heaters, coils, and panel radiators are priced as each.

Material Units: Wall-hung radiation of all types require supports. Indirect radiation requires enclosures, knob or chain dampers, access doors, special grilles, and finishes. All radiators usually require a supply control valve (manual or automatic), a balancing valve, on the return, or a radiator trap for steam radiation. Both water and steam radiation require automatic or manual air venting capability. If the radiator is down feed hot water, it should have a drain cock. Unit heaters, duct coils, and all terminal units would have almost identical trim (the size will vary) with the radiation. Manufacturers' catalogs showing all the options and accessories should be referred to while preparing the estimate.

Labor Units: The labor to install and pipe hydronic units should be based on the unit count (i.e., a certain number of each type per day). For fin tube and all baseboard, the linear footage of enclosure must also be considered as hanging so many feet per day of both back and front panels. For long runs of dummy enclosure in wall to wall installations, be sure to capture both the material and labor costs of the bare pipe concealed within the enclosure. Duct coils are installed by the tin knocker and piped by the pipefitter.

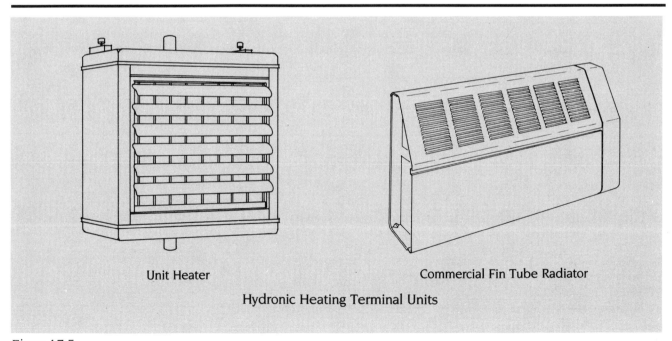

Unit Heater   Commercial Fin Tube Radiator

Hydronic Heating Terminal Units

Figure 17.5

Takeoff Procedure: On a hydronic radiation job the takeoff should begin at the radiation or heating units and proceed floor by floor, starting at the top. This procedure helps to familiarize the estimator with the building. The heating units can also be used later as targets for the piping takeoff. The units should be color coded to correspond with the piping color codes. All similar types of radiation can be totaled together for pricing and labor estimating. For example, if 42 cast-iron wall-hung radiators are specified for 1,260 S.F., this means 42 supply valves, 42 air vents, 84 wall brackets, and 42 traps are needed. The radiation itself is priced by the S.F. of radiation (E.D.R.). If 42 fin tube radiators, 1 row of slope top, 600 feet of enclosure, and 504 linear feet of element were specified, then 42 supply valves, 42 air vents, 84 element supports, 36 intermediate enclosure supports, and 84 access doors or valve enclosures are needed. If inside or outside corners are required, they should be listed separately.

Unit heaters, coils, and the related accessories are taken off and totaled on the summary sheets. These items may be priced in with fans and air-handling units by some manufacturers (or manufacturers' representatives) in order to gain an advantage over competitors who cannot quote a complete line. A well prepared estimator will have several sources of pricing and supply to compare individual prices with lump sum quotes. An example of this is shown in Figure 17.6.

Chillers and Hydronic Cooling

The central component of any chilled water air-conditioning system is the water chiller. Packaged water chillers (shown in Figure 17.7) are available in three basic designs: the reciprocating compressor, direct-expansion type; the centrifugal compressor, direct-expansion type; and the absorption type. Chillers and other air-conditioning apparatus are sized by the ton; that is, a ton of cooling which equals the melting rate of one ton of ice in a 24-hour period, or 12,000 Btu per hour. The three types of chillers vary significantly in their cooling power, as well as in their operation. The reciprocating compressor chiller, which generates cooling capacities in the range of 10 to 200 tons, is usually powered by an electric motor. The centrifugal compressor, which generates cooling capacities ranging from 100 to several thousand tons, is also commonly powered by an electric motor, but it may be designed for a steam-turbine drive as well. In some instances both of these types of chillers may be powered by internal combustion engines.

Absorption-type chillers provide cooling capacities ranging from 3 to 1,600 tons. Because it uses water as a refrigerant and lithium bromide or other salts as an absorbent, this system consumes about 10% of the electrical power required to operate the conventional reciprocating and centrifugal direct-expansion chillers. This low consumption advantage is particularly desirable in buildings where an electrical power failure triggers an emergency backup system, as in hospitals, data processing centers, electronic switching systems locations, and other buildings that must continue to function on auxiliary power. Absorption-type chillers are economically advantageous in areas where electric power is scarce or costly, where gas rates are low, or where waste or process steam or hot water is available during the cooling season. Solar power may also be used in some areas to generate the heat required for the absorption process.

The chillers themselves may be air or water cooled. Very small chillers are available with an air-cooled condenser built into the package. For larger systems

Means Forms

COST ANALYSIS

SHEET NO. CH-8

PROJECT **Office Building**

ESTIMATE NO.

ARCHITECT

DATE

TAKE OFF BY: QUANTITIES BY: **JJm**   PRICES BY: **FPW**   EXTENSIONS BY: **JJm**   CHECKED BY: **AHF**

		ABC Co.	USA Inc.	XYZ Assoc.	Individual misc. Quotes
Chiller	1	65000	60500 *	88000	no
Start & Service		YES	no	YES	1000
AH Units	6	25000	no	YES	23000
Vibr. Bases	6	no	no	YES	1000
Starters	1	YES	YES	YES	
Reheat Coil	1	200 *	no	YES	280
Roof Fan	1	no	no	1200 *	1220
FOB Job		YES	YES	YES	YES
Total		90200 –	60500 *—	89200 –	

			60500 –	Chiller
			1000 –	Service
			23000 –	Units
			1000 –	Bases
			200 –	Coil
			1200 –	Fan
	* Best Prices		# 86900	

Estimate Summary Sheet Used for Cost Comparison

Figure 17.6

166

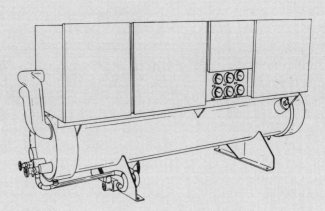

Reciprocating, Water Cooled Multiple Compressor, Semi Hermetic

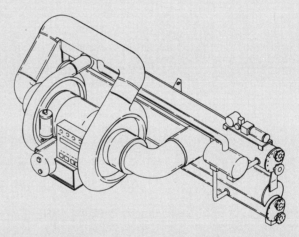

Centrifugal, Water Cooled, Hermetic

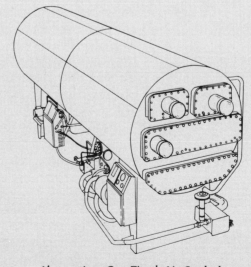

Absorption, Gas Fired, Air Cooled

Packaged Water Chillers

Figure 17.7

and for systems that may create too much noise for their location, the air-cooled condenser is installed at a distance from the chiller, and the two units are connected with refrigerant piping.

The basic process for air-conditioning systems using refrigerants as the cooling medium is the cooling and condensing back to liquid form of the refrigerant gas that was heated during the evaporation stage of the cycle. This condensation is achieved by cooling the gas with air, water, or a combination of both. Air-cooled condensers cool the refrigerant by blowing air directly across the refrigerant coil; evaporative condensers use the same method of cooling, with the addition of a spray of water over the coil to expedite the process.

The condenser for water- cooled chillers is piped into a remote water source, such as a cooling tower, pond, or river, via the "condenser water system." Completely packaged chiller systems may include built-in chilled water pumps, and all interconnecting piping, wiring, and controls. All of the components of the completely packaged unit are factory installed and tested prior to shipment to the installation site for connection to the chilled water and condenser water systems.

When water is used as the condensing medium and is abundant enough that recycling is not required, it may be piped to a drain after performing its cooling function and returned to its source. The source may be a river, a pond, or the ocean. If water supply is limited, expensive, or regulated by environmental restrictions, a water conserving or recycling system must be employed. Several types of systems may be installed to conform to these limitations. For example, a water regulating valve, a spray pond, a natural draft cooling tower, or a mechanical draft cooling tower (shown in Figure 17.8) may be used.

In very small cooling systems, a temperature controlled, water regulating valve may be used, provided that such a system is permitted by local environmental and/or building codes. The regulating valve system functions by allowing cooling water to flow when the condenser temperature rises, and, conversely, by stopping the flow as the temperature falls. The problem with this system, however, is that the heated condenser water cannot be recycled and is, therefore, wasted during the flow cycle.

The *spray pond* and *natural draft cooling tower systems*, although they are viable and available methods of cooling, are not commonly used for building air-conditioning. For reasons of water loss caused by excessive drift and the large amount of space required for their installation and operation, they are less desirable than the mechanical draft cooling tower method.

Mechanical draft cooling tower systems are classified in two basic designs: *induced draft* and *forced draft*. In an induced draft tower (as shown in Figure 17.8), a fan positioned at the top of the structure draws air upwards through the tower as the warm condenser water spills down. A cross-flow induced draft tower operates on the same principle, except that the air is drawn horizontally through the spill area from one side. The air is then discharged through a fan located on the opposite side. In a forced draft tower the fan is located at the bottom or the side of the structure. Air is forced by the fan into the water spill area, through the water, and then discharged at the top. All designs of mechanical draft towers are rated in tons of refrigeration; three gallons of condenser water per minute per ton is an approximate tower sizing method.

After the water has been cooled in the tower it passes through a heat exchanger, or condenser, in the refrigeration unit. Here, it again picks up heat and is pumped back to the cooling tower. The piping system is called the condenser water system. Figure 17.9 shows the layout of a typical condenser water system.

The actual process of cooling within the mechanical draft cooling tower takes place when air is moved across or counter to a stream of water falling through a system of baffles or "fill" to the tower basin. After the cooled water reaches the basin it is piped back to the condenser. Some of the droplets created by the fill are carried away by the moving air as "drift" and some of the droplets evaporate. This limited loss of water is to be expected as part of the operation of the tower system. Because of the loss of water by drift, evaporation, and bleed off, replenishment water must be added to the tower basin to maintain a predetermined level and to assure continuous operation of the system. To prevent scale buildup, algae, bacterial growth, or corrosion, tower water should be treated with chemicals or ozone applications. The materials used in constructing mechanical draft cooling towers include redwood (which is the most commonly employed material), other treated woods, various metals, plastics, concrete, or ceramic materials. The fill, which is the most important element in the tower's operation, may be manufactured from the same wide variety of materials used in the tower structure. Factory-assembled, prepackaged towers are available and are usually preferable to built-in place units, with multiple tower installations now being used for large systems.

The location of cooling towers is an important consideration for both practical and aesthetic reasons. They may be located outside of the building: on its roof or on the ground. If space permits, a cooling tower may be installed indoors

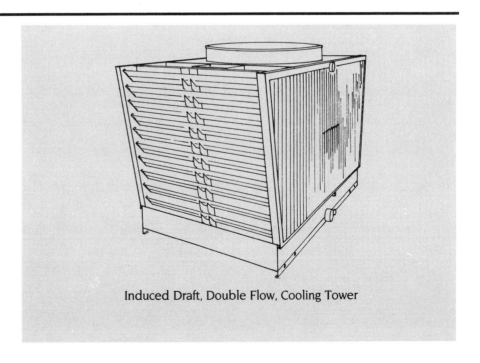

Induced Draft, Double Flow, Cooling Tower

Figure 17.8

by substituting centrifugal fans for the conventional noisy propeller type, adding air intake and exhaust ductwork. The tower discharge should not be directed into the prevailing wind, or towards doors, windows, and building air intakes. In general, common sense should be used when determining tower placement so that the noise, heat, and humidity the system creates do not interfere with building operation and comfort. The manufacturer's guidelines for installation should be strictly followed, especially those sections that address clearances for maximum airflow, maintenance, and future unit replacement.

Economical operation of a cooling tower system may be achieved through effective control and management of several critical aspects of its operation, including: careful monitoring of water treatment, selecting and maintaining the most efficient condensing temperature, and controlling water temperature with fan cycling.

A recent novel development in tower water system operation allows the tower to substitute for the water chiller under certain favorable climactic conditions. This new method cannot be implemented in all cases, but during periods when it can be employed, substantial savings result in the reduced cost of chiller operation.

Units of Measure: Chillers, cooling towers, air-cooled condensers, and any other major components of hydronic water chilling systems are taken off and recorded as each.

Material Units: Manufacturers' catalogs and quotations should be consulted for each estimate to confirm that all required components are included and how

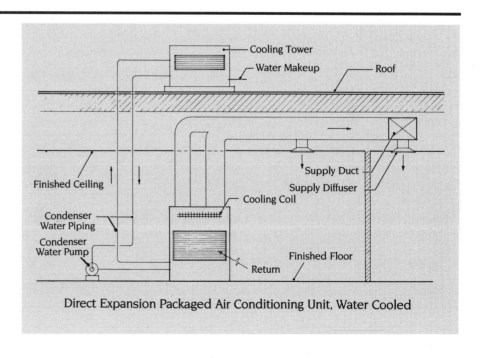

Direct Expansion Packaged Air Conditioning Unit, Water Cooled

Figure 17.9

each chiller or tower will be shipped (in one piece or broken down). Packaged chillers have built-in motor control panels and starters, whereas motor starters for cooling towers are purchased and shipped separately. If motor starting devices are to be furnished by the mechanical contractor and are not duplicated in the electrical specifications, the tower supplier's quotation should be checked and the starter cost added, if it has not been included. A successful bidder who is required to furnish motor starting equipment can save money by culling the starters from all of the equipment quotes and purchasing them as a package from an electrical outlet dealer. Packaged equipment, however, must be purchased with these controls built in.

The refrigerant charge and oil for centrifugal chillers will be quoted and shipped separately. The charging, startup, and warranty service must be estimated by an experienced technician, either in-house, by the manufacturer, or by a qualified subcontractor. This also applies to absorption units and their charges of lithium bromide, etc.

Labor Units: Because chillers, towers, or air-cooled condensers are large and cumbersome, they will always require mechanical assistance to be set into place. Placement can vary greatly, ranging from simple manhandling with skids or rollers (for smaller units) to the use of hoists and cranes (for larger sizes). Equipment location, regardless of size, will also dictate special hoisting apparatus. Cooling towers and air-cooled condensers may be installed on the roof of a building, indoors in one of the upper stories of a high rise building, in mechanical equipment rooms where the chillers and boilers are no longer in the traditional basement. Size and location will dictate whether chillers and towers should be shipped in one piece or knocked down for assembly in place. The manufacturer's quote usually includes assembly and installation.

The supplier's quotes for equipment, particularly for large shipments, should include job site delivery. Delivery may be F.O.B. the job, or F.O.B. the factory with freight allowed to the job or staging area.

Takeoff Procedure: All materials for air cooling should be taken off and recorded on a summary sheet where several manufacturers' quotations can be compared, as shown in Figure 17.6. The estimator can then compare choices, and can better determine the best price.

Air-Handling Equipment

Air-handling units consist of a filter section, a fan section, and a coil section on a common base. They are used to distribute clean, cooled, or heated air to the occupied building spaces. These units (some of which are shown in Figures 17.10, 17.11 and 17.12) are available in a wide range of capacities, from 200 cubic feet per minute to tens of thousands of cubic feet per minute. The units also vary in complexity of design and versatility of operation. Small units tend to have relatively simple coil and filter arrangements and modestly sized fan motor. Larger, more sophisticated units usually require remote placement. Because of the need to overcome losses caused by intake and supply ductwork, and by complex coil, filter, and damper configurations, the fan motor horsepower must be dramatically increased.

Small air-handling units may be located and mounted in a variety of settings and by different methods (as shown in Figure 17.10). They may be mounted on the floor or hung from walls or ceilings with no discharge ductwork required in the room they service. Small air-handling units require supply and return piping for heating and/or cooling. If these units are used for cooling, a drain connection is also required.

Determining the proper size, number, capacity, type, and configuration of coils in the unit is a prime consideration when selecting and/or designing an air-handling unit. As a general rule, the amount of air (in cubic feet per minute) to be handled by the unit determines the size and number of the various coils. Electric or hydronic coils are used for heating; chilled water or direct expansion coils are used for cooling. As the units increase in size and complexity, the coil configurations and arrangements become limitless. A simple heating and cooling unit, for example, may use the same coil for either hot or chilled water. A large unit usually demands different types of coils to perform many separate functions. In humid conditions, the air temperature may be intentionally lowered to remove moisture. In this case, a reheat coil is added

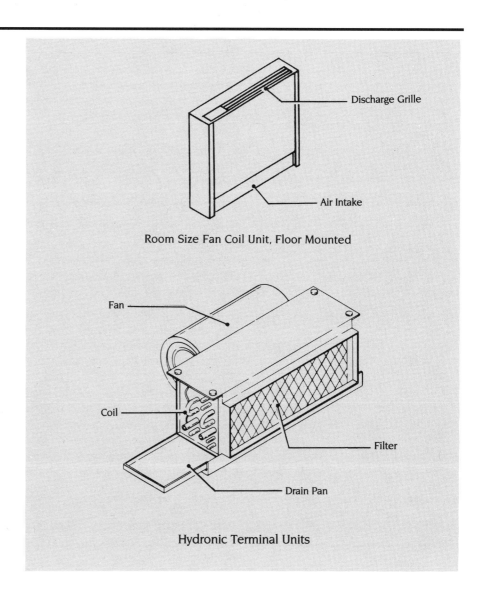

Room Size Fan Coil Unit, Floor Mounted

Hydronic Terminal Units

Figure 17.10

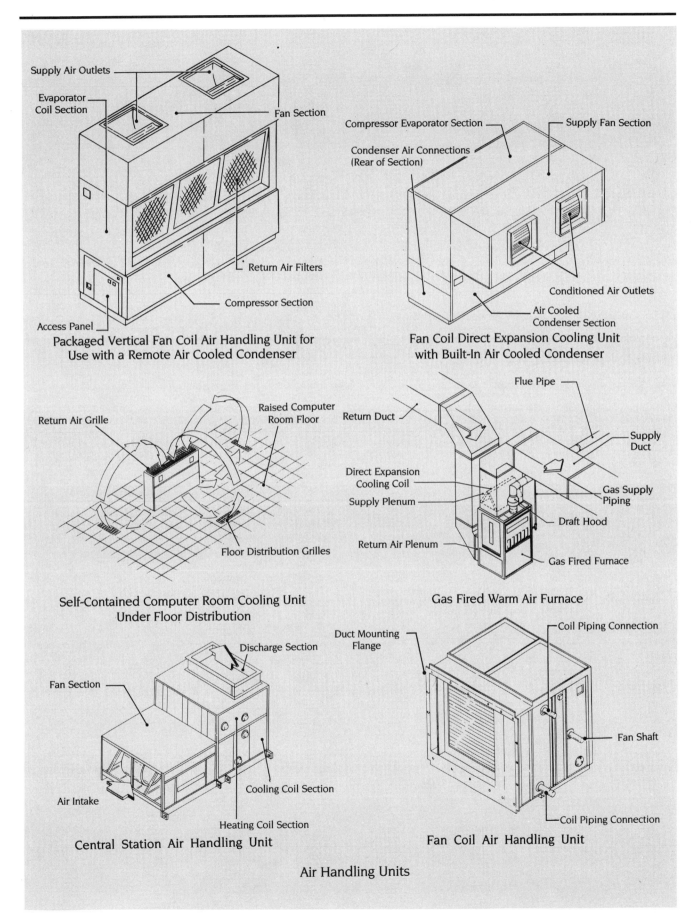

Supply Air Outlets

Evaporator Coil Section

Fan Section

Return Air Filters

Compressor Section

Access Panel

Packaged Vertical Fan Coil Air Handling Unit for
Use with a Remote Air Cooled Condenser

Compressor Evaporator Section

Condenser Air Connections
(Rear of Section)

Supply Fan Section

Conditioned Air Outlets

Air Cooled
Condenser Section

Fan Coil Direct Expansion Cooling Unit
with Built-In Air Cooled Condenser

Return Air Grille

Raised Computer
Room Floor

Floor Distribution Grilles

Self-Contained Computer Room Cooling Unit
Under Floor Distribution

Flue Pipe

Return Duct

Supply
Duct

Direct Expansion
Cooling Coil

Supply Plenum

Gas Supply
Piping

Draft Hood

Return Air Plenum

Gas Fired Furnace

Gas Fired Warm Air Furnace

Discharge Section

Fan Section

Cooling Coil Section

Air Intake

Heating Coil Section

Central Station Air Handling Unit

Duct Mounting
Flange

Coil Piping Connection

Fan Shaft

Coil Piping Connection

Fan Coil Air Handling Unit

Air Handling Units

Figure 17.11

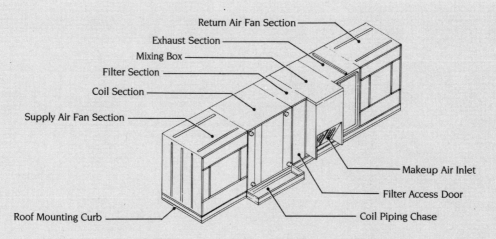

Central Station Air Handling Unit for Rooftop Location

Return Air Fan Section
Exhaust Section
Mixing Box
Filter Section
Coil Section
Supply Air Fan Section
Roof Mounting Curb
Makeup Air Inlet
Filter Access Door
Coil Piping Chase

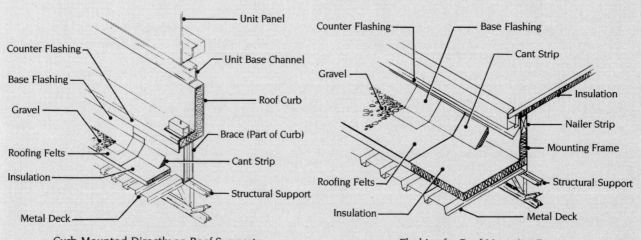

Curb Mounted Directly on Roof Supports

Unit Panel
Counter Flashing
Base Flashing
Gravel
Roofing Felts
Insulation
Metal Deck
Unit Base Channel
Roof Curb
Brace (Part of Curb)
Cant Strip
Structural Support

Flashing for Roof Mounting Frame

Counter Flashing
Gravel
Roofing Felts
Insulation
Base Flashing
Cant Strip
Insulation
Nailer Strip
Mounting Frame
Structural Support
Metal Deck

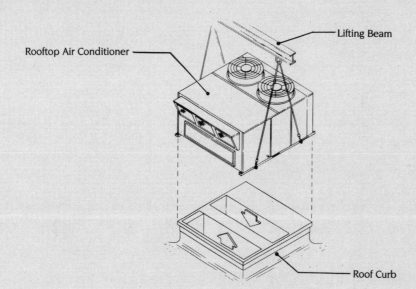

Installation of Rooftop Air Conditioner

Rooftop Air Conditioner
Lifting Beam
Roof Curb

Rooftop Air Handling Units and Mounting Details

Figure 17.12

174

to return the temperature to its desired level. Conversely, in dry conditions, a humidifier component is built into the unit. If outside air is introduced to the unit at subfreezing temperatures, then a preheat coil is placed in the outside air intake duct.

Certain precautions should be taken to prevent damage and to assure the efficiency of the unit's coils and other components. To protect the coil surfaces from accumulating dust and other airborne impurities, a filter section is a necessary addition to the unit. If a unit is designed to cool air, a drain pan must be included beneath the coiling section. This pan is then piped to an indirect drain to dispose of the unwanted condensation.

Another protection precaution involves the insulation of the fan coil casing internally. If this precaution is not taken to protect the cooling coil section and all other sections "down-stream," corrosive or rust-causing condensation will damage the casing and discharge ductwork. Insulating this casing also helps deaden the noise of the fan. Noise can also be controlled by the installation of flexible connections between the unit and its ductwork, mounting the unit on vibration-absorbing bases (if located on the floor), or suspending the unit from vibration absorbing hangers (if secured to the ceiling or wall).

Units of Measure: All air-handling units are taken off and recorded as each, separated by type for pricing and labor estimating.

Material Units: Material considerations in addition to the units themselves include curbs for roof-mounted units (shown in Figure 17.12), vibration mounts for suspended units, large base mounted equipment, and filters. Complete self-contained packaged units are available with built-in limit and operating controls. Central station fan coil units are not supplied with these controls or starters which must be supplied by the mechanical or electrical contractor. Job specifications and manufacturer's instruction should be carefully read for additional material required.

Labor Units: Air-handling equipment must be set in its designated place either manually or with mechanized equipment. A composite crew consisting of sheet metal workers and pipefitters usually handle large equipment on union jobs. After each unit has been assembled and set on its base, piping and duct connection must be made.

Takeoff Procedure: Air-handling equipment is taken off from the drawings and designated by equipment room or system. Like equipment should be listed together in one quote from various suppliers.

Fans and Gravity Ventilators

Fans are used to supply, circulate, or exhaust air for human comfort, safety, and health reasons. Fans may be exposed in the area being served or in a remote location, connected by ductwork to the served area. Fans may also be located outside of the building, on the roof or on side walls. In general, a fan consists of an electric motor and drive, blades, and a wheel or propeller. All of these parts are contained within an enclosure. The drive assembly may operate via belts and pulleys or may be directly connected to the motor. While the direct-drive type fan is less expensive, objectionable noise can result as the size or speed of the fan increases. Belt drive affords greater flexibility in speed and performance. Proper fan selection is important not only because of noise but also to avoid the feeling of air movement or drafts due to excess velocity. Fans are sized according to the cubic feet of air they can handle in one minute (CFM).

Fans are classified in two general groups, *centrifugal* and *axial-flow*. Centrifugal fans are further classified by the position of the blades on the fan wheel, either forward-curved or backward-curved. Axial-flow fans, where the air flows around the axis of the blade and through the impeller, are classified as propeller, vane-axial, and tube-axial. Some typical fans are shown in Figure 17.13.

Self-contained air-handling or air-conditioning units depend on a centrifugal fan for reasons of adaptability to duct configurations, and quiet and efficient operation. Air filters are used in air supply systems to protect the heating or cooling coils from dust or other particles picked up by the airflow.

The air-handling capacity or volume delivered by a fan may be varied by a motor speed control, outlet damper control, inlet vane control, or fan drive change. The most efficient method is a variable-speed motor, but this is also the most costly type of fan control.

Roof-mounted ventilators (shown in Figure 17.14) are designed to remove air from a building without the use of motor-driven fans. In some cases, this process is achieved by a rising of warm air and its displacement by denser or heavier cold air. Some ventilators, however, use the action of the wind to siphon air through the ventilator. Relief hoods may be used for exhaust air, as well as for makeup air intakes, through the use of dampers or self-acting shutters. Hoods and ventilators are usually constructed of galvanized steel or aluminum.

Gravity roof ventilators, which are not very efficient, are used in situations where rapid removal of stale air is not a factor. The cost of a motor-driven roof fan is two to three times that of a gravity ventilator.

Units of Measure: Fans and ventilators are taken off and priced as each.

Material Units: Roof-mounted fans and ventilators may be supplied (when specified) with self-flashing and sound-attenuating curbs. Fans will require motor control switches to be furnished by the mechanical or electrical contractor (check both mechanical and electrical specification for duplication). Dampers, shutters, and special finishes or linings may be required on a job by job basis. Large fans, whether supply or exhaust, will require vibration mounts or bases.

Labor Units: Fans and ventilators are installed by sheet-metal workers, even though the motor driven exhaust fan and the intake/exhaust hoods may be furnished by the mechanical contractor in accordance with local custom. Fans should be rigged or hoisted into place when other mechanical or roof mounted equipment is being placed, to take advantage of hoisting equipment availability.

Takeoff Procedure: Fans are taken off and recorded by type, size, or system. If they are being taken off by the trade that will also be purchasing the ventilators, the fans should be totaled on the same summary sheet. If not, there should be individual takeoff, summary, and pricing procedures. The sheet-metal subcontractor may be recording these fans for labor only, and the mechanical contractor for material cost only.

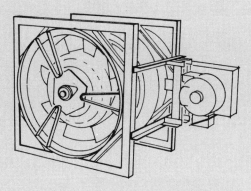

Axial Flow, Belt Drive, Centrifugal Fan

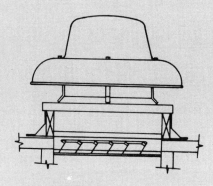

Centrifugal Roof Exhaust Fan

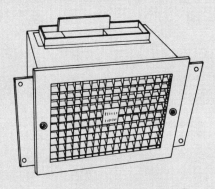

Ceiling Exhaust Fan

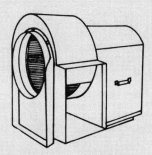

Belt Drive, Utility Set

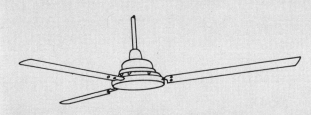

Paddle Blade Air Circulator

Belt Drive Propeller Fan with Shutter

Fans and Ventilators

Figure 17.13

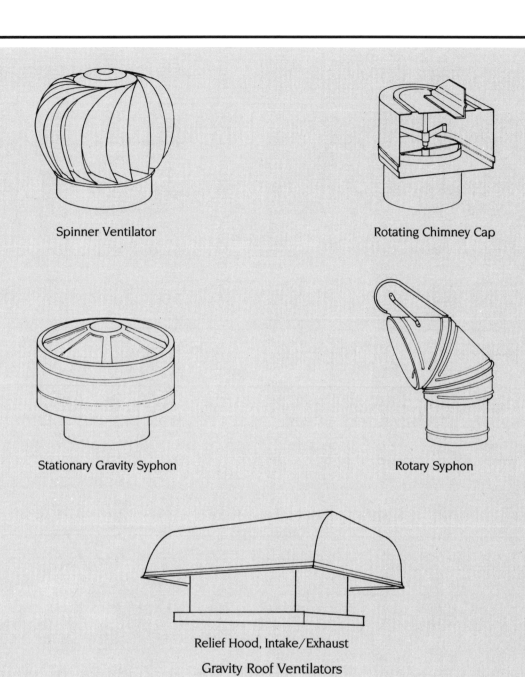

Spinner Ventilator

Rotating Chimney Cap

Stationary Gravity Syphon

Rotary Syphon

Relief Hood, Intake/Exhaust

Gravity Roof Ventilators

Figure 17.14

Chapter 18
DUCTWORK

Ductwork is the conduit or pipe which transports air through heating, ventilating, or air-conditioning systems. Although ductwork has traditionally been fabricated from sheets of galvanized steel, it is also made of aluminum sheets and fiberglass board. Stainless steel or one of several plastics are preferred materials for use in corrosive atmospheres.

Ductwork may be rectangular, round, or oval in shape, depending on requirements of size and application. Flexible or spiral-wound ductwork is also available for special installations. Fittings are available to accommodate changes in the direction, shape, and size of ductwork. Elbows, tees, transitions, reducers, and increasers come under this category. Typical ductwork systems depicting many of these items are shown in Figures 18.1 and 18.2.

Industry standards regulate the thickness or gauge of metal duct material. The basis for these regulations is the size of the ductwork and the pressure of the air contained. The appropriate weight for ductwork material is determined using these same standards. Based on this weight, the fabrication labor and installation labor may be calculated.

Units of Measure: Ductwork is taken off and recorded by the linear foot for each size. For metal ducts, the footage is then converted into pounds of metal and priced accordingly. The calculation for this conversion takes into account the gauge and weight per square foot of the metal being used. The table shown in Figure 18.3 can be used to obtain weight and area for any size and gauge of galvanized steel duct. The chart shown in Figure 18.4 can also be used to obtain weight per foot for various types of sheet metal. Both the table and the chart shown in these figures can be found in the reference section of *Means Mechanical Cost Data*. Smaller sized ducts and lighter gauges of metal will cost more per pound than longer and heavier ducts. Fiberglass and plastic ductwork is converted to square feet for pricing. Ductwork specialties (e.g., grilles, registers, diffusers, dampers, access doors, mixing boxes, etc., several of which are shown in Figure 18.5) are taken off and priced as each. Sound lining for ductwork is estimated by the square foot. Flexible connectors used at equipment connections and angle iron used for supports and special bracing are taken off and priced by the linear foot.

Material Units: The contractor's price per pound of metal ductwork should include a 10% to 15% allowance for waste, slips, and hangers. Fiberglass and plastic duct supports must be taken off and priced on an individual (as each) basis. Many other accessories are needed to complete a ductwork system. Supply and return "faces" include diffusers, registers, and grilles. Control devices include dampers and turning vanes.

Labor Units: Labor units for ductwork are unique in comparison to other mechanical trades in that the duct and fittings have to be fabricated as well as installed in place. Metal duct fabrication is normally done at the sheet-metal worker's shop and transported to the site for installation. Shop or fabrication labor may be estimated in one of two ways: by the average number of pounds of metal fabricated per man, per hour; or by the number of sheets converted to ductwork per man, per day. Installation labor can also be

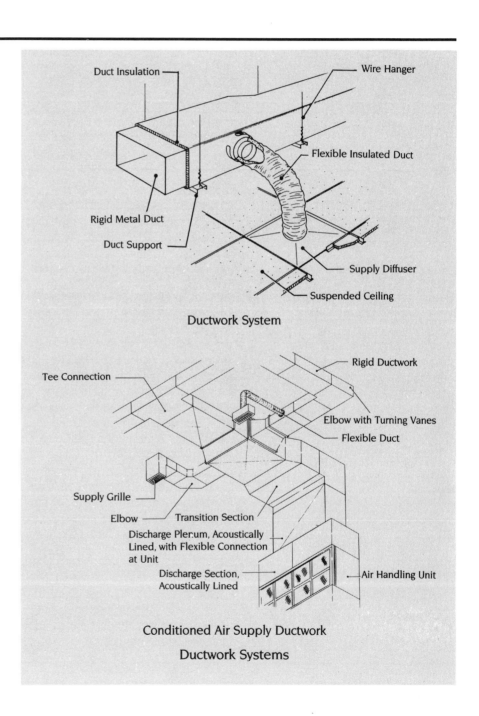

Ductwork System

Conditioned Air Supply Ductwork

Ductwork Systems

Figure 18.1

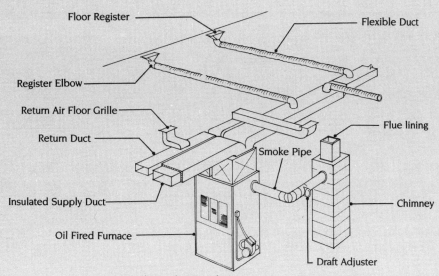

Floor Register — Flexible Duct
Register Elbow
Return Air Floor Grille
Return Duct
Flue lining
Smoke Pipe
Insulated Supply Duct
Chimney
Oil Fired Furnace
Draft Adjuster

Residential Forced Air Heating System

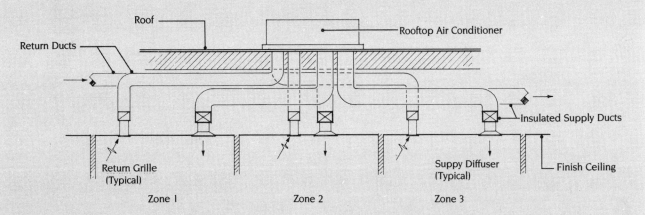

Roof — Rooftop Air Conditioner
Return Ducts
Insulated Supply Ducts
Return Grille (Typical)
Supply Diffuser (Typical)
Finish Ceiling
Zone 1   Zone 2   Zone 3

Rooftop Multizone Air Conditioning System

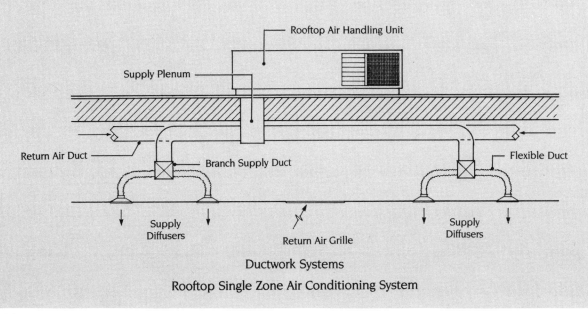

Rooftop Air Handling Unit
Supply Plenum
Return Air Duct
Branch Supply Duct
Flexible Duct
Supply Diffusers
Return Air Grille
Supply Diffusers

Ductwork Systems
Rooftop Single Zone Air Conditioning System

Figure 18.2

Table 8.4-009   Sheet Metal Calculator (Weight in Lb./Ft. of Length)

Gauge	26	24	22	20	18	16	Gauge	26	24	22	20	18	16
Wt.-Lb./S.F.	.906	1.156	1.406	1.656	2.156	2.656	Wt.-Lb./S.F.	.906	1.156	1.406	1.656	2.156	2.656
SMACNA Max. Dimension - Long Side		30"	54"	84"	85" Up		SMACNA Max. Dimension - Long Side		30"	54"	84"	85" Up	
Sum-2 Sides							Sum-2 Sides						
2	.3	.40	.50	.60	.80	.90	56	9.3	12.0	14.0	16.2	21.3	25.2
3	.5	.65	.80	.90	1.1	1.4	57	9.5	12.3	14.3	16.5	21.7	25.7
4	.7	.85	1.0	1.2	1.5	1.8	58	9.7	12.5	14.5	16.8	22.0	26.1
5	.8	1.1	1.3	1.5	1.9	2.3	59	9.8	12.7	14.8	17.1	22.4	26.6
6	1.	1.3	1.5	1.7	2.3	2.7	60	10.0	12.9	15.0	17.4	22.8	27.0
7	1.2	1.5	1.8	2.0	2.7	3.2	61	10.2	13.1	15.3	17.7	23.2	27.5
8	1.3	1.7	2.0	2.3	3.0	3.6	62	10.3	13.3	15.5	18.0	23.6	27.9
9	1.5	1.9	2.3	2.6	3.4	4.1	63	10.5	13.5	15.8	18.3	24.0	28.4
10	1.7	2.2	2.5	2.9	3.8	4.5	64	10.7	13.7	16.0	18.6	24.3	28.8
11	1.8	2.4	2.8	3.2	4.2	5.0	65	10.8	13.9	16.3	18.9	24.7	29.3
12	2.0	2.6	3.0	3.5	4.6	5.4	66	11.0	14.1	16.5	19.1	25.1	29.7
13	2.2	2.8	3.3	3.8	4.9	5.9	67	11.2	14.3	16.8	19.4	25.5	30.2
14	2.3	3.0	3.5	4.1	5.3	6.3	68	11.3	14.6	17.0	19.7	25.8	30.6
15	2.5	3.2	3.8	4.4	5.7	6.8	69	11.5	14.8	17.3	20.0	26.2	31.1
16	2.7	3.4	4.0	4.6	6.1	7.2	70	11.7	15.0	17.5	20.3	26.6	31.5
17	2.8	3.7	4.3	4.9	6.5	7.7	71	11.8	15.2	17.8	20.6	27.0	32.0
18	3.0	3.9	4.5	5.2	6.8	8.1	72	12.0	15.4	18.0	20.9	27.4	32.4
19	3.2	4.1	4.8	5.5	7.2	8.6	73	12.2	15.6	18.3	21.2	27.7	32.9
20	3.3	4.3	5.0	5.8	7.6	9.0	74	12.3	15.8	18.5	21.5	28.1	33.3
21	3.5	4.5	5.3	6.1	8.0	9.5	75	12.5	16.1	18.8	21.8	28.5	33.8
22	3.7	4.7	5.5	6.4	8.4	9.9	76	12.7	16.3	19.0	22.0	28.9	34.2
23	3.8	5.0	5.8	6.7	8.7	10.4	77	12.8	16.5	19.3	22.3	29.3	34.7
24	4.0	5.2	6.0	7.0	9.1	10.8	78	13.0	16.7	19.5	22.6	29.6	35.1
25	4.2	5.4	6.3	7.3	9.5	11.3	79	13.2	16.9	19.8	22.9	30.0	35.6
26	4.3	5.6	6.5	7.5	9.9	11.7	80	13.3	17.1	20.0	23.2	30.4	36.0
27	4.5	5.8	6.8	7.8	10.3	12.2	81	13.5	17.3	20.3	23.5	30.8	36.5
28	4.7	6.0	7.0	8.1	10.6	12.6	82	13.7	17.5	20.5	23.8	31.2	36.9
29	4.8	6.2	7.3	8.4	11.0	13.1	83	13.8	17.8	20.8	24.1	31.5	37.4
30	5.0	6.5	7.5	8.7	11.4	13.5	84	14.0	18.0	21.0	24.4	31.9	37.8
31	5.2	6.7	7.8	9.0	11.8	14.0	85	14.2	18.2	21.3	24.7	32.3	38.3
32	5.3	6.9	8.0	9.3	12.2	14.4	86	14.3	18.4	21.5	24.9	32.7	38.7
33	5.5	7.1	8.3	9.6	12.5	14.9	87	14.5	18.6	21.8	25.2	33.1	39.2
34	5.7	7.3	8.5	9.9	12.9	15.3	88	14.7	18.8	22.0	25.5	33.4	39.6
35	5.8	7.5	8.8	10.2	13.3	15.8	89	14.8	19.0	22.3	25.8	33.8	40.1
36	6.0	7.8	9.0	10.4	13.7	16.2	90	15.0	19.3	22.5	26.1	34.2	40.5
37	6.2	8.0	9.3	10.7	14.1	16.7	91	15.2	19.5	22.8	26.4	34.6	41.0
38	6.3	8.2	9.5	11.0	14.4	17.1	92	15.3	19.7	23.0	26.7	35.0	41.4
39	6.5	8.4	9.8	11.3	14.8	17.6	93	15.5	19.9	23.3	27.0	35.3	41.9
40	6.7	8.6	10.0	11.6	15.2	18.0	94	15.7	20.1	23.5	27.3	35.7	42.3
41	6.8	8.8	10.3	11.9	15.6	18.5	95	15.8	20.3	23.8	27.6	36.1	42.8
42	7.0	9.0	10.5	12.2	16.0	18.9	96	16.0	20.5	24.0	27.8	36.5	43.2
43	7.2	9.2	10.8	12.5	16.3	19.4	97	16.2	20.8	24.3	28.1	36.9	43.7
44	7.3	9.5	11.0	12.8	16.7	19.8	98	16.3	21.0	24.5	28.4	37.2	44.1
45	7.5	9.7	11.3	13.1	17.1	20.3	99	16.5	21.2	24.8	28.7	37.6	44.6
46	7.7	9.9	11.5	13.3	17.5	20.7	100	16.7	21.4	25.0	29.0	38.0	45.0
47	7.8	10.1	11.8	13.6	17.9	21.2	101	16.8	21.6	25.3	29.3	38.4	45.5
48	8.0	10.3	12.0	13.9	18.2	21.6	102	17.0	21.8	25.5	29.6	38.8	45.9
49	8.2	10.5	12.3*	14.2	18.6	22.1	103	17.2	22.0	25.8	29.9	39.1	46.4
50	8.3	10.7	12.5	14.5	19.0	22.5	104	17.3	22.3	26.0	30.2	39.5	46.8
51	8.5	11.0	12.8	14.8	19.4	23.0	105	17.5	22.5	26.3	30.5	39.9	47.3
52	8.7	11.2	13.0	15.1	19.8	23.4	106	17.7	22.7	26.5	30.7	40.3	47.7
53	8.8	11.4	13.3	15.4	20.1	23.9	107	17.8	22.9	26.8	31.0	40.7	48.2
54	9.0	11.6	13.5	15.7	20.5	24.3	108	18.0	23.1	27.0	31.3	41.0	48.6
55	9.2	11.8	13.8	16.0	20.9	24.8	109	18.2	23.3	27.3	31.6	41.4	49.1
							110	18.3	23.5	27.5	31.9	41.8	49.5

Example: If duct is 34" x 20" x 15' long, 34" is greater than 30" maximum, for 24 ga. so must be 22 ga. 34" + 20" = 54" going across from 54" find 13.5 lb. per foot. 13.5 x 15' = 202.5 lbs. For

S.F. of surface area 202.5 ÷ 1.406 = 144 S.F.
Note: figures include an allowance for scrap.

402

(Reprinted from Means Mechanical Cost Data 1991.)

Figure 18.3

estimated in one of two ways: by the number of pounds installed per man, per day; or by the linear feet of duct hung or installed per man, per day.

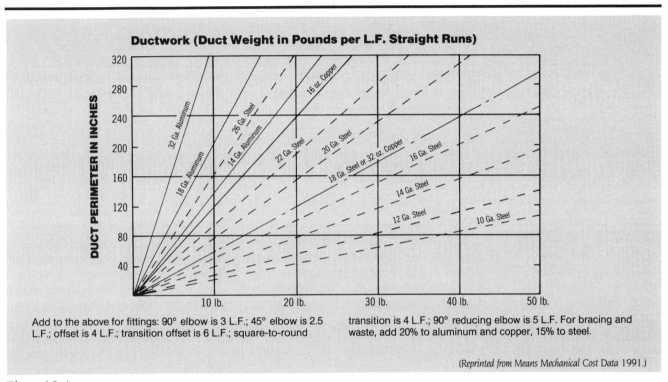

Add to the above for fittings: 90° elbow is 3 L.F.; 45° elbow is 2.5 L.F.; offset is 4 L.F.; transition offset is 6 L.F.; square-to-round

transition is 4 L.F.; 90° reducing elbow is 5 L.F. For bracing and waste, add 20% to aluminum and copper, 15% to steel.

(Reprinted from Means Mechanical Cost Data 1991.)

Figure 18.4

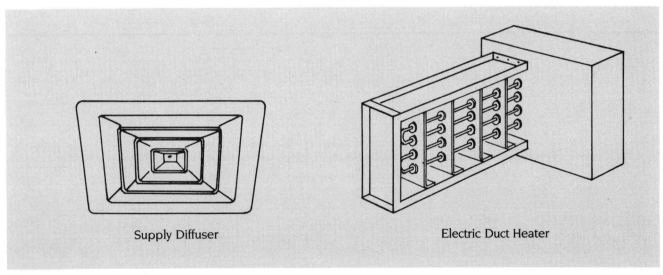

Supply Diffuser

Electric Duct Heater

Figure 18.5

Factors affecting fabrication and installation labor are listed below:

- Job conditions
- Height
- Methods of reinforcement and support
- Duct size and gauge
- The use of sound lining
- Sealing or taping the joint to prevent air leakage

For duct installation over 10' high, the percentage modifiers shown in Figure 18.6 (for field installation labor only) should be used.

The temperature control contractor is usually responsible for supplying the motorized dampers used in temperature control systems. These dampers must be installed, however, by the sheet-metal contractor. A schedule of dampers is supplied by the control company, identifying damper location and function, but final sizing to match the duct dimensions is the sheet-metal contractor's responsibility. Other labor concerns may include equipment usually furnished by the prime mechanical contractor but installed by the sheet-metal contractor (e.g., duct coils, induction boxes, mixing and terminal boxes, humidifiers, filters, fans, kitchen hoods, etc.).

Local labor practices may also require that a split crew of tin knockers and pipefitters install supply air-handling equipment if coils are involved. The designer may not be aware of such local practices, especially if the design firm is from another locality. The sheet-metal estimator should become familiar with all of the mechanical specifications to ensure that any "gray areas" of responsibility are properly defined, so as to avoid jurisdictional work stoppages during construction. Electrical and ceiling plans should be read in case ceiling trotter or lighting fixtures are to be integrated with air-handling systems. Architectural plans and specifications might contain information about louvers and wall or roof openings, which could affect the sheet-metal estimate as well. Again, be sure to check the job specifications carefully.

Height Modifications for Ductwork Installation	
Height Above Floor Level	Modification to Field Labor
10 to 15 feet	10 percent
15 to 20 feet	20 percent
20 to 25 feet	25 percent
25 to 30 feet	35 percent
31 feet and up	50 percent

Figure 18.6

Takeoff Procedure: The takeoff should begin at the source, the mechanical room, or wherever the supply units or exhaust fans are located. This will help the estimator to visualize from the start the number of systems and zones to be contended with. The number and location of ducts to be lined should be noted. Ductwork should be listed on a separate takeoff sheet, recorded by size and length. Each system should be listed individually, recording lined duct, insulated duct, and bare duct separately. A ductwork takeoff form, designed for this purpose, was shown in Figure 4.2.

Each system should be marked on the plans in a different colored pencil. Different materials should be listed separately, if, for example, galvanized and aluminum are both used in the same system. Ductwork should be measured straight through, rather than taking off each fitting. A 10% to 50% allowance for fittings must then be added to the ductwork totals. This percentage should be based on the fitting frequency indicated, the system complexity, and the estimator's own experience. The estimator with no company experience factor to refer to should take off and record each fitting, and then use a standard conversion factor for each. This data should be kept until the estimator's experience factors have been established.

After the ductwork has been taken off, the specialties (e.g., supply and return devices or faces, dampers, access doors, turning vanes, extractors, etc.) should be taken off. The footage totals can be divided by 4' or 8' lengths to obtain the number of field joints to be made up. The drawings should be reread at this point, with attention paid to the building perimeter. Items without duct connections, but which are the sheet-metal workers' responsibility to furnish or install, should be identified. For example, if louvers are built into a prefabricated panel wall, the sheet-metal contractor should check the architectural details and specifications for his involvement, as these items may have been ignored in the mechanical specifications. Other specially items might be propeller or wall fans, wall boxes, sleeves, drain pans, or radiator and convector casings.

Chapter 19
INSULATION

Thermal insulation is used in mechanical systems to prevent heat loss or gain and to provide a vapor barrier for piping and ductwork systems, boilers, tanks, chillers, heat exchangers, air-handling equipment casings, and the like. Most boilers, water heaters, chillers, and air-handling units are provided with insulated metal jackets and require little, if any, field insulation.

Insulation is available in rigid or flexible form and is manufactured from fiberglass, cellular glass, rock wool, polyurethane foam, closed cell polyethylene, flexible elastomeric, rigid calcium silicate, phenolic foam, or rigid urethane.

Insulation is produced in a variety of wall thicknesses and may be applied in layers for extreme temperatures. Rigid board or blocks, or flexible blanket are used for ductwork and equipment. Preformed sections are available for use with pipe.

The standard length for pipe insulation has always been 3'. This is still true for fiberglass and calcium silicate, but foam insulations (e.g., polyurethane, polyethylene, and urethane) are produced in 4' lengths. This means that 1/3 less butt joints will have to be made in straight runs of pipe. Flexible elastomeric insulation is shipped in 6' lengths.

Units of Measure: Pipe insulation is measured by the linear foot. Fittings, joints, etc., are measured as each. Ductwork, breeching, and equipment insulation, whether block, board, or blanket, are measured by the square foot.

Material Units: The materials required to apply and seal insulation will vary, depending on the type of insulation and exterior finish specified, (e.g., fire retardant, vapor barrier, weatherproof, etc.). For some insulation, these special finishes are applied at the factory, while others require field application.

Required materials may also include preformed fitting, valve, and flange covers unless the insulation is to be mitered or cut in the field to allow for fittings and valves. In addition, butt-joining strips, tape, wire, adhesives, stick clips, or welding pins, for adhering the insulation and protective jacketing, are required depending on the specified materials.

Labor Units: The simplest pipe covering installation method is slipping elastomeric insulation over straight sections of pipe or tubing, prior to pipe installation. This insulation can be formed around bends and elbows.

When the piping is already installed, the insulation must be slit lengthwise, placed, and both the butt joints and seams (at the slit) must be joined with a contact adhesive. This process requires more labor than does the simple slipping on method described above. Fittings, in this case, are covered by mitering or cutting a hole for a tee or a valve bonnet. Fiberglass insulation is already slit for placement around the pipe, and fittings are mitered. From straight sections, holes are cut for tees and valve bonnets using a knife. Elastomeric insulation usually does not require any additional finish.

A variety of jackets and fittings are available for all types of insulation, including roofing felt wired in place and preformed metal jackets used as closures where exposed to weather. Premolded fitting covers are available to give fittings and valves a finished appearance. Factory applied self-sealing jackets are an advantage of fiberglass insulation. Flanges, which are removed frequently, often require unique insulated metal jackets or boxes which are fabricated by the insulation contractor.

Calcium silicate insulation is often specified for higher temperatures (above 850°F). Installation of calcium silicate is more labor intensive than installation of fiberglass or elastomeric insulation. This rigid insulation is made in half sections for pipe and in 3' lengths. Multiple segments must be used in layers to achieve larger outside diameters. The segments are wired in place 9" on center, normally requiring two workers. Fittings for calcium silicate insulation are made by mitering sections using a saw. Flanges and valves require oversized sections cut to fit. These valve and fitting covers are wired in place and finished with a troweled coat of insulating cement. This type of insulation has a high waste factor, due to crumbling and breaking during cutting and ordinary use.

Special thicknesses of any insulation must be obtained by adding multiple layers of oversized insulation, and sealing and securing using the same method as was used for the base layer. This additional work should be included in the labor estimate.

Hangers and supports require special treatment for insulated piping systems. The pipe hanger or support must be placed outside the insulation, to allow for expansion and contraction in steam or hot water systems, and to maintain the vapor barrier in cold water systems. As a result, oversized supports should be used to fit the O.D. of the insulation, rather than the pipe.

For heated piping, preformed steel segments are welded to the bottom of the pipe at each point of support. These saddles are sized according to the insulation thickness plus the pipe O.D. For cold water (anti-sweat) systems, where no metal-to-metal contact between pipe and support can be tolerated, the hanger is oversized to allow for placement of the insulation. The insulation is then protected by a sheet-metal shield which matches the outer radius of the insulation.

For flexible insulation, which cannot support the weight of the pipe, a rigid insert is substituted for the lower section of insulation at each point of support. This insert may be formed from calcium silicate insulation, cork or even wood. The insulation jacket must enclose this insert within the vapor proof envelope. The labor for insulation around hangers should be estimated carefully. The insert, for example, might be installed by the pipe coverer, while the metal shield or protector is furnished by the pipefitter or plumber.

Duct and equipment insulation may be a wraparound blanket type insulation, wired in place and sealed with adhesive or tape. Rigid board is secured by the use of pins, secured to the duct exterior with mastic or spot welded into place. The insulation sheets are pressed onto these pins and secured with self locking washers.

The butt joints and seams are sealed with adhesive and tape. Block and segment insulation, when installed on round or irregular shapes, is wired on, and then covered with a chicken wire mesh and coated with a troweled application of insulation cement.

Labor units for installing and finishing insulation will vary from the simplest procedure to more complex and time-consuming processes. The plans and specifications should be read carefully to determine the type of insulation required. This will help assure that the method and sequence of installation labor is properly estimated.

Takeoff Procedure: Pipe insulation is recorded by size on a specialized takeoff sheet which is also used to record fittings and valves similar to the takeoff method for the pipe itself. A separate takeoff sheet should be used for each system, broken down by thickness and exterior finishes required. Fittings are converted to linear feet of pipe for pricing purposes, as shown in Figure 19.1.

Ductwork insulation is taken off in the same manner as the duct itself, then converted to square feet for pricing. Separate forms should be used to keep the ductwork insulation takeoff separate from the equipment insulation takeoff. The ductwork takeoff form (shown in Chapter 4, Figure 4.2), the piping schedule (shown in Chapter 4, Figure 4. 1), or the Quantity Sheet (shown in Chapter 8, Figure 8.1) might be employed for this purpose. In lieu of performing the takeoff, or to double check the takeoff, the estimator will often contact the sheet metal and/or piping contractors to obtain or verify quantities for ductwork insulation.

Conversion of Fittings into Equivalent Feet of Straight Pipe for Estimating Insulation:	
Fitting	Equivalent Pipe Footage
Elbows or Bends	Add 3 feet
Pair of Flanges	Add 3 feet
Reducing Couplings	Add 2 feet

Figure 19.1

Chapter 20
AUTOMATIC TEMPERATURE CONTROLS

Temperature control systems range from simple thermostats and boiler or furnace operating and limit controls to complex electronic HVAC controls integrated with energy management systems which are a part of the "intelligent building concept." To regulate indoor temperatures, the control system responds to indoor/outdoor temperature changes by opening, closing, and modulating valves and dampers (shown in Figure 20.1) in heating and cooling piping or ductwork. This includes, when necessary, resetting the temperature of the heating or cooling medium. Control systems also provide automatic temperature adjustments to correspond to occupied and unoccupied periods of building use. The temperature control system itself may be pneumatically, electrically, or electronically operated, or any combination thereof.

Units of Measure: Distances for wiring or tubing are recorded and priced by the linear foot. Control devices, such as controllers or sensors, are recorded as each. Complete temperature control systems are quoted as each, on either an "installed" or "supervised" basis.

Material Units: The experienced estimator should know which pieces of HVAC equipment to be furnished by the mechanical contractor will include temperature controls. Self-contained control valves will be furnished complete with built-in thermostats and sensing elements. A packaged air-conditioner, heat pump, warm-air furnace, or boiler will arrive with operating and limit controls installed and prewired. Remote thermostats, programmable controllers, or outdoor sensors should be handled as optional equipment for field installation and wiring, and should be ordered as each if not part of a package.

Depending on the project specifications, the temperature control system could be bid by a manufacturer as a turn-key installation. In this case, the temperature control bidder, or his representative, would assume complete responsibility for the installation.

Alternatively, the specifications may call for the temperature control manufacturer to bid the material only. Installation might then be performed by the mechanical or electrical subcontractor, or by a subcontractor specializing in control work. In this case, the installer must provide the necessary material to connect and hook up the control equipment. The control manufacturer will generally provide supervision for installation of his equipment.

Servicing is one important factor that should be considered with all temperature control systems, regardless of installation method. Most of the callbacks on an HVAC system are for adjustment or calibration of the temperature controls. For this reason, the proposed service agreement with the controls manufacturer should be carefully read.

Labor Units: Labor units for valve installation and sensing element wells, which are installed by the pipefitter, have already been covered in Chapter 13, "Piping." Control dampers and duct-mounted controllers are installed by the sheet-metal worker (see Chapter 18, "Ductwork"). Motorized operators for these dampers and valves are field installed by the control contractor. These operators may be either electric or pneumatic. If the system is pneumatic, the tubing will be installed by a pipefitter employed by the control contractor. If the system is electric or electronic, it should be wired by an electrician under the supervision of the control contractor.

Takeoff Procedure: Plans and specifications should be reviewed to determine placement and frequency of controls. The control specifications will dictate what has to be controlled and how. The drawings will indicate how many control points there should be and where they should be located.

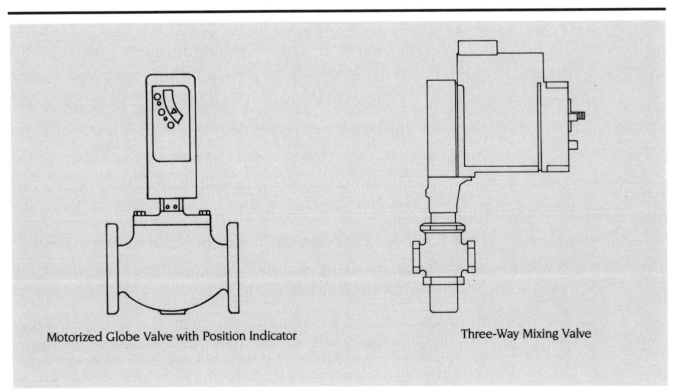

Motorized Globe Valve with Position Indicator   Three-Way Mixing Valve

Figure 20.1

After the specifications and the design sequence of operation have been read, a list should be made of control points. An average of 40' or 50' of tubing or wiring per control device should be anticipated. For shorter runs it may be better to scale off the actual distances. Some estimators may even prefer to make an actual tubing or wiring layout, rather than rely on averages. A wiring layout or control diagram is essential if the estimator has to obtain pricing from an electrician.

Temperature control systems can be so complex and specialized that the majority of mechanical contractors will subcontract out each job to a company that specializes in the particular installation and service of the type of system specified. If a subcontractor or several subcontractors are used, the mechanical contractor will be responsible for their installation. For this reason, supervision time should be included to handle this requirement.

Chapter 21

SUPPORTS, ANCHORS, AND GUIDES

Supports, anchors, and guides must be fabricated on site for the many occasions when there is no stock item available to meet the job requirement. Heat exchangers, tanks, coils, fan-coil units, and piping may, at some time or another, need job-fabricated supports.

Some relatively simple supports can be made from perforated channel metal framework and the accompanying fittings and brackets which bolt together. More substantial supports, however, must be fabricated from steel structural shapes (i.e., angles, tees, channels, and beams). Figure 21.1 shows the properties of the various steel shapes available.

The steel shapes are priced by the pound and measured or taken off by the linear foot. To convert linear feet into pounds, the estimator should refer to a handbook or stock list readily available from any steel distributor or warehouse. A table for estimating angle lengths is shown in Figure 21.2. If a handbook is not on hand, an approximation can be made from the fact that 12 cubic inches of steel weighs 3.4 lbs. This means that a piece of steel 1" x 1" and 12" long, a 1/4" plate 4" wide by 12" long, or a 2" x 2" x 1/4" angle 12" long each weigh approximately 3.4 lbs. The following example demonstrates the method of estimating a field-fabricated stand for a vertical-plate heat exchanger. The estimator first prepares a sketch indicating stock sizes and material lengths. The dimensions are shown in Figure 21.3.

Angle is stocked in 20' lengths but can be purchased in 5' increments. Therefore, the material for this stand, using 4" x 4" x 1/4" angle shape steel, would be the following:

Material: 30 L.F. 2 @ 6.6 lbs./ft. = 198 lbs. @ $.56=	$110.88
Fabricate: Cut to length, 8 pieces @ 1/4 hr. =	= $ 51.00
2 hrs. @ $25.50	= $ 25.50
Weld: 4' @ 4 L.F./hr. = 1 hour @ $25.50	= $ 16.00
Oxygen & Acetylene: 2 hours @ $8.00	= $ 4.90
Rental 300 AMP Arc Welder: 1 = hour @ $4.90	$208.28
TOTAL	

This estimate does not include overhead and profit. Priming and finish painting are to be accomplished by the painting contractor. Estimates for field fabricated items may be priced using the appropriate sections of *Means Mechanical Cost Data.*

Hot rolled structural steel is available in the following shapes.

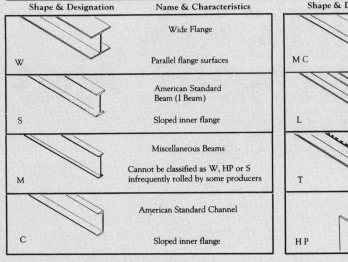

Shape & Designation	Name & Characteristics	Shape & Designation	Name & Characteristics
W	Wide Flange Parallel flange surfaces	M C	Miscellaneous Channel Infrequently rolled by some producers
S	American Standard Beam (I Beam) Sloped inner flange	L	Angle Equal or unequal legs, constant thickness
M	Miscellaneous Beams Cannot be classified as W, HP or S infrequently rolled by some producers	T	Structural Tee Cut from W, M or S on center of web
C	American Standard Channel Sloped inner flange	H P	Bearing Pile Parallel flanges and equal flange and web thickness

Common drawing designations follow.

Wide Flange
W 18 x 35 ←— Weight in Pounds Per Foot
└— Nominal Depth in Inches (Actual 17-3/4")

American Standard Beam
S 12 x 31.8
↑└— Weight in Pounds Per Foot
└— Depth in Inches

Miscellaneous Beam
M 8 x 6.5
↑└— Weight in Pounds Per Foot
└— Depth in Inches

American Standard Channel
C 8 x 11.5
↑└— Weight in Pounds Per Foot
└— Depth in Inches

Miscellaneous Channel
MC 8 x 22.8
↑└— Weight in Pounds Per Foot
└— Depth in Inches

Angle
↑ ┌— Length of One Leg in Inches
L 6 x 3-1/2 x 3/8 ←— Thickness of Each Leg in Inches
└— Length of Other Leg in Inches

Tee Cut From W16 x 100
WT 8 x 50
↑└— Weight in Pounds Per Foot
└— Nominal Depth in Inches (Actual 8-1/2")

Tee Cut From S 12 x 35
ST 6 x 17.5
↑└— Weight in Pounds Per Foot
└— Depth in Inches

Tee Cut From M 10 x 9
MT 5 x 4.5
↑└— Weight in Pounds Per Foot
└— Depth in Inches

Bearing Pile
HP 12 x 84
↑└— Weight in Pounds Per Foot
└— Nominal Depth in Inches (Actual 12-1/4")

Hot rolled structural shapes are generally available in the following ASTM specifications.

Steel Type	ASTM Designation	Minimum Yield Stress KSI	Characteristics
Carbon	A36	36	
	A529	42	
High-Strength Low Alloy	A441	50	Structural Manganese Vanadium Steel
	A572	42	Columbium-Vanadium Steel
		50	
		60	
		65	
Corrosion Resistant High Strength Low Alloy	A242	50	Corrosion Resistant
	A588	50	Corrosion Resistant to 4" Thick

Figure 21.1

Angles Equal Legs and Unequal Legs Properties for Designing			
Size and Thickness (in.)	Weight per Foot (lb.)	Size and Thickness (in.)	Weight per Foot (lb.)
L 9 x 4 x 5/8	26.3	L 4 x 4 x 3/4	18.5
9/16	23.8	5/8	15.7
1/2	21.3	1/2	12.8
L 8 x 8 x 1⅛	56.9	7/16	11.3
1	51.0	3/8	9.8
7/8	45.0	5/16	8.2
3/4	38.9	1/4	6.6
5/8	32.7	L 4 x 3½ x 5/8	14.7
9/16	29.6	1/2	11.9
1/2	26.4	7/16	10.6
L 8 x 6 x 1	44.2	3/8	9.1
7/8	39.1	5/16	7.7
3/4	33.8	1/4	6.2
5/8˙	28.5	L 3 x 2½ x 1/2	8.5
9/16	25.7	7/16	7.6
1/2	23.0	3/8	6.6
7/16	20.2	5/16	5.6
L 8 x 4 x 1	37.4	1/4	4.5
3/4	28.7	3/16	3.39
9/16	21.9	L 3 x 2 x 1/2	7.7
1/2	19.6	7/16	6.8
L 7 x 4 x 3/4	26.2	3/8	5.9
5/8	22.1	5/16	5.0
1/2	17.9	1/4	4.1
3/8	13.6	3/16	3.07
L 5 x 3½ x 3/4	19.8	L 2½ x 2½ x 1/2	7.7
5/8	16.8	3/8	5.9
1/2	13.6	5/16	5.0
7/16	12.0	1/4	4.1
3/8	10.4	3/16	3.07
5/16	8.7	L 2½ x 2 x 3/8	5.3
1/4	7.0	5/16	4.5
L 5 x 3 x 5/8	15.7	1/4	3.62
1/2	12.8	3/16	2.75
7/16	11.3	L 2 x 2 x 3/8	4.7
3/8	9.8	5/16	3.92
5/16	8.2	1/4	3.19
1/4	6.6	3/16	2.44
		1/8	1.65

Figure 21.2

Units of Measure: Piping supports, anchors, and guides may be listed as each according to pipe size. Equipment supports may be listed as each. These items may all be totaled, however, and the material priced by the pound.

Material Units: Costs for these field-fabricated items, in addition to the plate, rods, and shapes, include oxygen and acetylene for cutting and shaping, electrodes for welding, bolts, nuts, and other anchoring devices.

Labor Units: The following factors contribute to the labor units for supports, anchors, and guides:

- Physical unloading of the steel
- Cutting to desired lengths
- Relocation of the assembled units into place
- Responsibility for prime or finished coats of paint

Takeoff Procedure: From piping and equipment takeoffs, the estimator can determine where auxiliary steel is required. Mechanical sections and details, when shown, will also specify where supports, anchors, and guides are required. It is usually left to the estimator's judgment to size these fabrications, often estimated using rough sketches.

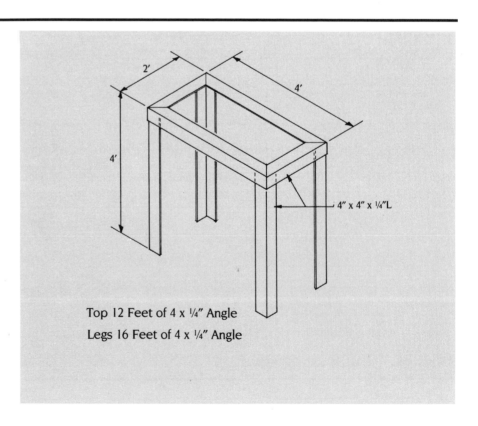

Top 12 Feet of 4 x ¼" Angle
Legs 16 Feet of 4 x ¼" Angle

Figure 21.3

Part III
SAMPLE
ESTIMATES

Chapter 22

USING MEANS MECHANICAL AND PLUMBING COST DATA

Users of *Means Mechanical Cost Data* and *Means Plumbing Cost Data* are chiefly interested in obtaining quick, reasonable, average prices for mechanical construction items. This is the primary purpose of the annual book–to eliminate guesswork when pricing unknowns. Many persons use the cost data, however, for bids, verification of quotations, or budgets, without being fully aware of how the prices are obtained and derived. Without this knowledge, this resource is not being used to fullest advantage. In addition to the basic cost data, the books also contain a wealth of information to aid the estimator, the contractor, the designer, and the owner to better plan and manage mechanical and plumbing construction projects. Productivity data is provided in order to assist with scheduling. National labor rates are analyzed. Tables and charts for location and time adjustments are included and help the estimator tailor the prices to a specific location. The costs in *Means Mechanical Cost Data* and *Means Plumbing Cost Data* consist of over 18,000 unit price line items, as well as prices for 3,000 mechanical systems. This information, organized according to the Construction Specifications Institute MASTERFORMAT Divisions, also provides an invaluable checklist to the construction professional to assure that all required items are included in a project.

Format and Data

The major portion of *Means Mechanical Cost Data* or *Means Plumbing Cost Data* is the Unit Price section, Section A. This is the primary source of unit cost data and is organized according to the CSI division index. This index was developed by representatives of all parties concerned with the building construction industry and has been accepted by the American Institute of Architects (AIA) the Associated General Contractors of America, Inc. (AGC), and the Construction Specifications Institute, Inc. (CSI). In *Means Mechanical Cost Data* and *Means Plumbing Cost Data*, relevant parts of other divisions are included along with Division 15–Mechanical. For example, items from Divisions 1, 2, 3, 4, 5, 6, 7, 8, 10, 11, 12, 13, 14, and 16 all appear in addition to the Division 15 entries.

CSI MASTERFORMAT Divisions:

Division 1 - General Requirements

Division 2 - Site Work

Division 3 - Concrete
Division 4 - Masonry
Division 5 - Metals
Division 6 - Wood & Plastics
Division 7 - Moisture-Thermal Control
Division 8 - Doors, Windows & Glass
Division 9 - Finishes
Division 10 - Specialties
Division 11 - Equipment
Division 12 - Furnishings
Division 13 - Special Construction
Division 14 - Conveying Systems
Division15 - Mechanical
Division 16 - Electrical

In addition to the 16 CSI divisions of the Unit Price section, Division 17, Square Foot and Cubic Foot Costs, presents consolidated data from over 11,300 actual reported construction projects and provides information based on total project costs as well as costs for major components.

Section B, Assemblies Cost Tables, contains over 3,000 costs for mechanical and appropriate related assemblies, or systems. Components of the systems are fully detailed and accompanied by illustrations.

Section C contains tables and reference charts. It also provides estimating procedures and explanations of cost development which support and supplement the unit price and systems cost data. Also in Section C are the City Cost Indexes, representing the compilation of construction data for 162 major U.S. and Canadian cities. Cost factors are given for each city, by trade, relative to the national average.

The prices presented in Means Mechanical Cost Data and Means Plumbing Cost Data are national averages. Material and equipment costs are developed through annual contact with manufacturers, dealers, distributors, and contractors throughout the United States and Canada. Means' staff of engineers is constantly updating prices and keeping abreast of changes and fluctuations within the industry. Labor rates are the national average of each trade as determined from union agreements from 30 major U.S. cities. Throughout the calendar year, as new wage agreements are negotiated, labor costs should be factored accordingly.

Following is a list of factors and assumptions on which the costs presented in Means Mechanical Cost Data and Means Plumbing Cost Data have been based:

- **Quality**: The costs are based on methods, materials, and workmanship in accordance with U.S. Government standards and represent good, sound construction practice.
- **Overtime**: The costs, as presented, include *no* allowance for overtime. If overtime or premium time is anticipated, labor costs must be factored accordingly.
- **Productivity**: The daily output and man-hour figures are based on an eight-hour workday, during daylight hours. The charts in Figures 11.7 and 11.8 shows that as the number of hours worked per day (over eight) increases, and as the days per week (over five) increase, production efficiency decreases (see Chapter 11, "Cost Control and Analysis").

- **Size of Project**: Costs in *Means Mechanical Cost Data* or *Means Plumbing Cost Data* are based on commercial and industrial buildings for which total project costs are $500,000 and up. Large residential projects are also included.
- **Local Factors**: Weather conditions, season of the year, local union restrictions, and unusual building code requirements can all have a significant impact on construction costs. The availability of a skilled labor force, sufficient materials, and even adequate energy and utilities will also affect costs. These factors vary in impact and are not necessarily dependent upon location. They must be reviewed for each project in every area.

In presenting prices in *Means Mechanical Cost Data* or *Means Plumbing Cost Data*, certain rounding rules are employed to make the numbers easy to use without significantly affecting accuracy. The rules are used consistently and are as follows:

Prices From	To	Rounded to nearest
$ 0.01	$ 5.00	$ 0.01
5.01	20.00	0.05
20.01	100.00	1.00
100.01	1,000.00	5.00
1,000.01	10,000.00	25.00
10,000.01	50,000.00	100.00
50,000.01	up	500.00

Section A– Unit Price Costs

The Unit Price section of *Means Mechanical Cost Data* or *Means Plumbing Cost Data* contains a great deal of information in addition to the unit cost for each construction component. Figure 22.1 is a typical page showing costs for fire valves. Note that prices are included for several types of valves, each in a wide range of size and capacity ratings. In addition, appropriate crews, workers, and productivity data are indicated. The information and cost data is broken down and itemized in this way to provide for the most detailed pricing possible. Both the unit price and the systems sections include detailed illustrations. The reference numbers enclosed in the squares refer the user to an appropriate assemblies section or reference table.

Within each individual line item, there is a description of the construction component, information regarding typical crews designated to perform the work and productivity shown as daily output and as man-hours. Costs are presented in two ways: "bare," or unburdened costs, and costs with markups for overhead and profit. Figure 22.2 is a graphic representation of how to use the Unit Price section as presented in *Means Plumbing Cost Data*.

Line Numbers

Every construction item in the Means unit price cost data books has a unique line number. This line number acts as an address so that each item can be quickly located and/or referenced. The numbering system is based on the CSI MASTERFORMAT classification by division. In Figure 22.2, note the bold number in reverse type, "153". This number represents the major subdivision, in this case "Plumbing Appliances", of the major CSI Division 15–Mechanical. Within each subdivision, the data is broken down into major classifications. These major classifications are listed alphabetically and are designated by

154 100	Fire Systems		CREW	DAILY OUTPUT	MAN-HOURS	UNIT	BARE COSTS MAT.	LABOR	EQUIP.	TOTAL	TOTAL INCL O&P
160 0050	For polished brass, add	B8.2-320				Ea.	30%				160
0060	For polished chrome, add						40%				
0080	Wheel handle, 300 lb., 1-½"	B8.2-390	1 Spri	12	.667		23.50	17.60		41.10	52
0090	2-½"		"	7	1.140		54	30		84	105
0100	For polished brass, add	C8.2-301					35%				
0110	For polished chrome, add						50%				
1000	Ball drip, automatic, rough brass, ½"	C8.2-302	1 Spri	20	.400		6	10.55		16.55	22
1010	¾"		"	20	.400		8	10.55		18.55	25
1100	Butterfly, 175 lb., sprinkler system, FM/UL, threaded, bronze										
1120	Slow close										
1150	1" size		1 Spri	19	.421	Ea.	57	11.10		68.10	79
1160	1-¼" size			15	.533		64	14.10		78.10	91
1170	1-½" size			13	.615		83.80	16.25		100.05	115
1180	2" size			11	.727		117	19.20		136.20	155
1190	2-½" size		Q-12	15	1.070		148	25		173	200
1230	For supervisory switch kit, all sizes										
1240	One circuit, add		1 Spri	48	.167	Ea.	56.60	4.40		61	69
1250	Two circuits, add		"	40	.200	"	56.60	5.30		61.90	70
1280	Quarter turn for trim										
1300	½" size		1 Spri	22	.364	Ea.	5.10	9.60		14.70	19.95
1310	¾" size			20	.400		7.60	10.55		18.15	24
1320	1" size			19	.421		9.90	11.10		21	28
1330	1-¼" size			15	.533		16.30	14.10		30.40	39
1340	1-½" size			13	.615		20.80	16.25		37.05	47
1350	2" size			11	.727		25.40	19.20		44.60	57
1400	Caps, polished brass with chain, ¾"						15			15	16.50
1420	1"						22.50			22.50	25
1440	1-½"						7.50			7.50	8.25
1460	2-½"						11			11	12.10
1480	3"						20.50			20.50	23
3000	Gate, hose, wheel handle, N.R.S., rough brass, 1-½"		1 Spri	12	.667		60	17.60		77.60	92
3040	2-½", 300 lb.		"	7	1.140		78	30		108	130
3080	For polished brass, add						40%				
3090	For polished chrome, add						50%				
3800	Hydrant, screw type, crank handle, brass										
3840	2-½" size		Q-12	11	1.450	Ea.	185	35		220	255
3880	For chrome, same price										
4200	Hydrolator, vent and draining, rough brass, 1-½"		1 Spri	12	.667		24.80	17.60		42.40	54
4220	2-½"		"	7	1.140		71.40	30		101.40	125
4280	For polished brass, add						50%				
4290	For polished chrome, add						90%				
5000	Pressure restricting, adjustable rough brass, 1-½"		1 Spri	12	.667		54	17.60		71.60	86
5020	2-½"		"	7	1.140		95.13	30		125.13	150
5080	For polished brass, add						30%				
5090	For polished chrome, add						45%				
6000	Roof manifold, horiz., brass, with valves & caps										
6040	2-½" x 2-½" x 4"		Q-12	4.80	3.330	Ea.	80	79		159	205
6060	2-½" x 2-½" x 6"			4.60	3.480		119	83		202	255
6080	2-½" x 2-½" x 2-½" x 4"			4.60	3.480		107	83		190	240
6090	2-½" x 2-½" x 2-½" x 6"			4.60	3.480		127	83		210	265
7000	Sprinkler line tester, cast brass						14.40			14.40	15.85
8000	Wye, leader line, ball type, swivel female x male x male										
8040	2-½" x 1-½" x 1-½" polished brass					Ea.	207			207	230
8060	2-½" x 1-½" x 1-½" polished chrome					"	215			215	235
170 0010	SPRINKLER SYSTEM COMPONENTS										170
0600	Accelerator		1 Spri	8	1	Ea.	237	26		263	300
0800	Air compressor for dry pipe system, automatic, complete	B8.2-110									
0820	280 gal. system capacity, ¾ HP		1 Spri	1.30	6.150	Ea.	560	160		720	860

177

(Reprinted from Means Plumbing Cost Data 1991.)

Figure 22.1

HOW TO USE UNIT PRICE PAGES

Important
Prices in this section are listed in two ways: as bare costs and as costs including overhead and profit of the installing contractor. In most cases, if the work is to be subcontracted, it is best for a general contractor to add an additional 10% to the figures found in the column titled **"TOTAL INCL. O&P"**.

Unit
The unit of measure listed here reflects the material being used in the line item. For example: Water coolers are expressed as each (Ea.).

Productivity
The daily output represents typical total daily amount of work that the designated crew will produce. Man-hours are a unit of measure for the labor involved in performing a task. To derive the total man-hours for a task, multiply the quantity of the item involved times the man-hour figure shown.

Line Number Determination
Each line item is identified by a unique ten-digit number.

MASTERFORMAT
Division
153 105 3400

.Subdivision

Mediumscope
153 105
153 **105** 3400.

Major Classification

153 105 **3400**

Individual Line Number

Description
The meaning of this line item shows a recessed water cooler will be installed by a Q-1 crew at a rate of 3.5 per day. (4.57 man-hours each)

| C8.1 403 | **Reference Number** |

These reference numbers refer to charts, tables, estimating data, cost derivations and other information which may be useful to the user of this book. These references may direct the reader to any section within the book.

Bare Costs are developed as follows for line no. **153-105-3400**
Mat. is **Bare Material Cost ($765)**
Labor for Crew Q1 = Man-hour Cost **($22.90)** × Man-hour Units **(4.570)** = **$105.00**
Equip. for Crew Q1 = Equip. Hour Cost **($0)** × Man-hour Units **(9.600)** = **$0**
Total = **Mat. Cost ($765)** + **Labor Cost ($105.00)** + **Equip. Cost ($0)** = **$870** each.
(**Note:** Where a Crew is indicated Equipment and Labor costs are derived from the Crew Tables. See example above.)

Total Costs Including O&P are developed as follows:
Mat. is **Bare Material Cost** + 10% = **$765** + **$76.50** = **$841.50**
Labor for Crew Q1 = Man-hour Cost **($34.17)** × Man-hour Units **(4.570)** = **$158.50**
Equip. for Crew Q1 = Equip. Hour Cost **($0)** × Man-hour Units **(4.570)** = **$0**
Total = **Mat. Cost ($841.50)** + **Labor Cost ($158.50)** + **Equip. Cost ($0)** = **$1,000.00**
(**Note:** Where a crew is indicated, Equipment and Labor costs are derived from the Crew Tables. See example above. **"Total Inc. O&P"** costs are rounded.)

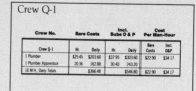

Crew Q-1

Crew No.	Bare Costs		Incl. Subs O & P		Cost Per Man-Hour	
Crew Q-1	Hr.	Daily	Hr.	Daily	Bare Costs	Incl. O&P
1 Plumber	$25.45	$203.60	$37.95	$303.60	$22.90	$34.17
1 Plumber Apprentice	20.36	162.88	30.40	243.20		
16 M.H., Daily Totals		$366.48		$546.80	$22.90	$34.17

---

153 | Plumbing Appliances

153 100 | Water Appliances

		CREW	DAILY OUTPUT	MAN-HOURS	UNIT	BARE COSTS				TOTAL INCL O&P
						MAT.	LABOR	EQUIP.	TOTAL	
0010	WATER COOLER									
0030	See line 153-105-9800 for rough-in, waste & vent									
0040	for all water coolers									
0100	Wall mounted, non-recessed									
0140	4 GPH	Q-1	4	4	Ea.	291	92		383	455
0180	8.2 GPH		4	4		347	92		439	520
0220	14.3 GPH		4	4		367	92		459	540
0260	16.1 GPH		4	4		390	92		482	565
0600	For hot and cold water, add					160			160	175
0640	For stainless steel cabinet, add					41.50			41.50	46
1040	14.3 GPH	Q-1	3.80	4.210		504	96		600	700
1240	For stainless steel cabinet, add					78			78	86
2600	Wheelchair type, 8 GPH	Q-1	4	4		875	92		967	1,100
3000	Simulated recessed, 8 GPH		4	4		416	92		508	595
3040	11.5 GPH		4	4		440	92		532	620
3200	For glass filler in addition to bubbler, add					69			69	76
3240	For stainless steel cabinet, add				Ea.	27			27	30
3300	Semi-recessed, 8.1 GPH	Q-1	4	4		468	92		560	650
3320	12 GPH	"	4	4		490	92		582	675
3340	For glass filler, add					69			69	76
3360	For stainless steel cabinet, add					27			27	30
3400	Full recessed, stainless steel, 8 GPH	Q-1	3.50	4.570		765	105		870	1,000
3420	11.5 GPH	"	3.50	4.570		825	105		930	1,075
3460	For glass filler, add					69			69	76
3600	For mounting can only					180			180	200
4600	Floor mounted, flush-to-wall									
4640	4 GPH	1 Plum	3	2.670	Ea.	312	68		380	445
4680	8.2 GPH		3	2.670		354	68		422	490
4720	14.3 GPH		3	2.670		370	68		438	510
4960	For hot and cold water, add					132			132	145
4980	For stainless steel cabinet, add					65.50			65.50	72
5040	14.3 GPH	1 Plum	2	4		540	100		640	745
5080	19.5 GPH	"	2	4		580	100		680	790
5120	For stainless steel cabinet, add					65.50			65.50	72
5600	Explosion Proof, 16 GPH	1 Plum	3	2.670		785	68		853	965
5640	For stainless steel cabinet, add					216			216	240

(Reprinted from Means Plumbing Cost Data 1991.)

Figure 22.2

bold type for both numbers and descriptions. Each item, or line, is further defined by an individual number. As shown in Figure 22.2, the full line number for each item consists of: a major CSI subdivision number–a major classification number–an item line number. Each full line number describes a unique construction element. For example, in Figure 22.1, the line number for a 2", slow close, bronze butterfly valve is 154-160-1180.

Line Description

Each line has a text description of the item for which costs are listed. The description may be self-contained and all inclusive or, if indented, the complete description for a line is dependent upon the information provided above. All indented items are delineations (by size, color, material, etc.) or breakdowns of previously described items. An index is provided in the back of *Means Mechanical Cost Data* and *Means Plumbing Cost Data* to aid in locating particular items.

Crew

For each construction element, (each line item), a minimum typical crew is designated as appropriate to perform the work. The crew may include one or more trades, foremen, craftsmen and helpers, and any equipment required for proper installation of the described item. If an individual trade installs the item using only hand tools, the smallest efficient number of tradesmen will be indicated (1 Plum, 2 Spri, etc.). Abbreviations for trades are shown in Figure 22.3. If more than one trade is required to install the item and/or if powered equipment is needed, a crew number will be designated (Q-19, Q-8, etc.). A complete listing of crews is presented in the Foreword pages of *Means Mechanical Cost Data* and *Means Plumbing Cost Data* (see Figure 22.4). On these pages, each crew is broken down into the following components:

1. Number and type of workers designated.
2. Number, size, and type of any equipment required.
3. Hourly labor costs listed two ways: "bare"–base rate including fringe benefits; and including installing contractor's overhead and profit - billing rate. (See Figure 22.3 from the inside back cover of *Means Mechanical Cost Data* or *Means Plumbing Cost Data* for labor rate information).
4. Daily equipment costs, based on the weekly equipment rental cost divided by 5, plus the hourly operating cost, times 8 hours. This cost is listed two ways: as a bare cost and with a 10% markup to cover handling and management costs.
5. Labor and equipment are broken down further into: cost per man-hour for labor, and cost per man-hour, for the equipment.
6. The total daily man-hours for the crew.
7. The total bare costs per day for the crew, including equipment.
8. The total daily cost of the crew including the installing contractor's overhead and profit.

The total daily cost of the required crew is used to calculate the unit installation cost for each item (for both bare costs and cost including overhead and profit).

The crew designation does not mean that this is the only crew that can perform the work. Crew size and content have been developed and chosen based on practical experience and feedback from contractors. These designations represent a labor and equipment makeup commonly found in the industry. The most appropriate crew for a given task is best determined based on particular project requirements. Unit costs may vary if crew sizes or content are significantly changed.

Installing Contractor's Overhead & Profit

Below are the **average** installing contractor's percentage mark-ups applied to base labor rates to arrive at typical billing rates.

Column A: Labor rates are based on union wages averaged for 30 major U.S. cities. Base rates including fringe benefits are listed hourly and daily. These figures are the sum of the wage rate and employer-paid fringe benefits such as vacation pay, employer-paid health and welfare costs, pension costs, plus appropriate training and industry advancement funds costs.

Column B: Workers' Compensation rates are the national average of state rates established for each trade.

Column C: Column C lists average fixed overhead figures for all trades. Included are Federal and State Unemployment costs set at 6.2%; Social Security Taxes (FICA) set at 7.65%; Builder's Risk Insurance costs set at 0.34%; and Public Liability costs set at 1.55%. All the percentages except those for Social Security Taxes vary from state to state as well as from company to company.

Column D and E: Percentages in Columns D and E are based on the presumption that the installing contractor has annual billing of $500,000 and up. Overhead percentages may increase with smaller annual billing. The overhead percentages for any given contractor may vary greatly and depend on a number of factors, such as the contractor's annual volume, engineering and logistical support costs, and staff requirements. The figures for overhead and profit will also vary depending on the type of job, the job location, and the prevailing economic conditions. All factors should be examined very carefully for each job.

Column F: Column F lists the total of columns B, C, D, and E.

Column G: Column G is Column A (hourly base labor rate) multiplied by the percentage in Column F (O&P percentage).

Column H: Column H is the total of Column A (hourly base labor rate) plus Column G (Total O&P).

Column I: Column I is Column H multiplied by eight hours.

		A		B	C	D	E	F		G	H		I
		Base Rate Incl. Fringes		Workers' Comp. Ins.	Average Fixed Over-head	Over-head	Profit	Total Overhead & Profit			Rate with O & P		
Abbr.	Trade	Hourly	Daily					%	Amount		Hourly	Daily	
Skwk	Skilled Workers Average (35 trades)	$22.65	$181.20	15.1%	15.7%	12.8%	10%	53.6%	$12.15		$34.80	$278.40	
	Helpers Average (5 trades)	17.10	136.80	16.2		13.0		54.9	9.40		26.50	212.00	
	Foremen Average, Inside (50¢ over trade)	23.15	185.20	15.1		12.8		53.6	12.40		35.55	284.40	
	Foremen Average, Outside ($2.00 over trade)	24.65	197.20	15.1		12.8		53.6	13.20		37.85	302.80	
Clab	Common Building Laborers	17.50	140.00	16.6		11.0		53.3	9.35		26.85	214.80	
Asbe	Asbestos Workers	24.70	197.60	13.5		16.0		55.2	13.65		38.35	306.80	
Boil	Boilermakers	25.05	200.40	9.3		16.0		51.0	12.80		37.85	302.80	
Bric	Bricklayers	22.75	182.00	13.7		11.0		50.4	11.45		34.20	273.60	
Brhe	Bricklayer Helpers	17.65	141.20	13.7		11.0		50.4	8.90		26.55	212.40	
Carp	Carpenters	22.00	176.00	16.6		11.0		53.3	11.75		33.75	270.00	
Cefi	Cement Finishers	21.65	173.20	9.6		11.0		46.3	10.00		31.65	253.20	
Elec	Electricians	25.15	201.20	6.0		16.0		47.7	12.00		37.15	297.20	
Elev	Elevator Constructors	25.35	202.80	7.7		16.0		49.4	12.50		37.85	302.80	
Eqhv	Equipment Operators, Crane or Shovel	23.30	186.40	10.4		14.0		50.1	11.65		34.95	279.60	
Eqmd	Equipment Operators, Medium Equipment	22.50	180.00	10.4		14.0		50.1	11.25		33.75	270.00	
Eqlt	Equipment Operators, Light Equipment	21.40	171.20	10.4		14.0		50.1	10.70		32.10	256.80	
Eqol	Equipment Operators, Oilers	19.20	153.60	10.4		14.0		50.1	9.60		28.80	230.40	
Eqmm	Equipment Operators, Master Mechanics	24.00	192.00	10.4		14.0		50.1	12.00		36.00	288.00	
Glaz	Glaziers	22.55	180.40	11.9		11.0		48.6	10.95		33.50	268.00	
Lath	Lathers	21.95	175.60	10.2		11.0		46.9	10.30		32.25	258.00	
Marb	Marble Setters	22.55	180.40	13.7		11.0		50.4	11.35		33.90	271.20	
Mill	Millwrights	22.95	183.60	9.9		11.0		46.6	10.70		33.65	269.20	
Mstz	Mosaic and Terrazzo Workers	22.10	176.80	8.3		11.0		45.0	9.95		32.05	256.40	
Pord	Painters, Ordinary	20.80	166.40	12.4		11.0		49.1	10.20		31.00	248.00	
Psst	Painters, Structural Steel	21.45	171.60	42.9		11.0		79.6	17.05		38.50	308.00	
Pape	Paper Hangers	20.80	166.40	12.4		11.0		49.1	10.20		31.00	248.00	
Pile	Pile Drivers	22.20	177.60	25.4		16.0		67.1	14.90		37.10	296.80	
Plas	Plasterers	21.80	174.40	13.4		11.0		50.1	10.90		32.70	261.60	
Plah	Plasterer Helpers	17.90	143.20	13.4		11.0		50.1	8.95		26.85	214.80	
Plum	Plumbers	25.45	203.60	7.5		16.0		49.2	12.50		37.95	303.60	
Rodm	Rodmen (Reinforcing)	23.90	191.20	27.6		14.0		67.3	16.10		40.00	320.00	
Rofc	Roofers, Composition	20.35	162.80	28.9		11.0		65.6	13.35		33.70	269.60	
Rots	Roofers, Tile & Slate	20.45	163.60	28.9		11.0		65.6	13.40		33.85	270.80	
Rohe	Roofer Helpers (Composition)	14.90	119.20	28.9		11.0		65.6	9.75		24.65	197.20	
Shee	Sheet Metal Workers	25.00	200.00	10.2		16.0		51.9	13.00		38.00	304.00	
Spri	Sprinkler Installers	26.40	211.20	7.7		16.0		49.4	13.05		39.45	315.60	
Stpi	Steamfitters or Pipefitters	25.50	204.00	7.5		16.0		49.2	12.55		38.05	304.40	
Ston	Stone Masons	22.65	181.20	13.7		11.0		50.4	11.40		34.05	272.40	
Sswk	Structural Steel Workers	24.10	192.80	35.4		14.0		75.1	18.10		42.20	337.60	
Tilf	Tile Layers (Floor)	22.15	177.20	8.3		11.0		45.0	9.95		32.10	256.80	
Tilh	Tile Layer Helpers	17.45	139.60	8.3		11.0		45.0	7.85		25.30	202.40	
Trlt	Truck Drivers, Light	18.10	144.80	13.5		11.0		50.2	9.10		27.20	217.60	
Trhv	Truck Drivers, Heavy	18.40	147.20	13.5		11.0		50.2	9.25		27.65	221.20	
Sswl	Welders, Structural Steel	24.10	192.80	35.4		14.0		75.1	18.10		42.20	337.60	
Wrck	*Wrecking	17.50	140.00	35.5		11.0		72.2	12.65		30.15	241.20	

*Not included in Averages.

(Reprinted from Means Plumbing Cost Data 1991.)

Figure 22.3

CREWS

Crew L-5	Hr.	Daily	Hr.	Daily	Bare Costs	Incl. O&P
1 Struc. Steel Foreman	$26.10	$208.80	$45.70	$365.60	$24.27	$41.66
5 Struc. Steel Workers	24.10	964.00	42.20	1688.00		
1 Equip. Oper. (crane)	23.30	186.40	34.95	279.60		
1 Hyd. Crane, 25 Ton		486.40		535.05	8.68	9.55
56 M.H., Daily Totals		$1845.60		$2868.25	$32.95	$51.21

Crew L-6	Hr.	Daily	Hr.	Daily	Bare Costs	Incl. O&P
1 Plumber	$25.45	$203.60	$37.95	$303.60	$25.35	$37.68
.5 Electrician	25.15	100.60	37.15	148.60		
12 M.H., Daily Totals		$304.20		$452.20	$25.35	$37.68

Crew L-7	Hr.	Daily	Hr.	Daily	Bare Costs	Incl. O&P
2 Carpenters	$22.00	$352.00	$33.75	$540.00	$21.16	$32.26
1 Building Laborer	17.50	140.00	26.85	214.80		
.5 Electrician	25.15	100.60	37.15	148.60		
28 M.H., Daily Totals		$592.60		$903.40	$21.16	$32.26

Crew L-8	Hr.	Daily	Hr.	Daily	Bare Costs	Incl. O&P
2 Carpenters	$22.00	$352.00	$33.75	$540.00	$22.69	$34.59
.5 Plumber	25.45	101.80	37.95	151.80		
20 M.H., Daily Totals		$453.80		$691.80	$22.69	$34.59

Crew L-9	Hr.	Daily	Hr.	Daily	Bare Costs	Incl. O&P
1 Labor Foreman (inside)	$18.00	$144.00	$27.60	$220.80	$19.92	$31.57
2 Building Laborers	17.50	280.00	26.85	429.60		
1 Struc. Steel Worker	24.10	192.80	42.20	337.60		
.5 Electrician	25.15	100.60	37.15	148.60		
36 M.H., Daily Totals		$717.40		$1136.60	$19.92	$31.57

Crew M-1	Hr.	Daily	Hr.	Daily	Bare Costs	Incl. O&P
3 Elevator Constructors	$25.35	$608.40	$37.85	$908.40	$24.08	$35.96
1 Elevator Apprentice	20.28	162.24	30.30	242.40		
Hand Tools		76.00		83.60	2.37	2.61
32 M.H., Daily Totals		$846.64		$1234.40	$26.45	$38.57

Crew M-2	Hr.	Daily	Hr.	Daily	Bare Costs	Incl. O&P
2 Millwrights	$22.95	$367.20	$33.65	$538.40	$22.95	$33.65
Power Tools		16.00		17.60	1.00	1.10
16 M.H., Daily Totals		$383.20		$556.00	$23.95	$34.75

Crew Q-1	Hr.	Daily	Hr.	Daily	Bare Costs	Incl. O&P
1 Plumber	$25.45	$203.60	$37.95	$303.60	$22.90	$34.17
1 Plumber Apprentice	20.36	162.88	30.40	243.20		
16 M.H., Daily Totals		$366.48		$546.80	$22.90	$34.17

Crew Q-2	Hr.	Daily	Hr.	Daily	Bare Costs	Incl. O&P
2 Plumbers	$25.45	$407.20	$37.95	$607.20	$23.75	$35.43
1 Plumber Apprentice	20.36	162.88	30.40	243.20		
24 M.H., Daily Totals		$570.08		$850.40	$23.75	$35.43

Crew Q-3	Hr.	Daily	Hr.	Daily	Bare Costs	Incl. O&P
1 Plumber Foreman (ins)	$25.95	$207.60	$38.70	$309.60	$24.30	$36.25
2 Plumbers	25.45	407.20	37.95	607.20		
1 Plumber Apprentice	20.36	162.88	30.40	243.20		
32 M.H., Daily Totals		$777.68		$1160.00	$24.30	$36.25

Crew Q-4	Hr.	Daily	Hr.	Daily	Bare Costs	Incl. O&P
1 Plumber Foreman (ins)	$25.95	$207.60	$38.70	$309.60	$24.30	$36.25
1 Plumber	25.45	203.60	37.95	303.60		
1 Welder (plumber)	25.45	203.60	37.95	303.60		
1 Plumber Apprentice	20.36	162.88	30.40	243.20		
1 Electric Welding Mach.		39.00		42.90	1.21	1.34
32 M.H., Daily Totals		$816.68		$1202.90	$25.51	$37.59

Crew Q-5	Hr.	Daily	Hr.	Daily	Bare Costs	Incl. O&P
1 Steamfitter	$25.50	$204.00	$38.05	$304.40	$22.95	$34.25
1 Steamfitter Apprentice	20.40	163.20	30.45	243.60		
16 M.H., Daily Totals		$367.20		$548.00	$22.95	$34.25

Crew Q-6	Hr.	Daily	Hr.	Daily	Bare Costs	Incl. O&P
2 Steamfitters	$25.50	$408.00	$38.05	$608.80	$23.80	$35.51
1 Steamfitter Apprentice	20.40	163.20	30.45	243.60		
24 M.H. Daily Totals		$571.20		$852.40	$23.80	$35.51

Crew Q-7	Hr.	Daily	Hr.	Daily	Bare Costs	Incl. O&P
1 Steamfitter Foreman (ins)	$26.00	$208.00	$38.80	$310.40	$24.35	$36.33
2 Steamfitters	25.50	408.00	38.05	608.80		
1 Steamfitter Apprentice	20.40	163.20	30.45	243.60		
32 M.H., Daily Totals		$779.20		$1162.80	$24.35	$36.33

Crew Q-8	Hr.	Daily	Hr.	Daily	Bare Costs	Incl. O&P
1 Steamfitter Foreman (ins)	$26.00	$208.00	$38.80	$310.40	$24.35	$36.33
1 Steamfitter	25.50	204.00	38.05	304.40		
1 Welder (steamfitter)	25.50	204.00	38.05	304.40		
1 Steamfitter Apprentice	20.40	163.20	30.45	243.60		
1 Electric Welding Mach.		39.00		42.90	1.21	1.34
32 M.H., Daily Totals		$818.20		$1205.70	$25.56	$37.67

Crew Q-9	Hr.	Daily	Hr.	Daily	Bare Costs	Incl. O&P
1 Sheet Metal Worker	$25.00	$200.00	$38.00	$304.00	$22.50	$34.20
1 Sheet Metal Apprentice	20.00	160.00	30.40	243.20		
16 M.H., Daily Totals		$360.00		$547.20	$22.50	$34.20

Crew Q-10	Hr.	Daily	Hr.	Daily	Bare Costs	Incl. O&P
2 Sheet Metal Workers	$25.00	$400.00	$38.00	$608.00	$23.33	$35.46
1 Sheet Metal Apprentice	20.00	160.00	30.40	243.20		
24 M.H., Daily Totals		$560.00		$851.20	$23.33	$35.46

Crew Q-11	Hr.	Daily	Hr.	Daily	Bare Costs	Incl. O&P
1 Sheet Metal Foreman (ins)	$25.50	$204.00	$38.75	$310.00	$23.87	$36.28
2 Sheet Metal Workers	25.00	400.00	38.00	608.00		
1 Sheet Metal Apprentice	20.00	160.00	30.40	243.20		
32 M.H., Daily Totals		$764.00		$1161.20	$23.87	$36.28

Crew Q-12	Hr.	Daily	Hr.	Daily	Bare Costs	Incl. O&P
1 Sprinkler Installer	$26.40	$211.20	$39.45	$315.60	$23.76	$35.50
1 Sprinkler Apprentice	21.12	168.96	31.55	252.40		
16 M.H., Daily Totals		$380.16		$568.00	$23.76	$35.50

Crew Q-13	Hr.	Daily	Hr.	Daily	Bare Costs	Incl. O&P
1 Sprinkler Foreman (ins)	$26.90	$215.20	$40.20	$321.60	$25.20	$37.66
2 Sprinkler Installers	26.40	422.40	39.45	631.20		
1 Sprinkler Apprentice	21.12	168.96	31.55	252.40		
32 M.H., Daily Totals		$806.56		$1205.20	$25.20	$37.66

(Reprinted from Means Plumbing Cost Data 1991.)

Figure 22.4

Figure 22.5 is a page from Division 15 of *Means Plumbing Cost Data*. This page lists the equipment costs used in the presentation and calculation of the crew costs and unit price data. Rental costs are shown as daily, weekly, and monthly rates. The Hourly Operating Cost represents the cost of fuel, lubrication, and routine maintenance. Equipment costs used in the crews are calculated as follows:

Line Number:	016-420-7800
Equipment:	Electric Welder, 300 Amp.
Rent per week:	$125.00
Hourly Operating Cost:	$ 1.75

$$\frac{\text{Weekly rental}}{5 \text{ Days}} + (\text{Hourly Oper. Cost x 8 hrs/day}) = \text{Daily Equipment Costs}$$

$$\frac{125}{5} + (\$1.75 \times 8) = \$39.00 \text{ per day}$$

Note: Crew operating labor is not included.

Units

The unit column (see Figures 22.1 and 22.2) defines the component for which the costs have been calculated. It is this "unit" on which Unit Price Estimating is based. The units as used represent standard estimating and quantity takeoff procedures. However, the estimator should always check to be sure that the units taken off are the same as those priced. A list of standard abbreviations is included at the back of *Means Mechanical Cost Data* and *Means Plumbing Cost Data*.

Bare Costs

The four columns listed under "Bare Costs," – "Material," "Labor," "Equipment," and "Total" represent the actual cost of construction items to the contractor. In other words, bare costs are those which do not include the overhead and profit of the installing contractor, whether it is a subcontractor or a general contracting company using its own crews.

Material: Material costs are based on the national average contractor purchase price delivered to the job site. Delivered costs are assumed to be within a 20-mile radius of metropolitan areas. No sales tax is included in the material prices because of variations from state to state.

The prices are based on quantities that would normally be purchased for complete buildings or projects costing $500,000 and up. Prices for small quantities must be adjusted accordingly. If more current costs for materials are available for the appropriate location, it is recommended that adjustments be made to the unit costs to reflect any cost difference.

Labor: Labor costs are calculated by multiplying the "Bare Labor Cost" per man-hour times the number of man-hours, from the "Man-Hours" column. The "Bare" labor rate is determined by adding the base rate plus fringe benefits. The base rate is the actual hourly wage of a worker used in figuring payroll. It is from this figure that employee deductions are taken (Federal withholding, FICA, State withholding). Fringe benefits include all employer-paid benefits, above and beyond the payroll amount (employer-paid health, vacation pay, pension, profit sharing). The "Bare Labor Cost" is, therefore, the actual amount that the contractor must pay directly for construction workers. Figure 22.3 shows labor rates for the 35 construction trades plus skilled worker, helper,

016 400 | Equipment Rental

			UNIT	HOURLY OPER. COST.	RENT PER DAY	RENT PER WEEK	RENT PER MONTH	CREW EQUIPMENT COST	
420	4600	1-½″, 83 GPM	Ea.	.27	33	100	300	22.15	420
	4700	2″, 120 GPM		.27	35	105	315	23.15	
	4800	3″, 300 GPM		.55	44	130	390	30.40	
	4900	4″, 560 GPM		1	59	180	535	44	
	5000	6″, 1590 GPM		4.65	165	500	1,500	137.20	
	5100	Diaphragm pump, gas, single, 1-½″ diameter		.43	16	48	145	13.05	
	5200	2″ diameter		.45	18	54	160	14.40	
	5300	3″ diameter		.60	31	94	280	23.60	
	5400	Double, 4″ diameter		1.40	65	195	585	50.20	
	5500	Trash pump, self-priming, gas, 2″ diameter		.95	27	80	240	23.60	
	5600	Diesel, 4″ diameter		1.50	81	245	735	61	
	5650	Diesel, 6″ diameter		4.10	140	420	1,275	116.80	
	5700	Salamanders, L.P. gas fired, 100,000 B.T.U.		.60	12	36	110	12	
	5800	Saw, chain, gas engine, 18″ long		.40	28	84	250	20	
	5900	36″ long		.95	57	170	510	41.60	
	5950	60″ long		.96	55	165	495	40.70	
	6000	Masonry, table mounted, 14″ diameter, 5 H.P.		1.50	33	100	300	32	
	6100	Circular, hand held, electric, 7″ diameter		.15	11	34	100	8	
	6200	12″ diameter		.25	22	66	200	15.20	
	6350	Torch, cutting, acetylene-oxygen, 150′ hose		6.40	19	58	175	62.80	
	6360	Hourly operating cost includes tips and gas		6.75	16.65	50	150	64	
	6410	Toilet, portable chemical			8.30	25	75	5	
	6420	Recycle flush type			10.80	31	93	6.20	
	6430	Toilet, fresh water flush, garden hose,			16.20	49	145	9.80	
	6440	Hoisted, non-flush, for high rise			8.65	26	78	5.20	
	6450	Toilet, trailers, minimum			20	60	180	12	
	6460	Maximum			81	245	730	49	
	6500	Trailers, platform, flush deck, 2 axle, 25 ton capacity		1.10	125	380	1,150	84.80	
	6600	40 ton capacity		1.45	200	595	1,785	130.60	
	6700	3 axle, 50 ton capacity		2.45	220	660	1,975	151.60	
	6800	75 ton capacity		3.10	320	970	2,900	218.80	
	7100	Truck, pickup, ¾ ton, 2 wheel drive		9.05	58	175	525	107.40	
	7200	4 wheel drive		10.25	60	175	530	117	
	7300	Tractor, 4 x 2, 30 ton capacity, 195 H.P.		9.05	325	980	2,950	268.40	
	7410	250 H.P.		12.50	360	1,075	3,225	315	
	7700	Welder, electric, 200 amp		.80	18	54	160	17.20	
	7800	300 amp		1.75	42	125	375	39	
	7900	Gas engine, 200 amp		4.05	35	105	315	53.40	
	8000	300 amp		4.30	57	170	510	68.40	
	8100	Wheelbarrow, any size			6.65	20	60	4	
460	0010	**LIFTING & HOISTING EQUIPMENT RENTAL**							460
	0100	without operators							
	0200	Crane, climbing, 106′ jib, 6000 lb. capacity, 410 FPM	Ea.	22.45	1,075	3,225	9,675	824.60	
	0300	101′ jib, 10,250 lb. capacity, 270 FPM	″	29.95	1,375	4,100	12,300	1,059	
	0400	Tower, static, 130′ high, 106′ jib,							
	0500	6200 lb. capacity at 400 FPM	Ea.	48.15	1,275	3,800	11,400	1,145	
	0600	Crawler, cable, ½ C.Y., 15 tons at 12′ radius		15.80	415	1,250	3,750	376.40	
	0700	¾ C.Y., 20 tons at 12′ radius		16.40	450	1,350	4,050	401.20	
	0800	1 C.Y., 25 tons at 12′ radius		17.60	490	1,475	4,425	435.80	
	0900	1-½ C.Y., 40 tons at 12′ radius		25.50	760	2,275	6,825	659	
	1000	2 C.Y., 50 tons at 12′ radius		29.75	865	2,600	7,800	758	
	1100	3 C.Y., 75 tons at 12′ radius		36.40	965	2,900	8,700	871.20	
	1200	100 ton capacity, standard boom		35.30	1,350	4,025	12,100	1,087	
	1300	165 ton capacity, standard boom		55.65	2,150	6,500	19,500	1,745	
	1400	200 ton capacity, 150′ boom		105	2,350	7,075	21,200	2,255	
	1500	450′ boom		118	2,950	8,850	26,500	2,714	
	1600	Truck mounted, cable operated, 6 x 4, 20 tons at 10′ radius		11.55	590	1,800	5,400	452.40	
	1700	25 tons at 10′ radius		17.60	980	2,950	8,850	730.80	

9

(Reprinted from Means Plumbing Cost Data 1991.)

Figure 22.5

and foreman averages. These rates are the averages of union wage agreements effective January 1 of the current year from 30 major cities in the United States. The "Bare Labor Cost" for each trade, as used in *Means Mechanical Cost Data* and *Means Plumbing Cost Data*, is shown in column "A" as the base rate including fringes. Refer to the "Crew" column to determine what rate is used to calculate the "Bare Labor Cost" for a particular line item.

Equipment: Equipment costs are calculated by multiplying the "Bare Equipment Cost" per man-hour, from the appropriate "Crew" listing, times the man-hours in the "Man-Hours" column. The calculation of the equipment portion of installation costs is outlined earlier in this chapter.

Total Bare Costs

This column simply represents the arithmetic sum of the bare material, labor, and equipment costs. This total is the average cost to the contractor for the particular item of construction, supplied and installed, or "in place." No overhead and/or profit is included.

Total Including Overhead and Profit

This column represents the total cost of an item including the installation contractor's overhead and profit. The installing contractor could be either the prime mechanical contractor or a subcontractor. If these costs are used for an item to be installed by a subcontractor, the prime mechanical contractor should include an additional percentage (usually 10% to 20%) to cover the expenses of supervision and management. Consideration must be given also to sub-subcontractors who often appear in mechanical contracting. An example might be an electrical sub to the Temperature Control Subcontractor who, of course, is a sub to the HVAC contractor. Each must have their own overhead and profit markup.

The costs in the "Total Including Overhead and Profit" are the arithmetical sum of the following three calculations:

- Bare Material Cost plus 10%
- Labor Cost, including overhead and profit, per man-hour times the number of man-hours
- Equipment Costs, including overhead and profit, per man-hour times the number of man-hours

The Labor and Equipment Costs, including overhead and profit are found in the appropriate crew listings. The overhead and profit percentage factor for Labor is obtained from Column F in Figure 22.3. The overhead and profit for Equipment is 10% of "Bare" cost.

Labor costs are increased by percentages for overhead and profit, depending on trade as shown in Figure 22.3. The resulting rates are listed in the right-hand columns of the same figure. Note that the percentage increase for overhead and profit for the average skilled worker is 53.6% of the base rate. The following items are included in the increase for overhead and profit, as shown in Figure 22.3.

Workers' Compensation and Employer's Liability: Workers' Compensation and Employer's Liability Insurance rates vary from state to state and are tied into the construction trade safety records in that particular state. Rates also vary by trade according to the hazard involved (see Figure 22.6, average insurance rates as of January, 1991). The proper authorities will most likely keep the contractor well informed of the rates and obligations.

State and Federal Unemployment Insurance: The employer's tax rate is adjusted by a merit-rating system according to the number of former employees applying for benefits. Contractors who find it possible to offer a maximum of steady employment can enjoy a reduction in the unemployment tax rate.

Employer-Paid Social Security (FICA): The tax rate is adjusted annually by the federal government. It is a percentage of an employee's salary up to a maximum annual contribution.

Builder's Risk and Public Liability: These insurance rates vary according to the trades involved and the state in which the work is done.

Overhead: The column listed as "Overhead" provides percentages to be added for office or operating overhead. This is the cost of doing business. The percentages are presented as national averages by trade as shown in Figure 22.3. Note that the operating overhead costs are applied to labor only in *Means Mechanical Cost Data* and *Means Plumbing Cost Data*.

Profit: This percentage is the fee added by the contractor to offer both a return on investment and an allowance to cover the risk involved in the type of construction being bid. The profit percentage may vary from 4% on large, straightforward projects to as much as 25% on smaller, high-risk jobs. Profit percentages are directly affected by economic conditions, the expected number of bidders, and the estimated risk involved in the project. For estimating purposes, *Means Mechanical Cost Data* and *Means Plumbing Cost Data* assume 10% (applied to labor) as a reasonable average profit factor.

Square Foot and Cubic Foot Costs

Division 17 in *Means Mechanical Cost Data* and *Means Plumbing Cost Data* has been developed to facilitate the preparation of rapid preliminary budget estimates. The cost figures in this division are derived from more than 11,300 actual building projects contained in the Means data bank of construction costs and include the contractor's overhead and profit. The prices shown *do not* include architectural fees or land costs. The files are updated each year with costs for new projects. In no case are all subdivisions of a project listed.

These projects were located throughout the United States and reflect differences in square foot and cubic foot costs due to both the variations in labor and material costs, and the differences in the owners' requirements. For instance, a bank in a large city would have different features and costs than one in a rural area. This is true of all the different types of buildings analyzed. All individual cost items were computed and tabulated separately. Thus, the sum of the median figures for Plumbing, HVAC, and Electrical will not normally add up to the total Mechanical and Electrical costs arrived at by separate analysis and tabulation of the projects.

The data and prices presented on a Division 17 page (as shown in Figure 22.7) are listed both as square foot or cubic foot costs and as a percentage of total costs. Each category tabulates the data in a similar manner. The median, or middle figure, is listed. This means that 50% of all projects had lower costs, and 50% had higher costs than the median figure. Figures in the "1/4" column indicate that 25% of the projects had lower costs and 75% had higher costs. Similarly, figures in the "3/4" column indicate that 75% had lower costs and 25% of the projects had higher costs.

The costs and figures represent all projects and do not take into account project size. As a rule, larger buildings (of the same type and relative location)

Table 10.2-201 Insurance Rates by Trade

The table below tabulates the national averages for Workers' Compensation insurance rates by trade and type of building. The average "Insurance Rate" is multiplied by the "% of Building Cost" for each trade. This produces the "Workers' Compensation Cost" by % of total labor cost, to be added for each trade by building type to determine the weighted average Workers' Compensation rate for the building types analyzed.

Trade	Insurance Rate (% of Labor Cost)		% of Building Cost			Workers' Compensation Cost		
	Range	Average	Office Bldgs.	Schools & Apts.	Mfg.	Office Bldgs.	Schools & Apts.	Mfg.
Excavation, Grading, etc.	4.7% to 27.2%	10.4%	4.8%	4.9%	4.5%	.50%	.51%	.47%
Piles & Foundations	5.0 to 54.8	25.4	7.1	5.2	8.7	1.80	1.32	2.21
Concrete	5.0 to 44.3	15.3	5.0	14.8	3.7	.77	2.26	.57
Masonry	4.0 to 44.2	13.7	6.9	7.5	1.9	.95	1.03	.26
Structural Steel	5.0 to 162.5	35.4	10.7	3.9	17.6	3.79	1.38	6.23
Miscellaneous & Ornamental Metals	3.9 to 22.4	10.8	2.8	4.0	3.6	.30	.43	.39
Carpentry & Millwork	5.0 to 43.4	16.6	3.7	4.0	0.5	.61	.66	.08
Metal or Composition Siding	5.0 to 34.2	13.7	2.3	0.3	4.3	.32	.04	.59
Roofing	5.0 to 62.5	28.9	2.3	2.6	3.1	.66	.75	.90
Doors & Hardware	4.7 to 20.9	9.6	0.9	1.4	0.4	.09	.13	.04
Sash & Glazing	4.1 to 23.4	11.9	3.5	4.0	1.0	.42	.48	.12
Lath & Plaster	5.0 to 38.5	13.4	3.3	6.9	0.8	.44	.92	.11
Tile, Marble & Floors	3.2 to 23.1	8.3	2.6	3.0	0.5	.22	.25	.04
Acoustical Ceilings	3.7 to 26.9	10.2	2.4	0.2	0.3	.24	.02	.03
Painting	4.6 to 44.2	12.4	1.5	1.6	1.6	.19	.20	.20
Interior Partitions	5.0 to 43.4	16.6	3.9	4.3	4.4	.65	.71	.73
Miscellaneous Items	2.2 to 139.7	15.1	5.2	3.7	9.7	.79	.56	1.46
Elevators	2.5 to 16.2	7.7	2.1	1.1	2.2	.16	.08	.17
Sprinklers	2.2 to 15.1	7.7	0.5	—	2.0	.04	—	.15
Plumbing	2.7 to 16.0	7.5	4.9	7.2	5.2	.37	.54	.39
Heat., Vent., Air Conditioning	4.1 to 25.5	10.2	13.5	11.0	12.9	1.38	1.12	1.32
Electrical	2.4 to 12.9	6.0	10.1	8.4	11.1	.61	.50	.67
Total	2.2% to 162.5%	—	100.0%	100.0%	100.0%	15.30%	13.89%	17.13%
				Overall Weighted Average			15.44%	

Table 10.2-202 Insurance Rates by States

The table below lists the weighted average Workers' Compensation base rate for each state with a factor comparing this with the national average of 15.1%.

State	Weighted Average	Factor	State	Weighted Average	Factor	State	Weighted Average	Factor
Alabama	13.2%	87	Kentucky	13.2%	87	North Dakota	13.5%	89
Alaska	22.2	147	Louisiana	12.8	85	Ohio	13.4	89
Arizona	15.8	105	Maine	24.1	160	Oklahoma	12.8	85
Arkansas	11.7	77	Maryland	15.4	102	Oregon	27.9	185
California	16.5	109	Massachusetts	25.3	168	Pennsylvania	15.2	101
Colorado	22.1	146	Michigan	15.1	100	Rhode Island	19.2	127
Connecticut	23.5	156	Minnesota	25.2	167	South Carolina	10.6	70
Delaware	12.1	80	Mississippi	10.0	66	South Dakota	10.3	68
District of Columbia	21.0	139	Missouri	8.2	54	Tennessee	10.5	70
Florida	30.4	201	Montana	37.5	248	Texas	19.2	127
Georgia	13.9	92	Nebraska	9.5	63	Utah	9.7	64
Hawaii	17.6	117	Nevada	16.2	107	Vermont	10.7	71
Idaho	12.8	85	New Hampshire	19.0	126	Virginia	9.1	60
Illinois	22.6	150	New Jersey	7.9	52	Washington	13.3	88
Indiana	6.5	43	New Mexico	18.5	123	West Virginia	10.1	67
Iowa	13.1	87	New York	11.7	77	Wisconsin	13.9	92
Kansas	9.5	63	North Carolina	7.3	48	Wyoming	5.5	36
			Weighted Average for U.S. is 15.4% of payroll = 100					

Rates in the following table are the base or manual costs per $100 of payroll for Workers' Compensation in each state. Rates are usually applied to straight time wages only and not to premium time wages and bonuses.

The weighted average skilled worker rate for 35 trades is 15.1%. For bidding purposes, apply the full value of Workers'
366

Compensation directly to total labor costs, or if labor is 32%, materials 48% and overhead and profit 20% of total cost, carry 32/80 x 15.1% = 6.0% of cost (before overhead and profit) into overhead. Rates vary not only from state to state but also with the experience rating of the contractor.

Rates are the most current available at the time of publication.

(Reprinted from Means Mechanical Cost Data 1991.)

Figure 22.6

171 | S.F., C.F. and % of Total Costs

171 000	S.F. & C.F. Costs	UNIT	UNIT COSTS			% OF TOTAL			
			¼	MEDIAN	¾	¼	MEDIAN	¾	
570 0010	**MEDICAL OFFICES**	S.F.	57.20	71.90	87.45				**570**
0020	Total project costs	C.F.	4.44	5.95	7.90				
2720	Plumbing	S.F.	3.49	5.20	7.10	5.70%	6.90%	9.20%	
2770	Heating, ventilating, air conditioning		4.04	6.10	7.90	6.50%	8%	10.40%	
2900	Electrical		4.83	6.95	9.30	7.60%	9.70%	11.70%	
3100	Total: Mechanical & Electrical		11.30	16.05	21.05	17.30%	22.40%	27.10%	
590 0010	**MOTELS**		41.75	55.85	75.85				**590**
0020	Total project costs	C.F.	3.61	5.55	8.50				
2720	Plumbing	S.F.	3.76	4.70	5.70	9.40%	10.60%	12.50%	
2770	Heating, ventilating, air conditioning		1.99	3.42	4.99	4.90%	5.60%	8.20%	
2900	Electrical		3.48	4.27	5.50	7%	8.10%	10.40%	
3100	Total: Mechanical & Electrical		8.25	10.85	14.50	17.90%	23.10%	26.20%	
9000	Per rental unit, total cost	Unit	14,900	28,400	38,800				
9500	Total: Mechanical & Electrical	"	4,025	5,700	6,150				
600 0010	**NURSING HOMES**	S.F.	57.40	75.75	92.15				**600**
0020	Total project costs	C.F.	4.58	6	8				
2720	Plumbing	S.F.	5.10	6.25	9.10	9.30%	10.30%	13.30%	
2770	Heating, ventilating, air conditioning		5.20	7.25	9.30	9.20%	11.40%	11.80%	
2900	Electrical		5.75	7.25	9.45	9.70%	11%	12.80%	
3100	Total: Mechanical & Electrical		13.40	17.90	26.95	22.30%	28.30%	33.30%	
9000	Per bed or person, total cost	Bed	22,200	28,900	36,700				
610 0010	**OFFICES Low-Rise (1 to 4 story)**	S.F.	47.10	60.35	79.90				**610**
0020	Total project costs	C.F.	3.49	4.86	6.55				
2720	Plumbing	S.F.	1.79	2.70	3.86	3.60%	4.50%	6%	
2770	Heating, ventilating, air conditioning		3.81	5.35	7.84	7.20%	10.40%	11.90%	
2900	Electrical		3.98	5.50	7.55	7.40%	9.50%	11%	
3100	Total: Mechanical & Electrical		8.25	12.25	17.90	14.90%	20.80%	26.80%	
620 0010	**OFFICES Mid-Rise (5 to 10 story)**		52.95	65.05	87.25				**620**
0020	Total project costs	C.F.	3.63	4.69	6.60				
2720	Plumbing	S.F.	1.60	2.43	3.50	2.80%	3.60%	4.50%	
2770	Heating, ventilating, air conditioning		3.94	5.65	9	7.60%	9.30%	11%	
2900	Electrical		3.36	4.81	7.40	6.50%	8%	10%	
3100	Total: Mechanical & Electrical		9.50	12.20	20.35	16.70%	20.50%	25.70%	
630 0010	**OFFICES High-Rise (11 to 20 story)**		62.70	79.25	98.30				**630**
0020	Total project costs	C.F.	4.01	5.70	8.10				
2900	Electrical	S.F.	3.13	4.56	7	5.80%	7%	10.50%	
3100	Total: Mechanical & Electrical		11.45	14.55	24.20	17.20%	21.40%	29.40%	
640 0010	**POLICE STATIONS**		78.40	101	129				**640**
0020	Total project costs	C.F.	5.75	7.45	9.90				
2720	Plumbing	S.F.	4.28	6.25	10.50	5.60%	6.80%	10.70%	
2770	Heating, ventilating, air conditioning		6.30	8.50	12.10	7%	10.50%	11.90%	
2900	Electrical		7.85	12.50	15.80	9.40%	11.70%	14.70%	
3100	Total: Mechanical & Electrical		21.25	26.55	34.70	22.60%	27.50%	33.10%	
650 0010	**POST OFFICES**		63.85	76.20	99.55				**650**
0020	Total project costs	C.F.	3.52	4.60	5.45				
2720	Plumbing	S.F.	2.72	3.44	4.36	4.20%	5.30%	5.60%	
2770	Heating, ventilating, air conditioning		3.93	5.25	8	6.60%	8%	9.80%	
2900	Electrical		5	7.05	8.35	7.40%	9.40%	11%	
3100	Total: Mechanical & Electrical		11	1.51	21	16.50%	21.40%	26.30%	
660 0010	**POWER PLANTS**		375	555	850				**660**
0020	Total project costs	C.F.	10.20	18.85	52.10				
2900	Electrical	S.F.	22.05	57.65	93.80	9.20%	12.30%	18.40%	
8100	Total: Mechanical & Electrical		66.55	143	322	28.80%	32.50%	52.60%	
670 0010	**RELIGIOUS EDUCATION**		47.65	56.30	69.80				**670**
0020	Total project costs	C.F.	2.81	3.97	5.25				
2720	Plumbing	S.F.	2.02	2.94	4.06	3.90%	4.90%	6.90%	
2770	Heating, ventilating, air conditioning	"	4.64	5.50	7.25	8.10%	9.90%	11.20%	

For expanded coverage of these items see *Means Square Foot Cost Data 1991*

283

Figure 22.7

will cost less to build per square foot than similar buildings of a smaller size. This cost difference is due to economies of scale as well as a lower exterior envelope-to-floor area ratio. A conversion is necessary to adjust project costs based on size relative to the norm. See Figure 23.3.

There are two stages of project development when square foot cost estimates are most useful. The first is during the conceptual stage when few, if any, details are available. At this time, square foot costs are appropriate for ballpark budget purposes. As soon as details become available in the project design, the square foot approach should be discontinued and the project priced more accurately. After the estimate is completed, square foot costs can be used again–this time for verification and as a check against gross errors.

When using the figures in Division 17, it is recommended that the median cost column be consulted for preliminary figures if no additional information is available. When costs have been converted for location (see City Cost Indexes) the median numbers (as shown in Figure 22.7) should provide a fairly accurate base figure. This figure should then be adjusted according to the estimator's experience, local economic conditions, code requirements, and the owner's particular project requirements. There is no need to factor the percentage figures, as these should remain relatively constant from city to city.

Repair and Remodeling

Cost figures in *Means Mechanical Cost Data* and *Means Plumbing Cost Data* are based on new construction utilizing the most cost-effective combination of labor, equipment, and material. The work is scheduled in the proper sequence to allow the various trades to accomplish their tasks in an efficient manner. Figure 22.8 (from Division 010 of *Means Mechanical Cost Data*) shows factors that can be used to adjust figures in other sections of the book for repair and remodeling projects. For expanded coverage, see *Means Repair and Remodeling Cost Data*.

Section B ~ Assemblies Cost Tables

Means' assemblies data are divided into twelve "uniformat" divisions, which organize the components of construction into logical groupings. The Systems or Assemblies approach was devised to provide quick and easy methods for estimating even when only preliminary design data are available. The groupings, or systems, are presented in such a way so that the estimator can substitute one system for another. This is extremely useful when adapting to budget, design, or other considerations. Figure 22.9 shows how the data is presented in the Assemblies section.

Each system is illustrated and accompanied by a detailed description. The book lists the components and sizes of each system. Each individual component is found in the Unit Price section.

Quantity

A unit of measure is established for each assembly. For example, sprinkler systems are measured by the square foot of floor area; plumbing fixture systems are measured by "each;" HVAC systems are measured by the square foot of

010 000	Overhead		CREW	DAILY OUTPUT	MAN-HOURS	UNIT	BARE COSTS				TOTAL INCL O&P	
							MAT.	LABOR	EQUIP.	TOTAL		
012	0011	**CONSTRUCTION COST INDEX** For 162 major U.S. and										**012**
	0020	Canadian cities, total cost, min. (Greensboro, NC)				%					78.60%	
	0050	Average **C13.1 - 100**									100%	
	0100	Maximum (Anchorage, AK)									127%	
020	0010	**CONTINGENCIES** Allowance to add at conceptual stage				Project					15%	**020**
	0050	Schematic stage									10%	
	0100	Preliminary working drawing stage									7%	
	0150	Final working drawing stage									2%	
022	0010	**CONTRACTOR EQUIPMENT** See division 016 **C10.3 - 300**										**022**
024	0010	**CREWS** For building construction, see How To Use This Book										**024**
028	0010	**ENGINEERING FEES** Educational planning consultant, minimum				Project					.50%	**028**
	0100	Maximum				"					2.50%	
	0200	Electrical, minimum **C10.1 - 103**				Contrct					4.10%	
	0300	Maximum									10.10%	
	0400	Elevator & conveying systems, minimum									2.50%	
	0500	Maximum									5%	
	0600	Food service & kitchen equipment, minimum									8%	
	0700	Maximum									12%	
	1000	Mechanical (plumbing & HVAC), minimum									4.10%	
	1100	Maximum									10.10%	
032	0010	**FACTORS** To be added to construction costs for particular job **C18.1 - 100**										**032**
	0200											
	0500	Cut & patch to match existing construction, add, minimum				Costs	2%	3%				
	0550	Maximum					5%	9%				
	0800	Dust protection, add, minimum					1%	2%				
	0850	Maximum					4%	11%				
	1100	Equipment usage curtailment, add, minimum					1%	1%				
	1150	Maximum					3%	10%				
	1400	Material handling & storage limitation, add, minimum					1%	1%				
	1450	Maximum					6%	7%				
	1700	Protection of existing work, add, minimum					2%	2%				
	1750	Maximum					5%	7%				
	2000	Shift work requirements, add, minimum						5%				
	2050	Maximum						30%				
	2300	Temporary shoring and bracing, add, minimum					2%	5%				
	2350	Maximum					5%	12%				
034	0010	**FIELD OFFICE EXPENSE**										**034**
	0100	Office equipment rental, average				Month	135				148.50	
	0120	Office supplies, average				"	250				275	
	0125	Office trailer rental, see division 015-904										
	0140	Telephone bill; avg. bill/month incl. long dist.				Month	225				247.50	
	0160	Field office lights & HVAC				"	78				85.80	
038	0011	**HISTORICAL COST INDEXES** Back to 1947										**038**
040	0010	**INSURANCE** Builders risk, standard, minimum				Job					.22%	**040**
	0050	Maximum									.59%	
	0200	All-risk type, minimum **C10.1 - 301**									.25%	
	0250	Maximum									.62%	
	0400	Contractor's equipment floater, minimum				Value					.50%	
	0450	Maximum				"					1.50%	
	0600	Public liability, average				Job					1.55%	
	0800	Workers' compensation & employer's liability, average										
	0850	by trade, carpentry, general **C10.2 - 200**				Payroll		16.64%				
	1000	Electrical						5.97%				
	1150	Insulation						13.48%				
	1450	Plumbing						7.45%				

For expanded coverage of these items see *Means Building Construction Cost Data 1991*

1

(Reprinted from Means Mechanical Cost Data 1991.)

Figure 22.8

HOW TO USE
ASSEMBLIES COST TABLES

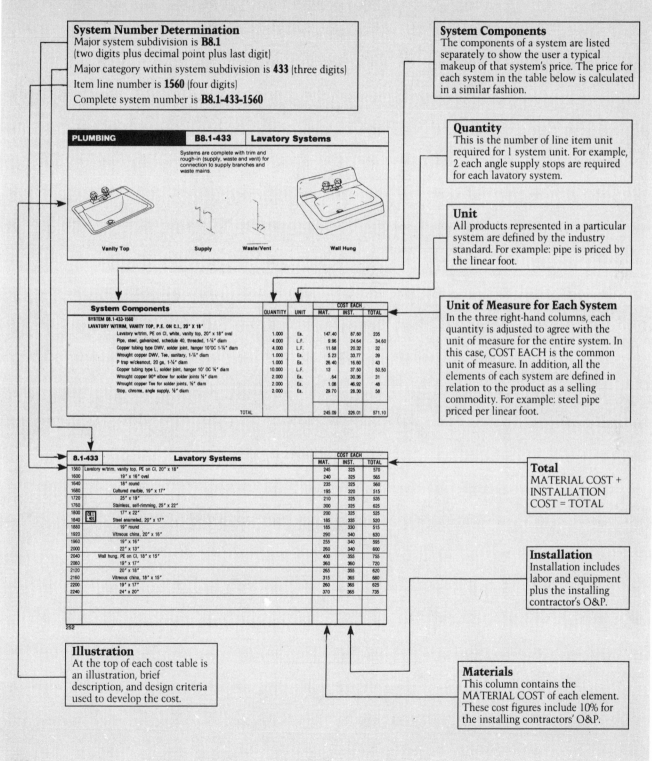

System Number Determination

Major system subdivision is **B8.1**
(two digits plus decimal point plus last digit)

Major category within system subdivision is **433** (three digits)

Item line number is **1560** (four digits)

Complete system number is **B8.1-433-1560**

System Components
The components of a system are listed separately to show the user a typical makeup of that system's price. The price for each system in the table below is calculated in a similar fashion.

Quantity
This is the number of line item unit required for 1 system unit. For example, 2 each angle supply stops are required for each lavatory system.

Unit
All products represented in a particular system are defined by the industry standard. For example: pipe is priced by the linear foot.

Unit of Measure for Each System
In the three right-hand columns, each quantity is adjusted to agree with the unit of measure for the entire system. In this case, COST EACH is the common unit of measure. In addition, all the elements of each system are defined in relation to the product as a selling commodity. For example: steel pipe priced per linear foot.

Total
MATERIAL COST + INSTALLATION COST = TOTAL

Installation
Installation includes labor and equipment plus the installing contractor's O&P.

Materials
This column contains the MATERIAL COST of each element. These cost figures include 10% for the installing contractors' O&P.

Illustration
At the top of each cost table is an illustration, brief description, and design criteria used to develop the cost.

PLUMBING	B8.1-433	Lavatory Systems

Systems are complete with trim and rough-in (supply, waste and vent) for connection to supply branches and waste mains.

Vanity Top   Supply   Waste/Vent   Wall Hung

System Components	QUANTITY	UNIT	MAT.	INST.	TOTAL
SYSTEM B8.1-433-1560					
LAVATORY W/TRIM, VANITY TOP, P.E. ON C.I., 20" X 18"					
Lavatory w/trim, PE on CI, white, vanity top, 20" x 18" oval	1.000	Ea.	147.40	87.60	235
Pipe, steel, galvanized, schedule 40, threaded, 1-¼" diam	4.000	L.F.	9.96	24.64	34.60
Copper tubing type DWV, solder joint, hanger 10'OC 1-¼" diam	4.000	L.F.	11.68	20.32	32
Wrought copper DWV, Tee, sanitary, 1-¼" diam	1.000	Ea.	5.23	33.77	39
P trap w/cleanout, 20 ga, 1-¼" diam	1.000	Ea.	26.40	16.60	43
Copper tubing type L, solder joint, hanger 10' OC ½" diam	10.000	L.F.	13	37.50	50.50
Wrought copper 90° elbow for solder joints ½" diam	2.000	Ea.	.64	30.36	31
Wrought copper Tee for solder joints, ½" diam	2.000	Ea.	1.08	46.92	48
Stop, chrome, angle supply, ½" diam	2.000	Ea.	29.70	28.30	58
TOTAL			245.09	326.01	571.10

8.1-433	Lavatory Systems	MAT.	INST.	TOTAL
1560	Lavatory w/trim, vanity top, PE on CI, 20" x 18"	245	325	570
1600	19" x 16" oval	240	325	565
1640	18" round	235	325	560
1680	Cultured marble, 19" x 17"	195	320	515
1720	25" x 19"	210	325	535
1760	Stainless, self-rimming, 25" x 22"	300	325	625
1800	17" x 22"	200	325	525
1840	Steel enameled, 20" x 17"	185	335	520
1880	19" round	185	330	515
1920	Vitreous china, 20" x 16"	290	340	630
1960	19" x 16"	255	340	595
2000	22" x 13"	260	340	600
2040	Wall hung, PE on CI, 18" x 15"	400	355	755
2080	19" x 17"	360	360	720
2120	20" x 18"	265	355	620
2160	Vitreous china, 18" x 15"	315	365	680
2200	19" x 17"	260	365	625
2240	24" x 20"	370	365	735

252

238

(Reprinted from Means Plumbing Cost Data 1991.)

Figure 22.9

floor area. Within each system, the components are measured by industry standard, using the same units as in the Unit Price section.

Material

The cost of each component in the Material column is the "Bare Material Cost," plus 10% handling, for the unit and quantity as defined in the "Quantity" column.

Installation

Installation costs as listed in the Assemblies pages contain both labor and equipment costs. The labor rate includes the "Bare Labor Costs" plus the installing contractor's overhead and profit. These rates are shown in Figure 22.3. The equipment rate is the "Bare Equipment Cost," plus 10%.

Section C - Estimating References

Throughout the Unit Price and Assemblies sections are reference numbers highlighted with bold squares. These numbers serve as footnotes, referring the reader to illustrations, charts, and estimating reference tables in Section C, as well as to related information in Sections A and B. Figure 22.10 shows an example reference number (for plumbing fixtures) as it appears on a Unit Price page. Figure 22.11 shows the corresponding reference page from Section B. The development of unit costs for many items is explained in these reference tables. Design criteria for many types of mechanical systems are also included to aid the designer/estimator in making appropriate choices.

City Cost Indexes

The unit prices in *Means Mechanical Cost Data* and *Means Plumbing Cost Data* are national averages. When they are to be applied to a particular location, these prices must be adjusted to local conditions. Means has developed the City Cost Indexes for just that purpose. Section D of *Means Mechanical Cost Data* and *Means Plumbing Cost Data* contains tables of indexes for 162 U.S. and Canadian cities based on a 30 major city average of 100. The figures are broken down into material and installation for all trades, as shown in Figure 22.12. Please note that for each city there is a weighted average based on total project costs. This average is based on the relative contribution of each division to the construction process as a whole.

In addition to adjusting the figures in *Means Mechanical Cost Data* for particular locations, the City Cost Index can also be used to adjust costs from one city to another. For example, the price of the mechanical work for a particular building type is known for City A. In order to budget the costs of the same building type in City B, the following calculation can be made:

$$\frac{\text{City B Index}}{\text{City A Index}} \text{ x City A Cost } = \text{City B Cost}$$

While City Cost Indexes provide a means to adjust prices for location, the Historical Cost Index, (also included in *Means Mechanical Cost Data* and *Means Plumbing Cost Data* and shown in Figure 22.13) provides a means to adjust for time. Using the same principle as above, a time-adjustment factor can be calculated:

$$\frac{\text{Index for Year X}}{\text{Index for Year Y}} \text{ x Time-adjustment Factor}$$

This time-adjustment factor can be used to determine the budget costs for a particular building type in Year X, based on costs for a similar building type known from Year Y. Used together, the two indexes allow for cost adjustments

152 100	Fixtures		CREW	DAILY OUTPUT	MAN-HOURS	UNIT	BARE COSTS				TOTAL INCL O&P	
							MAT.	LABOR	EQUIP.	TOTAL		
168	6980	Rough-in, supply, waste and vent	Q-1	1.99	8.040	Ea.	107.85	185		292.85	395	168
172	0010	**WASH CENTER** Prefabricated, stainless steel, semirecessed										172
	0050	Lavatory, storage cabinet, mirror, light & switch, electric										
	0060	outlet, towel dispenser, waste receptacle & trim										
	0100	Foot water valve, cup & soap dispenser,16″ W x 54-¾″ H	Q-1	8	2	Ea.	900	46		946	1,050	
	0200	Handicap, wrist blade handles, 17″ W x 66-½″ H		8	2		800	46		846	950	
	0220	20″ W x 67-⅜″ H		8	2		1,050	46		1,096	1,225	
	0300	Push button metering & thermostatic mixing valves										
	0320	Handicap 17″ W x 27-½″ H	Q-1	8	2	Ea.	880	46		926	1,025	
	0400	Rough-in, supply, waste and vent	"	2.10	7.620	"	65.36	175		240.36	330	
176	0010	**WASH FOUNTAINS** Rigging not included										176
	1900	Group, foot control										
	2000	Precast terrazzo, circular, 36″ diam., 5 or 6 persons	Q-2	3	8	Ea.	1,190	190		1,380	1,600	
	2100	54″ diameter for 8 or 10 persons		2.50	9.600		1,480	230		1,710	1,975	
	2400	Semi-circular, 36″ diam. for 3 persons		3	8		1,090	190		1,280	1,475	
	2420	36″ diam. for 3 persons in wheelchairs		3	8		1,760	190		1,950	2,225	
	2500	54″ diam. for 4 or 5 persons		2.50	9.600		1,325	230		1,555	1,800	
	2520	54″ diam. for 4 persons in wheelchairs		2.50	9.600		1,900	230		2,130	2,425	
	2700	Quarter circle (corner), 54″ for 3 persons		3.50	6.860		1,285	165		1,450	1,650	
	2720	54″ diam. for 3 persons in wheelchairs		3.50	6.860		1,845	165		2,010	2,275	
	3000	Stainless steel, circular, 36″ diameter		3.50	6.860		1,330	165		1,495	1,700	
	3100	54″ diameter		2.80	8.570		1,720	205		1,925	2,200	
	3400	Semi-circular, 36″ diameter		3.50	6.860		1,160	165		1,325	1,525	
	3500	54″ diameter		2.80	8.570		1,485	205		1,690	1,925	
	5000	Thermoplastic, pre-assembled, circular, 36″ diameter		6	4		1,065	95		1,160	1,325	
	5100	54″ diameter		4	6		1,235	145		1,380	1,575	
	5400	Semi-circular, 36″ diameter		6	4		980	95		1,075	1,225	
	5600	54″ diameter		4	6		1,195	145		1,340	1,525	
	5700	Rough-in, supply, waste and vent for above wash fountains	Q-1	1.82	8.790		120.35	200		320.35	435	
	6200	Duo for small washrooms, stainless steel		2	8		647	185		832	985	
	6400	Bowl with backsplash		2	8		745	185		930	1,100	
	6500	Rough-in, supply, waste & vent for duo fountains		2.02	7.920		59.91	180		239.91	335	
180	0010	**WATER CLOSETS**										180
	0020	For seats, see 152-164										
	0150	Tank type, vitreous china, incl. seat, supply pipe w/stop										
	0200	Wall hung, one piece	Q-1	5.30	3.020	Ea.	495	69		564	650	
	0400	Two piece, close coupled		5.30	3.020		339	69		408	475	
	0960	For rough-in, supply, waste, vent and carrier		2.60	6.150		147.31	140		287.31	370	
	1000	Floor mounted, one piece		5.30	3.020		418	69		487	565	
	1020	One piece, low profile		5.30	3.020		666	69		735	835	
	1050	One piece combination		5.30	3.020		650	69		719	820	
	1100	Two piece, close coupled, water saver		5.30	3.020		138	69		207	255	
	1150	With wall outlet		5.30	3.020		245	69		314	375	
	1200	With 18″ high bowl		5.30	3.020		236	69		305	365	
	1960	For color, add					30%					
	1961	For designer colors and trim add					55%					
	1980	For rough-in, supply, waste and vent	Q-1	1.94	8.250		85.65	190		275.65	375	
	3000	Bowl only, with flush valve, seat										
	3100	Wall hung	Q-1	5.80	2.760	Ea.	275	63		338	395	
	3150	Hospital type, slotted rim for bed pan										
	3160	Elongated bowl	Q-1	5.80	2.760	Ea.	336	63		399	465	
	3200	For rough-in, supply, waste and vent, single WC		2.05	7.800		156.94	180		336.94	440	
	3300	Floor mounted		5.80	2.760		275	63		338	395	
	3350	With wall outlet		5.80	2.760		315	63		378	440	
	3360	Hospital type, slotted rim for bed pan										
	3370	Elongated bowl	Q-1	5	3.200	Ea.	278	73		351	415	

159

(Reprinted from Means Plumbing Cost Data 1991.)

Figure 22.10

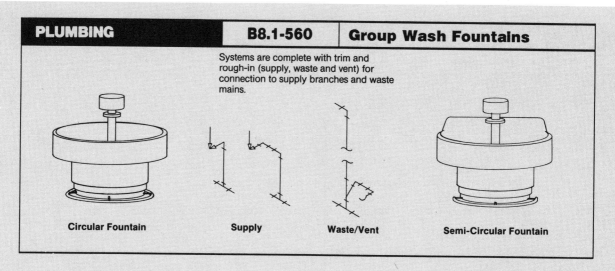

Systems are complete with trim and rough-in (supply, waste and vent) for connection to supply branches and waste mains.

Circular Fountain   **Supply**   **Waste/Vent**   **Semi-Circular Fountain**

System Components	QUANTITY	UNIT	COST EACH MAT.	COST EACH INST.	COST EACH TOTAL
SYSTEM 08.1-560-1760					
GROUP WASH FOUNTAIN, PRECAST TERRAZZO					
CIRCULAR, 36" DIAMETER					
Wash fountain, group, precast terrazzo, foot control 36" diam	1.000	Ea.	1,309	291	1,600
Copper tubing type DWV, solder joint, hanger 10'OC, 2" diam	10.000	L.F.	48.20	68.80	117
P trap, standard, copper, 2" diam	1.000	Ea.	47.30	20.70	68
Wrought copper, Tee, sanitary, 2" diam	1.000	Ea.	8.91	43.09	52
Copper tubing type L, solder joint, hanger 10' OC ½" diam	20.000	L.F.	26	75	101
Wrought copper 90° elbow for solder joints ½" diam	3.000	Ea.	.96	45.54	46.50
Wrought copper Tee for solder joints, ½" diam	2.000	Ea.	1.08	46.92	48
TOTAL			1,441.45	591.05	2,032.50

8.1-560	Group Wash Fountain Systems	COST EACH MAT.	COST EACH INST.	COST EACH TOTAL
1740	Group wash fountain, precast terrazzo			
1760	Circular, 36" diameter	1,450	590	2,040
1800	54" diameter	1,750	645	2,395
1840	Semi-circular, 36" diameter	1,325	575	1,900
1880	54" diameter	1,600	645	2,245
1960	Stainless steel, circular, 36" diameter	1,600	535	2,135
2000	54" diameter	2,025	610	2,635
2040	Semi-circular, 36" diameter	1,400	550	1,950
2080	54" diameter	1,775	590	2,365
2160	Thermoplastic, circular, 36" diameter	1,300	455	1,755
2200	54" diameter	1,500	515	2,015
2240	Semi-circular, 36" diameter	1,200	445	1,645
2280	54" diameter	1,450	510	1,960

(For 1800/1840 rows: box labeled C8.1 -401)

261

(Reprinted from Means Plumbing Cost Data 1991.)

Figure 22.11

from one city during a given year to another city in another year (the present or otherwise). For example, an office building built in San Francisco in 1974 originally cost $1,000,000. How much will a similar building cost in Phoenix in 1987? Adjustment factors are developed as shown above using data from Figures 22.12 and 22.13:

$$\frac{\text{Phoenix Index}}{\text{San Francisco Index}} = \frac{91.1}{124.9} = 0.73$$

$$\frac{1987 \text{ Index}}{1974 \text{ Index}} = \frac{200.7}{94.7} = 2.12$$

Original cost x location adjustment x time adjustment = Proposed new cost $1,000,000 x 0.73 x 2.12 = $1,547,600

CITY COST INDEXES

Table 1

DIVISION		BIRMINGHAM (ALABAMA)			HUNTSVILLE			MOBILE			MONTGOMERY			ANCHORAGE (ALASKA)			PHOENIX (ARIZONA)		
		MAT.	INST.	TOTAL	MAT.	INST.	TOTAL	MAT.	INST.	TOTAL	MAT.	INST.	TOTAL	MAT.	INST.	TOTAL	MAT.	INST.	TOTAL
2	SITE WORK	100.0	89.2	95.2	119.5	87.0	105.0	122.5	86.2	106.4	91.9	85.4	89.0	158.9	126.5	144.5	92.4	93.9	93.1
3.1	FORMWORK	97.7	70.2	76.4	103.3	61.0	70.5	106.6	72.0	79.8	112.5	64.2	75.1	124.4	136.2	133.6	108.6	85.3	90.5
3.2	REINFORCING	94.5	69.8	84.5	95.7	62.8	82.4	82.9	69.9	77.6	82.9	69.8	77.6	117.7	130.6	123.0	110.4	89.8	102.1
3.3	CAST IN PLACE CONC.	89.2	91.2	90.4	101.8	89.5	94.3	99.9	92.5	95.4	101.1	88.9	93.7	225.7	111.2	155.8	105.4	92.2	97.4
3	CONCRETE	92.1	81.1	85.1	100.8	76.0	85.0	97.4	82.5	87.9	99.3	77.6	85.5	181.3	122.7	144.1	107.2	89.3	95.8
4	MASONRY	81.6	75.2	76.7	88.4	62.1	68.3	93.8	75.4	79.7	86.5	50.9	59.2	149.9	133.3	137.2	93.5	77.1	81.0
5	METALS	95.6	76.9	89.0	100.0	72.1	90.2	93.4	77.7	87.9	95.7	76.9	89.1	116.3	122.4	118.4	99.1	90.5	96.1
6	WOOD & PLASTICS	92.1	71.5	80.6	107.5	64.0	83.3	91.9	74.2	82.0	101.7	68.0	83.0	117.9	132.6	126.1	99.3	83.8	90.7
7	MOISTURE PROTECTION	84.5	59.0	76.4	92.1	57.2	81.0	87.3	60.6	78.8	88.6	57.4	78.7	102.6	137.0	113.5	92.6	83.7	89.8
8	DOORS, WINDOWS, GLASS	90.7	70.6	80.3	100.9	58.9	79.2	98.4	71.5	84.5	98.0	65.3	81.1	128.5	125.8	127.1	103.0	81.5	91.9
9.1	LATH & PLASTER	95.8	67.9	74.7	91.1	66.2	72.3	91.8	77.8	81.2	108.3	66.7	76.7	120.3	137.8	133.5	93.4	89.1	90.2
9.2	DRYWALL	100.8	70.7	86.8	108.6	63.3	87.5	92.8	74.4	84.2	101.0	68.0	85.6	121.9	134.5	127.8	90.6	83.9	87.5
9.5	ACOUSTICAL WORK	97.7	70.7	83.1	100.1	63.0	80.0	93.1	73.2	82.3	93.1	66.8	78.9	123.9	133.8	129.2	103.6	82.7	92.3
9.6	FLOORING	112.0	71.4	101.1	97.5	62.1	88.0	114.0	76.8	104.1	100.9	45.1	86.0	117.3	130.2	120.7	93.2	86.1	91.3
9.8	PAINTING	104.2	66.7	74.5	110.6	65.4	74.7	121.4	75.7	85.2	119.6	74.4	83.8	123.1	137.8	134.8	96.3	79.1	82.6
9	FINISHES	103.3	69.2	85.1	105.3	64.1	83.3	100.4	75.2	87.0	102.4	68.4	84.3	121.1	135.5	128.8	92.8	82.6	87.4
10-14	TOTAL DIV. 10-14	100.0	76.3	93.1	100.0	75.5	92.9	100.0	80.3	94.3	100.0	74.9	92.7	100.0	133.8	109.8	100.0	89.8	97.0
15	MECHANICAL	96.5	69.6	83.1	99.3	70.0	84.7	97.3	72.2	84.8	99.0	67.4	83.3	107.2	123.6	115.4	98.4	84.6	91.5
16	ELECTRICAL	94.9	70.0	77.8	92.6	69.1	76.4	90.5*	73.6	78.9	91.6	59.5	69.5	107.5	133.0	125.1	105.3	76.8	85.7
1-16	WEIGHTED AVERAGE	95.0	74.2	83.8	100.2	69.4	83.8	97.8	76.2	86.3	96.9	67.4	81.1	125.4	128.4	127.0	99.1	84.1	91.1

Table 2

DIVISION		TUCSON (ARIZONA)			FORT SMITH (ARKANSAS)			LITTLE ROCK			ANAHEIM (CALIFORNIA)			BAKERSFIELD			FRESNO		
		MAT.	INST.	TOTAL	MAT.	INST.	TOTAL	MAT.	INST.	TOTAL	MAT.	INST.	TOTAL	MAT.	INST.	TOTAL	MAT.	INST.	TOTAL
2	SITE WORK .	110.2	96.1	103.9	99.9	89.4	95.2	106.8	91.6	100.0	104.6	112.6	108.2	97.0	111.3	103.4	94.7	120.2	106.0
3.1	FORMWORK	109.1	85.1	90.5	111.2	63.8	74.4	103.4	63.6	72.6	104.6	122.6	118.6	124.8	122.6	123.1	110.1	122.3	119.5
3.2	REINFORCING	95.0	89.8	92.9	124.5	63.9	99.9	117.7	59.1	94.0	99.3	129.3	111.4	96.0	129.3	109.5	106.5	129.3	115.7
3.3	CAST IN PLACE CONC.	105.5	96.8	100.2	90.4	90.0	90.1	98.4	90.4	93.5	109.3	109.9	109.6	103.2	109.9	107.3	92.9	107.9	102.1
3	CONCRETE	103.8	91.6	96.1	102.2	77.4	86.5	103.7	77.2	86.9	106.1	116.5	112.7	105.8	116.6	112.7	99.4	115.4	109.6
4	MASONRY	92.3	77.1	80.7	95.3	70.3	76.1	88.8	70.3	74.6	108.8	129.4	124.6	100.9	112.7	109.9	119.7	110.4	112.6
5	METALS	90.8	92.1	91.3	96.5	73.2	88.3	106.2	70.2	93.5	99.1	121.6	107.0	99.3	121.8	107.2	94.9	123.1	104.8
6	WOOD & PLASTICS	106.2	83.3	93.5	107.2	65.0	83.7	94.9	65.0	78.3	96.2	117.6	108.1	95.2	117.6	107.7	96.5	119.4	109.4
7	MOISTURE PROTECTION	105.5	73.5	95.3	84.8	61.8	77.5	84.3	61.8	77.2	107.9	129.2	114.6	84.9	114.7	94.4	107.5	109.7	108.2
8	DOORS, WINDOWS, GLASS	88.1	81.5	84.7	92.7	58.0	74.7	95.1	58.1	76.0	93.3	122.1	108.2	99.9	116.1	108.3	101.1	118.0	109.9
9.1	LATH & PLASTER	109.0	86.0	91.5	92.9	70.4	75.8	98.4	70.4	77.2	97.2	131.2	122.9	92.2	99.4	97.7	102.1	114.0	111.1
9.2	DRYWALL	82.1	83.9	82.9	95.0	63.7	80.4	114.8	63.7	91.0	97.4	122.4	109.0	97.9	112.0	104.5	98.9	118.8	108.2
9.5	ACOUSTICAL WORK	113.6	82.7	96.9	83.7	63.7	72.9	83.7	63.7	72.9	81.4	118.2	101.3	93.2	118.2	106.8	96.7	120.2	109.4
9.6	FLOORING	110.0	82.1	102.5	89.5	70.9	84.5	88.7	70.9	84.0	116.9	137.6	122.4	111.7	100.3	108.7	88.7	99.1	91.5
9.8	PAINTING	98.5	78.1	82.3	111.0	49.1	61.9	104.7	62.0	70.8	108.3	117.9	115.9	120.1	119.7	119.8	107.9	98.0	100.0
9	FINISHES	93.1	81.8	87.1	94.5	59.6	75.9	105.1	64.0	83.2	101.6	122.1	112.5	102.8	113.6	108.5	97.5	110.0	104.2
10-14	TOTAL DIV. 10-14	100.0	89.0	96.8	100.0	72.4	92.0	100.0	73.0	92.1	100.0	126.6	107.7	100.0	124.1	107.0	100.0	143.9	112.7
15	MECHANICAL	98.7	88.7	93.8	97.2	62.9	80.2	96.8	66.7	81.8	96.8	120.5	108.6	95.0	95.3	95.2	92.8	111.7	102.2
16	ELECTRICAL	103.2	81.4	88.2	100.1	68.6	78.4	94.2	71.9	78.8	99.5	117.1	111.6	107.1	100.0	102.2	110.6	94.7	99.6
1-16	WEIGHTED AVERAGE	99.0	85.6	91.9	97.2	69.3	82.3	98.9	70.7	83.8	101.0	120.9	111.7	99.1	110.1	105.0	99.6	112.9	106.7

Table 3 — CALIFORNIA

DIVISION		LOS ANGELES			OXNARD			RIVERSIDE			SACRAMENTO			SAN DIEGO			SAN FRANCISCO		
		MAT.	INST.	TOTAL	MAT.	INST.	TOTAL	MAT.	INST.	TOTAL	MAT.	INST.	TOTAL	MAT.	INST.	TOTAL	MAT.	INST.	TOTAL
2	SITE WORK	97.8	116.0	105.9	101.9	104.6	103.1	98.7	111.6	104.5	86.4	105.1	94.7	95.3	108.3	101.1	102.1	116.4	108.5
3.1	FORMWORK	111.4	123.0	120.4	98.5	123.1	117.6	114.0	122.6	120.7	110.1	125.9	122.3	105.2	123.1	119.1	103.2	136.2	128.8
3.2	REINFORCING	86.6	129.3	103.9	99.3	129.3	111.4	124.5	129.3	126.4	99.3	129.3	111.4	117.6	129.3	122.3	123.1	129.3	125.6
3.3	CAST IN PLACE CONC.	96.7	112.9	106.6	102.3	110.5	107.3	102.3	110.2	107.1	115.8	107.1	110.5	99.4	104.7	102.6	100.3	117.6	110.8
3	CONCRETE	97.4	118.3	110.7	100.9	117.1	111.2	109.6	116.7	114.1	111.0	116.4	114.4	104.6	114.1	110.6	106.0	125.9	118.6
4	MASONRY	108.7	129.4	124.5	100.8	123.6	118.3	105.2	114.9	112.7	103.3	112.0	109.9	110.4	109.6	109.8	126.4	149.9	144.4
5	METALS	101.6	122.5	109.0	105.2	121.8	111.1	99.1	121.6	107.0	111.1	123.0	115.3	99.1	120.6	106.6	103.9	126.1	111.7
6	WOOD & PLASTICS	99.6	118.6	110.2	92.7	118.5	107.1	94.7	117.6	107.4	78.3	124.0	103.7	96.1	118.2	108.4	93.0	135.1	116.4
7	MOISTURE PROTECTION	103.9	131.7	112.7	90.2	128.8	102.4	90.7	124.7	101.5	85.3	119.6	96.2	94.5	110.0	99.4	100.4	134.8	111.3
8	DOORS, WINDOWS, GLASS	102.6	122.1	112.7	102.6	122.1	112.7	103.2	122.1	112.9	91.8	118.5	105.7	107.4	122.2	115.1	113.5	132.7	123.5
9.1	LATH & PLASTER	96.3	131.2	122.7	97.6	121.0	115.4	97.6	123.6	117.3	99.1	125.6	119.2	101.9	109.6	107.7	101.5	143.9	133.6
9.2	DRYWALL	89.1	122.4	104.6	98.7	119.7	108.5	94.8	122.4	107.6	97.4	124.3	109.9	99.9	118.8	108.7	81.0	138.0	107.6
9.5	ACOUSTICAL WORK	98.9	118.2	109.4	87.5	118.2	104.1	87.5	118.2	104.1	85.6	124.8	106.9	100.9	118.9	110.7	100.9	136.8	120.3
9.6	FLOORING	96.2	137.6	107.3	95.9	137.6	107.1	95.9	137.6	107.1	86.0	126.8	96.9	98.2	128.7	106.4	107.0	136.1	114.8
9.8	PAINTING	83.9	127.0	118.1	92.1	112.9	108.6	100.7	117.9	114.4	112.2	128.0	124.8	91.4	126.4	119.2	102.1	143.7	135.2
9	FINISHES	91.1	125.3	109.3	96.5	118.6	108.3	95.1	121.6	109.2	95.5	125.9	111.7	98.8	121.6	110.9	91.0	140.1	117.1
10-14	TOTAL DIV. 10-14	100.0	126.9	107.8	100.0	126.5	107.7	100.0	126.4	107.6	100.0	145.6	113.2	100.0	124.6	107.1	100.0	151.6	114.9
15	MECHANICAL	97.6	124.0	110.7	98.5	120.1	109.3	96.6	123.3	109.9	97.9	117.0	107.4	102.8	121.4	112.1	101.0	172.3	136.5
16	ELECTRICAL	101.9	126.4	118.8	99.5	112.7	108.6	99.0	120.9	114.1	110.6	90.8	97.0	105.7	103.1	103.9	108.0	148.6	136.0
1-16	WEIGHTED AVERAGE	99.3	123.7	112.3	99.4	118.9	109.8	99.6	119.7	110.3	99.8	115.1	108.0	101.7	115.0	108.8	103.3	143.7	124.9

383

(Reprinted from Means Mechanical Cost Data 1991.)

Figure 22.12

222

Historical Cost Indexes

The table below lists both the Means City Cost Index based on Jan. 1, 1975 = 100 as well as the computed value of an index based on January 1, 1991 costs. Since the Jan. 1, 1991 figure is estimated, space is left to write in the actual index figures as they become available thru either the quarterly "Means Construction Cost Indexes" or as printed in the "Engineering News-Record". To compute the actual index based on Jan. 1, 1991 = 100, divide the Quarterly City Cost Index for a particular year by the actual Jan. 1, 1991 Quarterly City Cost Index. Space has been left to advance the index figures as the year progresses.

Year	"Quarterly City Cost Index" Jan. 1, 1975 = 100		Current Index Based on Jan. 1, 1991 = 100		Year	"Quarterly City Cost Index" Jan. 1, 1975 = 100	Current Index Based on Jan. 1, 1991 = 100		Year	"Quarterly City Cost Index" Jan. 1, 1975 = 100	Current Index Based on Jan. 1, 1991 = 100	
	Est.	Actual	Est.	Actual		Actual	Est.	Actual		Actual	Est.	Actual
Oct. 1991					July 1978	122.4	56.0		July 1962	46.2	21.1	
July 1991					1977	113.3	51.9		1961	45.4	20.8	
April 1991					1976	107.3	49.1		1960	45.0	20.6	
Jan. 1991	218.5		100.0	100.0	1975	102.6	47.0		1959	44.2	20.2	
July 1990		215.9	98.8		1974	94.7	43.3		1958	43.0	19.7	
1989		210.9	96.5		1973	86.3	39.5		1957	42.2	19.3	
1988		205.7	94.1		1972	79.7	36.5		1956	40.4	18.5	
1987		200.7	91.9		1971	73.5	33.6		1955	38.1	17.4	
1986		192.8	88.2		1970	65.8	30.1		1954	36.7	16.8	
1985		189.1	86.5		1969	61.6	28.2		1953	36.2	16.6	
1984		187.6	85.9		1968	56.9	26.0		1952	35.3	16.2	
1983		183.5	84.0		1967	53.9	24.7		1951	34.4	15.7	
1982		174.3	79.8		1966	51.9	23.8		1950	31.4	14.4	
1981		160.2	73.3		1965	49.7	22.7		1949	30.4	13.9	
1980		144.0	65.9		1964	48.6	22.2		1948	30.4	13.9	
1979		132.3	60.5		1963	47.3	21.6		1947	27.6	12.6	

City Cost Indexes

Tabulated on the following pages are average construction cost indexes for 162 major U.S. and Canadian cities. Index figures for both material and installation are based on the 30 major city average of 100 and represent the cost relationship as of July 1, 1990. The index for each division is computed from representative material and labor quantities for that division. The weighted average for each city is a weighted total of the components listed above it, but does not include relative productivity between trades or cities.

The material index for the weighted average includes about 100 basic construction materials with appropriate quantities of each material to represent typical "average" building construction projects.

The installation index for the weighted average includes the contribution of about 30 construction trades with their representative man-days in proportion to the material items installed. Also included in the installation costs are the representative equipment costs for those items requiring equipment.

Since each division of the book contains many different items, any particular item multiplied by the particular city index may give incorrect results. However, when all the book costs for a particular division are summarized and then factored, the result should be very close to the actual costs for that particular division for that city.

If a project has a preponderance of materials from any particular division (say structural steel), then the weighted average index should be adjusted in proportion to the value of the factor for that division.

Adjustments to Costs

Time Adjustment using the Historical Cost Indexes:

$$\frac{\text{Index for Year A}}{\text{Index for Year B}} \text{ X Cost in Year B} = \text{Cost in Year A}$$

Location Adjustment using the City Cost Indexes:

$$\frac{\text{Index for City A}}{\text{Index for City B}} \text{ X Cost in City B} = \text{Cost in City A}$$

Adjustment from the National Average:

$$\text{National Average Cost X } \frac{\text{Index for City A}}{100} = \text{Cost in City A}$$

Note: The City Cost Indexes for Canada can be used to convert U.S. national averages to local costs in Canadian dollars.

(Reprinted from Means Mechanical Cost Data 1991.)

Figure 22.13

Chapter 23

SQUARE FOOT AND SYSTEMS ESTIMATING EXAMPLES

Often the contractor is faced with the need to develop a preliminary estimate for a project. The estimate may be for the entire project or for only a portion of the work which is often the case for the mechanical contractor. This chapter contains two complete sample estimates for the mechanical portion of an office building: a *square foot estimate* and a *systems estimate*. In these step-by-step examples, forms are filled out and calculations made according to the techniques described in Part 1, "The Estimating Process." All cost and reference tables are from the annual *Means Mechanical Cost Data* or *Means Plumbing Cost Data*.

Project Description

A mechanical contractor has been invited by a familiar general contractor to submit a budget estimate on an office building. The project is a three-story office building with a penthouse and garage. Area calculations for this building are shown in Figure 23.1.

For purposes of illustration, a budget estimate will first be developed as a square foot estimate based on the bare minimum of information supplied.ˑ A sketch of the building concept with overall dimensions is shown in Figure 23.1.

Square Foot Estimating Example

There are two occasions when square foot cost estimates are useful. The first is during the conceptual stage when few, if any, details are available. At this time, square foot costs make a useful starting point for ballpark budget purposes. The second instance where square foot costs are used is after bids are received. Square foot costs may be used at this time to check on the accuracy or competitiveness of the bids. As soon as details become available in the project design, the square foot approach should be discontinued and the project priced by its particular assemblies or unit costs.

The dimensions of the project (based on Figure 23.1) and the mathematical procedures to obtain the areas of the office building and the parking garage are shown in Figure 23.2.

Once the areas of the proposed building are known, size modification calculations can be made as shown in Figure 23.3 using data from *Means Mechanical Cost Data* or *Means Plumbing Cost Data*. The appropriate building type is located and its typical size provides the denominator for the modifier equation. Dividing through produces the factors of 6.8 for the office and .12 for the garage. It will be seen that 6.8 is off the horizontal scale to the right and .12 is off the scale to the left. In both cases use the maximum value indicated on the vertical scale at its intersection. In other words, the modifier for any size factor less than .5 will be 1.1 and for any size factor greater than 3.5 will be .90 as the example problem illustrates.

Square foot costs are then found in Division 17 of *Means Mechanical Cost Data* or *Means Plumbing Cost Data*, a portion of which appears in Figure 23.4, or taken from your own cost records. These square foot costs are multiplied by the appropriate modified areas, and costs for plumbing, heating, ventilating, and

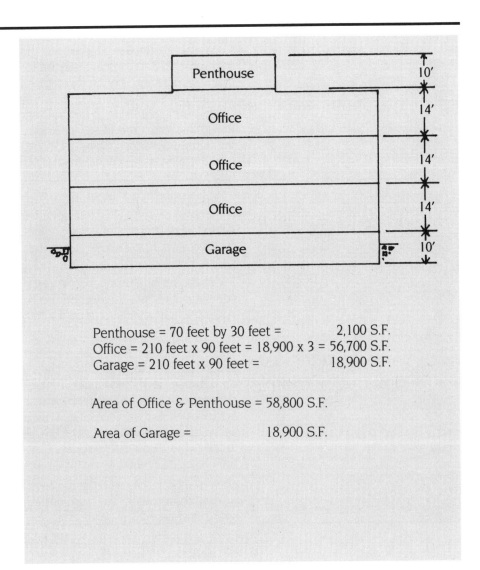

Penthouse = 70 feet by 30 feet = 2,100 S.F.
Office = 210 feet x 90 feet = 18,900 x 3 = 56,700 S.F.
Garage = 210 feet x 90 feet = 18,900 S.F.

Area of Office & Penthouse = 58,800 S.F.

Area of Garage = 18,900 S.F.

Figure 23.1

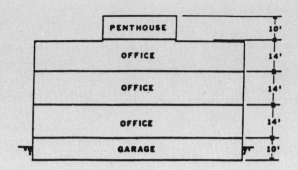

```
Penthouse =  70 feet x 30 feet =                     2,100 S.F.
Office    = 210 feet x 90 feet = 18,900 x 3 =  56,700 S.F.
Garage    = 210 feet x 90 feet =                    18,900 S.F.


Area of Office & Penthouse =          A_O =  58,800 S.F.
Area of Garage             =          A_G =  18,900 S.F.
```

SIZE MODIFICATION

Office $\dfrac{58,800}{8,600}$ = (6.8) modifier = (.9) multiplier (M_O)

Garage $\dfrac{18,900}{163,000}$ = (.12) modifier = (1.1) multiplier (M_G)

COST

    Office
171-610-2720 Plumbing $2.70 x (M_O) .9 x (A_O) 58,800 S.F. = $142,884
        -2770 HVAC $5.35 x (M_O) .9 x (A_O) 58,800 S.F. = $283,122
        -2900 Electrical $5.50 x (M_O) .9 x (A_O) 58,800 S.F. = $291,060

    Garage
171-410-2720 Plumbing $.62 x (M_G) 1.1 x (A_G) 18,900 S.F. = 12,890
        -2900 Electrical $1.09 x (M_G) 1.1 x (A_G) 18,900 S.F. = 22,661

                Total Plumbing $155,774
                Total HVAC $283,122
                Total Mechanical $438,896
        (Geographic correction) $438,896 x 1.112 = $488,052
                Total Electrical $313,721
        (Geographic correction) $313,721 x 1.232 = $386,504

Figure 23.2

Table 14.1-102 Square Foot Project Size Modifier

One factor that affects the S.F. cost of a particular building is the size. In general, for buildings built to the same specifications in the same locality, the larger building will have the lower S.F. Cost. This is due mainly to the decreasing contribution of the exterior walls plus the economy of scale usually achievable in larger buildings. The Area Conversion Scale shown below will give a factor to convert costs for the typical size building to an adjusted cost for the particular project.

The Square Foot Base Size lists the median costs, most typical project size in our accumulated data and the range in size of the projects.

The Size Factor for your project is determined by dividing your project area in S.F. by the typical project size for the particular Building Type. With this factor, enter the Area Conversion Scale at the appropriate Size Factor and determine the appropriate cost multiplier for your building size.

Example: Determine the cost per S.F. for a 100,000 S.F. Mid-rise apartment building.

$$\frac{\text{Proposed building area} = 100,000 \text{ S.F.}}{\text{Typical size from below} = 50,000 \text{ S.F.}} = 2.00$$

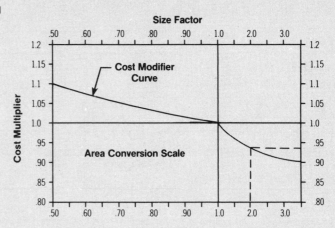

Enter Area Conversion scale at 2.0, intersect curve, read horizontally the appropriate cost multiplier of 0.94. Size adjusted cost becomes 0.94 x $55.40=$52.08 based on national average costs.

Note: For Size Factors less than .50, the Cost Multiplier is 1.1
      For Size Factors greater than 3.5, the Cost Multiplier is .90

Square Foot Base Size							
Building Type	**Median Cost per S.F.**	**Typical Size Gross S.F.**	**Typical Range Gross S.F.**	**Building Type**	**Median Cost per S.F.**	**Typical Size Gross S.F.**	**Typical Range Gross S.F.**
Apartments, Low Rise	$ 44.30	21,000	9,700 - 37,200	Jails	$130.00	13,700	7,500 - 28,000
Apartments, Mid Rise	55.40	50,000	32,000 - 100,000	Libraries	79.15	12,000	7,000 - 31,000
Apartments, High Rise	64.35	310,000	100,000 - 650,000	Medical Clinics	76.50	7,200	4,200 - 15,700
Auditoriums	75.80	25,000	7,600 - 39,000	Medical Offices	71.90	6,000	4,000 - 15,000
Auto Sales	47.05	20,000	10,800 - 28,600	Motels	55.85	27,000	15,800 - 51,000
Banks	103.00	4,200	2,500 - 7,500	Nursing Homes	75.75	23,000	15,000 - 37,000
Churches	67.40	9,000	5,300 - 13,200	Offices, Low Rise	60.35	8,600	4,700 - 19,000
Clubs, Country	66.05	6,500	4,500 - 15,000	Offices, Mid Rise	65.05	52,000	31,300 - 83,100
Clubs, Social	65.05	10,000	6,000 - 13,500	Offices, High Rise	79.25	260,000	151,000 - 468,000
Clubs, YMCA	69.80	28,300	12,800 - 39,400	Police Stations	101.00	10,500	4,000 - 19,000
Colleges (Class)	89.75	50,000	23,500 - 98,500	Post Offices	76.20	12,400	6,800 - 30,000
Colleges (Science Lab)	109.00	45,600	16,600 - 80,000	Power Plants	555.00	7,500	1,000 - 20,000
College (Student Union)	96.95	33,400	16,000 - 85,000	Religious Education	56.30	9,000	6,000 - 12,000
Community Center	70.65	9,400	5,300 - 16,700	Research	103.00	19,000	6,300 - 45,000
Court Houses	94.00	32,400	17,800 - 106,000	Restaurants	89.75	4,400	2,800 - 6,000
Dept. Stores	41.55	90,000	44,000 - 122,000	Retail Stores	43.60	7,200	4,000 - 17,600
Dormitories, Low Rise	68.20	24,500	13,400 - 40,000	Schools, Elementary	65.40	41,000	24,500 - 55,000
Dormitories, Mid Rise	86.35	55,600	36,100 - 90,000	Schools, Jr. High	65.95	92,000	52,000 - 119,000
Factories	36.35	26,400	12,900 - 50,000	Schools, Sr. High	66.10	101,000	50,500 - 175,000
Fire Stations	73.20	5,800	4,000 - 8,700	Schools, Vocational	62.85	37,000	20,500 - 82,000
Fraternity Houses	64.10	12,500	8,200 - 14,800	Sports Arenas	51.40	15,000	5,000 - 40,000
Funeral Homes	64.25	7,800	4,500 - 11,000	Supermarkets	43.35	20,000	12,000 - 30,000
Garages, Commercial	48.45	9,300	5,000 - 13,600	Swimming Pools	74.15	13,000	7,800 - 22,000
Garages, Municipal	53.50	8,300	4,500 - 12,600	Telephone Exchange	119.00	4,500	1,200 - 10,600
Garages, Parking	22.50	163,000	76,400 - 225,300	Terminals, Bus	57.25	11,400	6,300 - 16,500
Gymnasiums	62.60	19,200	11,600 - 41,000	Theaters	61.35	10,500	8,800 - 17,500
Hospitals	123.00	55,000	27,200 - 125,000	Town Halls	72.25	10,800	4,800 - 23,400
House (Elderly)	62.00	37,000	21,000 - 66,000	Warehouses	28.55	25,000	8,000 - 72,000
Housing (Public)	52.65	36,000	14,400 - 74,400	Warehouse & Office	32.95	25,000	8,000 - 72,000
Ice Rinks	49.95	29,000	27,200 - 33,600				

(Reprinted from Means Mechanical Cost Data 1991.)

Figure 23.3

air conditioning are determined. The total mechanical cost is then modified for geographic area (see Chapter 22, "Using *Means Mechanical Cost Data* and *Means Plumbing Cost Data,*" Figure 22.12).

Systems Estimating Example

The next phase of the estimating process, after the preliminary square foot estimate, may be to complete a systems estimate. However, the systems estimate, like the square foot estimate, is not a substitute for a *unit price estimate*. A systems estimate is most often prepared during the conceptual stage of a project when certain parameters are known, but the building plans are not yet completed. This preliminary estimate may help the designer keep the project within the owner's budget. The sketch and dimensions of the proposed building would be identical to that used for the square foot estimate; however, in addition to the areas, the occupied volume is calculated (see Figure 23.5).

Once the physical area and occupied volume have been defined, the number of occupants needs to be determined. Occupancy is based on the net floor area. Figures 23.6 and 23.7 can be used to determine the occupancy requirements. The floor area ratio table in Figure 23.6 indicates a net to gross ratio for offices as 75% on a per occupied floor basis.

Therefore, 18,900 S.F. (area of each office floor as shown in Figure 23.5) is reduced to 14,175 S.F. net area. This net or reduced area has excluded the space occupied by stairwells, corridors, mechanical rooms, etc. In a commercial building, the net area might be referred to as the "leasable area." The occupancy determinations table in Figure 23.7 indicates the occupancy recommendations of four major building code authorities for several building classifications. All agree that for office buildings, the maximum density is one person per 100 S.F. Based on this calculation, the net floor area of 14,175 S.F. allows a maximum of 142 persons per floor. Without any data to the contrary, a male: female ratio of 50:50 should be assumed. With the basic requirements of the building established, the next step is to develop costs using the systems estimating method.

In a systems estimate, individual components are grouped together into systems that reflect the way buildings are constructed. The grouping of many components into a single system allows the estimator to compare various systems and select the one best suited to accommodate costs, usage, compatibility, and any special requirements for the particular project.

Prior to beginning the estimate, all pertinent data must be assembled and analyzed. A typical preprinted form to record the available data is shown in Figure 23.8.

In a systems estimate the Uniformat grouping of 12 construction divisions is used, instead of the 16 Construction Specifications Institute MASTERFORMAT divisions used for unit price estimates. The mechanical data is found in Division 8 of the Uniformat data. (See Chapter 2 for more information on the Uniformat Divisions.)

It is desirable that the building dimensions, the owner's/occupants' requirements, local building codes, existing utilities, local labor sources, and anticipated economic conditions be known. With this information and the inclusion of any contingencies, a reasonable estimate may be arrived at, which can be updated as more design parameters are developed. For this example, the proposed building is a regional office building of good quality for an insurance company located in the Boston area.

171 000		S.F. & C.F. Costs	UNIT	UNIT COSTS			% OF TOTAL			
				¼	MEDIAN	¾	¼	MEDIAN	¾	
360	2720	Plumbing	S.F.	3.49	5.45	7.40	5.90%	7.30%	9.50%	
	2770	Heating, ventilating, air conditioning		2.99	4.80	7.65	4.80%	7.30%	9.20%	
	2900	Electrical		4.09	6.80	9.75	7.20%	9.60%	11.90%	
	3100	Total: Mechanical & Electrical		10.35	16	22.10	17.50%	22.60%	27.50%	
370	0010	FRATERNITY HOUSES And Sorority Houses	↓	53.70	64.10	71.50				**370**
	0020	Total project costs	C.F.	5.15	6.25	6.90				
	2720	Plumbing	S.F.	4.05	4.74	6.30	5.90%	8%	10.80%	
	2900	Electrical		3.54	4.66	8.45	6.50%	8.80%	10.40%	
	3100	Total: Mechanical & Electrical	↓	10.05	14.25	17.15	14.60%	20.70%	24.20%	
380	0010	FUNERAL HOMES	S.F.	51.10	64.25	94.15				**380**
	0020	Total project costs	C.F.	3.61	5.20	6.45				
	2720	Plumbing	S.F.	2.02	2.81	3.05	4.10%	4.40%	4.70%	
	2770	Heating, ventilating, air conditioning		4.50	4.55	5.45	7%	9.20%	10.40%	
	2900	Electrical		3.38	4.58	6.55	5.90%	7.70%	11%	
	3100	Total: Mechanical & Electrical		9.25	12.25	14.15	17.80%	20.80%	27.10%	
390	0010	GARAGES, COMMERCIAL (service)	↓	30.90	48.45	63.90				**390**
	0020	Total project costs	C.F.	1.97	2.89	4.16				
	2720	Plumbing	S.F.	1.98	3.12	6.25	4.90%	7.40%	11%	
	2730	Heating & ventilating		2.91	4	4.83	5.20%	7.20%	9.50%	
	2900	Electrical		2.71	4.49	6.30	7%	9%	11.10%	
	3100	Total: Mechanical & Electrical		6.40	11.35	16.30	15.70%	21.90%	27.80%	
400	0010	GARAGES, MUNICIPAL (repair)	↓	39	53.50	81				**400**
	0020	Total project costs	C.F.	2.44	3.40	4.55				
	2720	Plumbing	S.F.	2.06	3.87	6.35	4.10%	6.90%	8.70%	
	2730	Heating & ventilating		2.58	4.62	6.95	6.10%	7.80%	11.60%	
	2900	Electrical		2.97	4.65	6.80	6.60%	8.10%	10.50%	
	3100	Total: Mechanical & Electrical		7.80	15.50	21.90	17%	24.40%	32.40%	
410	0010	GARAGES, PARKING	↓	17.75	22.50	37.35				**410**
	0020	Total project costs	C.F.	1.53	2.03	3.34				
	2720	Plumbing	S.F.	.32	.62	.90	2.10%	2.80%	3.80%	
	2900	Electrical		.75	1.09	1.76	4.20%	5.20%	6.50%	
	3100	Total: Mechanical & Electrical	↓	1.29	1.74	2.60	7.10%	8.30%	9.50%	
	9000	Per car, total cost	Car	5,675	7,725	10,900				
	9500	Total: Mechanical & Electrical	Car	405	610	910				

	2720	Plumbing	S.F.	5.10	6.25	9.10	9.30%	10.30%	13.30%	
	2770	Heating, ventilating, air conditioning		5.20	7.25	9.30	9.20%	11.40%	11.80%	
	2900	Electrical		5.75	7.25	9.45	9.70%	11%	12.80%	
	3100	Total: Mechanical & Electrical	↓	13.40	17.90	26.95	22.30%	28.30%	33.30%	
	9000	Per bed or person, total cost	Bed	22,200	28,900	36,700				
610	0010	OFFICES Low-Rise (1 to 4 story)	S.F.	47.10	60.35	79.90				**610**
	0020	Total project costs	C.F.	3.49	4.86	6.55				
	2720	Plumbing	S.F.	1.79	2.70	3.86	3.60%	4.50%	6%	
	2770	Heating, ventilating, air conditioning		3.81	5.35	7.84	7.20%	10.40%	11.90%	
	2900	Electrical		3.98	5.50	7.55	7.40%	9.50%	11%	
	3100	Total: Mechanical & Electrical		8.25	12.25	17.90	14.90%	20.80%	26.80%	

For expanded coverage of these items see *Means Square Foot Cost Data 1991*

283

Figure 23.4

A systems estimate for the mechanical trades is accomplished by obtaining all the available data, including physical area, function, occupancy, etc., and selecting the appropriate mechanical systems to cover all the requirements for such a building. The figures and other pertinent information to compile systems and their costs should come from the estimator's own experience and records, or from a reliable cost data source such as *Means Mechanical Cost Data*, which contains 24 plumbing, 8 fire protection, and 25 heating and cooling systems. Some of these systems are priced by the square foot area of a building

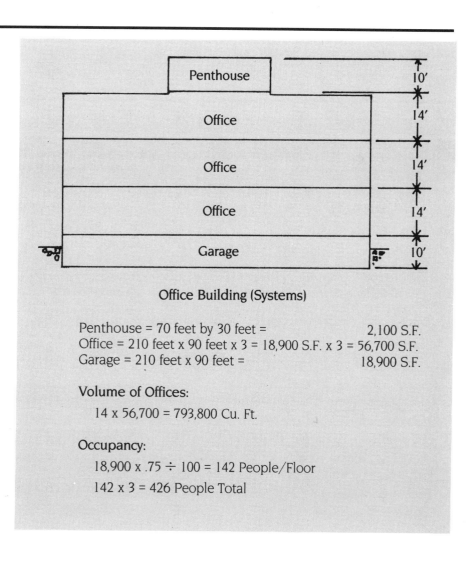

Office Building (Systems)

Penthouse = 70 feet by 30 feet =   2,100 S.F.
Office = 210 feet x 90 feet x 3 = 18,900 S.F. x 3 = 56,700 S.F.
Garage = 210 feet x 90 feet =   18,900 S.F.

Volume of Offices:

    14 x 56,700 = 793,800 Cu. Ft.

Occupancy:

    18,900 x .75 ÷ 100 = 142 People/Floor

    142 x 3 = 426 People Total

Figure 23.5

(e.g., sprinkler systems), others are priced by the required number of units (e.g., bathrooms, water heaters, etc.). All systems costs include an allowance for overhead and profit.

Plumbing

The table shown in Figure 23.9, "Minimum Plumbing Fixture Requirements," from *Means Plumbing Cost Data* is the basis for determining the number of plumbing fixtures required depending on building use and the number of occupants. It has been determined (See Figure 23.3) that this office building will house 142 persons per floor, with a 50:50 male/female ratio.

Based on this determined occupancy of 142 persons per floor, the minimum plumbing fixture requirements mandate the following fixture count for business offices:

			Floor Area Ratios Commonly Used Gross-to-Net Area and Net-to-Gross Area Ratios Expressed in % for Various Building Types		
Building Type	Gross to Net Ratio	Net to Gross Ratio	Building Type	Gross to Net Ratio	Net to Gross Ratio
Apartment	156	64	School Buildings (campus type)		
Bank	140	72	Administrative	150	67
Church	142	70	Auditorium	142	70
Courthouse	162	61	Biology	161	62
Department Store	123	81	Chemistry	170	59
Garage	118	85	Classroom	152	66
Hospital	183	55	Dining Hall	138	72
Hotel	158	63	Dormitory	154	65
Laboratory	171	58	Engineering	164	61
Library	132	76	Fraternity	160	63
Office	135	75	Gymnasium	142	70
Restaurant	141	70	Science	167	60
Warehouse	108	93	Service	120	83
			Student Union	172	59

The gross area of a building is the total floor area based on outside dimensions. The net area of a building is the usable floor area for the function intended and excludes such items as stairways, corridors and mechanical rooms. In the case of commerical building, it might be considered as the "leaseable area."

Figure 23.6

Fixtures	Men	Women	Handicapped	Misc	Total Per Floor
water closets	2	3	1	—	6
lavatories	3	3	1	—	7
urinals	2	—	—	—	2
water coolers	—	—	1	2	3
service sink	—	—	—	1	1

As shown in Figure 23.9, 1/3 of the water closets can be replaced with urinals. This would change the required three men's water closets to two with one urinal. A second urinal has been arbitrarily added based on an estimate of what an occupant would require. Three handicapped fixtures per floor have also been included in the estimate to conform with local and national codes.

In addition to the fixtures and appliances (water coolers), the plumbing estimate would not be complete without water heaters and roof drains. The roof area is 18,900 S.F. and the building height is 52 feet (see Figure 23.10). An additional four feet is allowed to bury the storm drain. With these measurements, the

Occupancy Determinations				
		S.F. Required Per Person*		
Description		BOCA	SSBC	UBC
Assembly Areas	Fixed Seats	**	6	7
	Movable Seats		15	15
	Concentrated	7		
	Unconcentrated	15		
	Standing Space	3		
Educational	Unclassified			
	Classrooms	20	40	20
	Shop Areas	50	100	50
Institutional	Unclassified		125	
	In-Patient Areas	240		
	Sleeping Areas	120		
Mercantile	Basement	30	30	20
	Ground Floor	30	30	30
	Upper Floors	60	60	50
Office		100	100	100

*The occupancy load for assembly area with fixed seats shall be determined by the number of fixed seats installed.
BOCA = Building Officials & Code Administrators
SSBC = Southern Standard Building Code
UBC = Uniform Building Code

Figure 23.7

233

Means Forms

ASSEMBLY NUMBER	DESCRIPTION	QTY	UNIT	TOTAL COST		COST PER S.F.
				UNIT	TOTAL	
8.0	**Mechanical System**					
9.0	**Electrical**					

Page 4 of 6

Figure 23.8

Table 8.1-401 Minimum Plumbing Fixture Requirements

TYPE OF BUILDING/USE	WATER CLOSETS		URINALS		LAVATORIES		BATHTUBS OR SHOWERS		DRINKING FOUNTAIN	OTHER
	Persons	Fixtures	Persons	Fixtures	Persons	Fixtures	Persons	Fixtures	Fixtures	Fixtures
Assembly Halls Auditoriums Theater Public assembly	1-100 101-200 201-400	1 2 3	1-200 201-400 401-600	1 2 3	1-200 201-400 401-750	1 2 3			1 for each 1000 persons	1 service sink
	Over 400 add 1 fixt. for ea. 500 men: 1 fixt. for ea. 300 women		Over 600 add 1 fixture for each 300 men		Over 750 add 1 fixture for each 500 persons					
Assembly Public Worship	300 men 150 women	1 1	300 men	1	men women	1 1			1	
Dormitories	Men: 1 for each 10 persons Women: 1 for each 8 persons		1 for each 25 men, over 150 add 1 fixture for each 50 men		1 for ea. 12 persons 1 separate dental lav. for each 50 persons recom.		1 for ea. 8 persons For women add 1 additional for each 30. Over 150 persons add 1 for each 20.		1 for each 75 persons	Laundry trays 1 for each 50 serv. sink 1 for ea. 100
Dwellings Apartments and homes	1 fixture for each unit				1 fixture for each unit		1 fixture for each unit			
Hospitals Indiv. Room Ward Waiting room	8 persons	1 1 1			10 persons	1 1 1	20 persons	1 1	1 for 100 patients	1 service sink per floor
Industrial Mfg. plants Warehouses	1-10 11-25 26-50 51-75 76-100	1 2 3 4 5	0-30 31-80 81-160 161-240	1 2 3 4	1-100 over 100	1 for ea. 10 1 for ea. 15	1 Shower for each 15 persons subject to excessive heat or occupational hazard		1 for each 75 persons	
	1 fixture for each additional 30 persons									
Public Buildings Businesses Offices	1-15 16-35 36-55 56-80 81-110 111-150	1 2 3 4 5 6	Urinals may be provided in place of water closets but may not replace more than 1/3 required number of men's water closets		1-15 16-35 36-60 61-90 91-125	1 2 3 4 5			1 for each 75 persons	1 service sink per floor
	1 fixture for ea. additional 40 persons				1 fixture for ea. additional 45 persons					
Schools Elementary	1 for ea. 30 boys 1 for ea. 25 girls		1 for ea. 25 boys		1 for ea. 35 boys 1 for ea. 35 girls		For gym or pool shower room 1/5 of a class		1 for each 40 pupils	
Schools Secondary	1 for ea. 40 boys 1 for ea. 30 girls		1 for ea. 25 boys		1 for ea. 40 boys 1 for ea. 40 girls		For gym or pool shower room 1/5 of a class		1 for each 50 pupils	

(Reprinted from Means Plumbing Cost Data 1991.)

Figure 23.9

roof drain system can be sized using the chart shown in Figure 23.10 from *Means Plumbing Cost Data*. Six 4" drains are capable of handling up to 20,760 S.F.

Hot water consumption rates are determined by using the table shown in Figure 23.11 from *Means Plumbing Cost Data*. This table shows that an office building has a minimum hourly demand of 0.4 gallons per person. The office building in this estimating example, having an occupancy of 142 persons per floor or 426 total, requires a 170 gallon minimum hourly demand (gallons per hour-GPH) water heater or series of heaters.

All the plumbing fixtures and appliances should be listed by categories for pricing on a plumbing estimate worksheet, as shown in Figure 23.12. Using the assemblies cost tables shown in Figure 23.13 through 23.19 from *Means Plumbing Cost Data*, a systems table number is assigned for each fixture. The line number for pricing is also recorded. The lavatories, for example, can be priced using Figure 23.13. Vanity top oval lavatories, 19" x 16", are selected and priced according to line 1600. The same procedure is followed for the remaining fixtures and appliances. The complete plumbing systems estimate is shown in Figure 23.20.

To compensate for materials not included in the systems groupings certain percentages must be added to the plumbing system. These percentages can be found in table C8.1-031 from *Means Plumbing Cost Data* (reproduced in Figure 23.21). The first of these is for "water control" and covers water meter, backflow preventer, shock absorbers, and vacuum breakers. An additional 10-15% should be added to the subtotal for these items.

Roof Drain Systems		
Pipe Diameter	Max. S.F. Roof Area	Gallons per Min.
2"	544	23
3"	1,610	67
4"	3,460	144
5"	6,280	261
6"	10,200	424
8"	22,000	913

Design Assumptions: Vertical conductor size is based on a maximum rate of rainfall of 4" per hour. To convert roof area to other rates multiply "Max S.F. Roof Area" shown by four and divide the result by desired rate. The answer is the local roof area that may be handled by the indicated pipe diameter.

Basic cost is for roof drain, 10' of vertical leader and 10' of horizontal, plus connection to the main.

Figure 23.10

Table 8.1-101 Hot Water Consumption Rates

Type of Building	Size Factor	Maximum Hourly Demand	Average Day Demand
Apartment Dwellings	No. of Apartments: Up to 20 21 to 50 51 to 75 76 to 100 101 to 200 201 up	 12.0 Gal. per apt. 10.0 Gal. per apt. 8.5 Gal. per apt. 7.0 Gal. per apt. 6.0 Gal. per apt. 5.0 Gal. per apt.	 42.0 Gal. per apt. 40.0 Gal. per apt. 38.0 Gal. per apt. 37.0 Gal. per apt. 36.0 Gal. per apt. 35.0 Gal. per apt.
Dormitories	Men Women	3.8 Gal. per man 5.0 Gal. per woman	13.1 Gal. per man 12.3 Gal. per woman
Hospitals	Per bed	23 Gal. per patient	90 Gal. per patient
Hotels	Single room with bath Double room with bath	17 Gal. per unit 27 Gal. per unit	50 Gal. per unit 80 Gal. per unit
Motels	No. of units: Up to 20 21 to 100 101 Up	 6.0 Gal. per unit 5.0 Gal. per unit 4.0 Gal. per unit	 20.0 Gal. per unit 14.0 Gal. per unit 10.0 Gal. per unit
Nursing Homes		4.5 Gal. per bed	18.4 Gal. per bed
Office buildings		0.4 Gal. per person	1.0 Gal. per person
Restaurants	Full meal type Drive-in snack type	1.5 Gal./max. meals/hr. 0.7 Gal./max. meals/hr.	2.4 Gal. per meal 0.7 Gal. per meal
Schools	Elementary Secondary & High	0.6 Gal. per student 1.0 Gal. per student	0.6 Gal. per student 1.8 Gal. per student

For evaluation purposes, recovery rate and storage capacity are inversely proportional. Water heaters should be sized so that the maximum hourly demand anticipated can be met in addition to allowance for the heat loss from the pipes and storage tank.

Table 8.1-102 Fixture Demands in Gallons Per Fixture Per Hour

Table below is based on 140°F final temperature except for dishwashers in public places (*) where 180°F water is mandatory.

Fixture	Apartment House	Club	Gym	Hospital	Hotel	Indust. Plant	Office	Private Home	School
Bathtubs	20	20	30	20	20			20	
Dishwashers, automatic*	15	50-150		50-150	50-200	20-100		15	20-100
Kitchen sink	10	20		20	30	20	20	10	20
Laundry, stationary tubs	20	28		28	28			20	
Laundry, automatic wash	75	75		100	150			75	
Private lavatory	2	2	2	2	2	2	2	2	2
Public lavatory	4	6	8	6	8	12	6		15
Showers	30	150	225	75	75	225	30	30	225
Service sink	20	20		20	30	20	20	15	20
Demand factor	0.30	0.30	0.40	0.25	0.25	0.40	0.30	0.30	0.40
Storage capacity factor	1.25	0.90	1.00	0.60	0.80	1.00	2.00	0.70	1.00

To obtain the probable maximum demand multiply the total demands for the fixtures (gal./fixture/hour) by the demand factor. The heater should have a heating capacity in gallons per hour equal to this maximum. The storage tank should have a capacity in gallons equal to the probable maximum demand multiplied by the storage capacity factor.

(Reprinted from Means Plumbing Cost Data 1991.)

Figure 23.11

Pipe and fittings is the next category and includes the interconnecting piping (mains) between the several systems. For pipe and fittings, 30 to 60% should be added to the subtotal. This percentage depends on the building design. For example, a building similar in size to the office building in this example, but having toilet rooms at each of the far ends of the building, would have a much higher percentage than the 30% indicated for this plumbing estimate. In extreme cases, this percentage might reach 100%.

Quality complexity is the last additive. It should be applied to fire protection, HVAC, and plumbing estimates to compensate for quality/complexity over and above the basic level assumed in the Means assemblies prices. Percentages for quality/complexity are 0-5% for economy installation, 5-15% for good quality medium complexity installation, and 15-25% for above average quality, and complexity. Construction and estimating experience is necessary to determine the proper percentages to be added. Keeping cost records of completed projects will contribute to this experience factor.

Fire Protection

Fire protection systems are usually composed of sprinkler systems, hose standpipes, or combinations thereof. System classifications according to hazard and design have been developed by the National Fire Protection Association (NFPA) and are discussed in Chapter 16, "Fire Protection." Based on these criteria, a firehose standpipe and sprinkler system for the example office building can be estimated.

Depending on the system's design, the firehose standpipe could be taken off and priced as part of the sprinkler system or the two could be taken off and priced separately. For this estimate, the firehose standpipe system and the sprinkler system are taken off and priced separately.

Sprinkler System: The system classification for an office building is "Light Hazard Occupancy." A conventional "wet type" distribution is selected for this heated building, and the designer has chosen Schedule #40 black steel pipe and threaded fittings as the piping material.

Using the previously determined area data (18,900 S.F. per floor and 14 L.F. floor heights), and the cost data from Means Plumbing Cost Data, shown in Figure 23.22a through 23.22c, the mathematical procedures shown in Figure 23.23 can be carried out. Line number 8.2-110-0620 (from Figure 23.22a) indicates a cost per square foot for a light hazard, steel pipe installation for the first floor. This price is for a 10,000 S.F. floor, which is closer to the 18,900 S.F. than the next line number which prices a 50,000 S.F. floor area. For the second and third floors, which do not require the alarm valve, water motor, and other sprinkler systems specialties, a square foot cost is selected. Line number 8.2-110-0740 indicates 1.02 cents per square foot. These prices for the first and upper floors are transferred to the fire protection worksheet shown in Figure 23.23.

Fire Standpipe System: To estimate the cost of a fire standpipe system for the proposed office building, the system classification and type should first be determined. Table 8.2-302 from Means Plumbing Cost Data (shown in Figure 23.24) indicates the several choices based on the NFPA 14 basic standpipe design. The building is Class III type usage, which allows a system to be operated by either occupants or by fire department personnel. A wet pipe system, a typical system for an office building, is selected. Two standpipes, at the building stairwells, will adequately cover the areas to be protected. The pipe size is 4", because the building height is less than 100'.

COST ANALYSIS

PROJECT		Plumbing Systems Worksheet			SHEET NO.			
ARCHITECT					ESTIMATE NO.			
					DATE			

TAKE OFF BY: QUANTITIES BY: PRICES BY: EXTENSIONS BY: CHECKED BY:

Fixture	SOURCE/DIMENSIONS Table Number	Qty.		Unit Cost	Total Costs	Calculations
Bathtubs						
Drinking Fountain						
Kitchen Sink						
Laundry Sink						
Lavatory		21				
Service Sink		3				
Urinal		6				
Water Cooler		9				
Water Closet		18				
WashFount. Group						
Water Heater		1				
Roof Drains		6				
Subtotal						
Water Control						
Pipe & Fittings						
Quality Complexity						
Total						

Figure 23.12

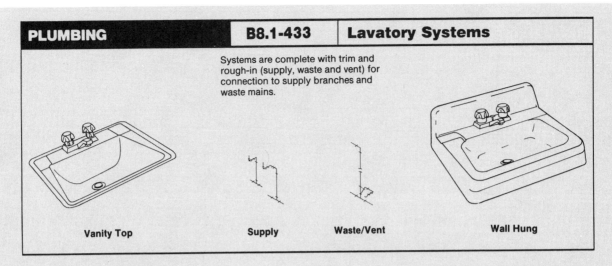

Systems are complete with trim and rough-in (supply, waste and vent) for connection to supply branches and waste mains.

Vanity Top **Supply** **Waste/Vent** **Wall Hung**

System Components	QUANTITY	UNIT	COST EACH MAT.	COST EACH INST.	COST EACH TOTAL
SYSTEM 08.1-433-1560					
LAVATORY W/TRIM, VANITY TOP, P.E. ON C.I., 20" X 18"					
Lavatory w/trim, PE on CI, white, vanity top, 20" x 18" oval	1.000	Ea.	147.40	87.60	235
Pipe, steel, galvanized, schedule 40, threaded, 1-¼" diam	4.000	L.F.	9.96	24.64	34.60
Copper tubing type DWV, solder joint, hanger 10'OC 1-¼" diam	4.000	L.F.	11.68	20.32	32
Wrought copper DWV, Tee, sanitary, 1-¼" diam	1.000	Ea.	5.23	33.77	39
P trap w/cleanout, 20 ga, 1-¼" diam	1.000	Ea.	26.40	16.60	43
Copper tubing type L, solder joint, hanger 10' OC ½" diam	10.000	L.F.	13	37.50	50.50
Wrought copper 90° elbow for solder joints ½" diam	2.000	Ea.	.64	30.36	31
Wrought copper Tee for solder joints, ½" diam	2.000	Ea.	1.08	46.92	48
Stop, chrome, angle supply, ½" diam	2.000	Ea.	29.70	28.30	58
TOTAL			245.09	326.01	571.10

8.1-433	Lavatory Systems	COST EACH MAT.	COST EACH INST.	COST EACH TOTAL
1560	Lavatory w/trim, vanity top, PE on CI, 20" x 18" *	245	325	570
1600	19" x 16" oval	240	325	565
1640	18" round	235	325	560
1680	Cultured marble, 19" x 17"	195	320	515
1720	25" x 19"	210	325	535
1760	Stainless, self-rimming, 25" x 22"	300	325	625
1800	17" x 22"	200	325	525
1840	C8.1 -401 Steel enameled, 20" x 17"	185	335	520
1880	19" round	185	330	515
1920	Vitreous china, 20" x 16"	290	340	630
1960	19" x 16"	255	340	595
2000	22" x 13"	260	340	600
2040	Wall hung, PE on CI, 18" x 15"	400	355	755
2080	19" x 17"	360	360	720
2120	20" x 18"	265	355	620
2160	Vitreous china, 18" x 15"	315	365	680
2200	19" x 17"	260	365	625
2240	24" x 20"	370	365	735
*	Vanity Top built in by others			

(Reprinted from Means Plumbing Cost Data 1991.)

Figure 23.13

240

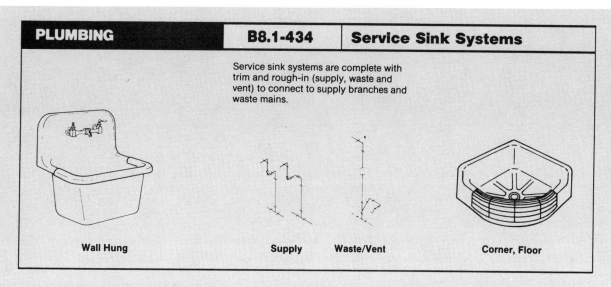

PLUMBING	B8.1-434	Service Sink Systems

Service sink systems are complete with trim and rough-in (supply, waste and vent) to connect to supply branches and waste mains.

Wall Hung   **Supply**   **Waste/Vent**   **Corner, Floor**

System Components	QUANTITY	UNIT	MAT.	INST.	TOTAL
SYSTEM 08.1-434-4260					
SERVICE SINK, PE ON CI, CORNER FLOOR, 28"X28", W/RIM GUARD & TRIM					
Service sink, corner floor, PE on CI, 28" x 28", w/rim guard & trim	1.000	Ea.	462	123	585
Copper tubing type DWV, solder joint, hanger 10'OC 3" diam	6.000	L.F.	48.90	56.70	105.60
Copper tubing type DWV, solder joint, hanger 10'OC 2" diam	4.000	L.F.	19.28	27.52	46.80
Wrought copper DWV, Tee, sanitary, 3" diam	1.000	Ea.	17.49	78.51	96
P trap with cleanout & slip joint, copper 3" diam	1.000	Ea.	60.50	27.50	88
Copper tubing, type L, solder joints, hangers 10' OC, ½" diam	10.000	L.F.	13	37.50	50.50
Wrought copper 90° elbow for solder joints ½" diam	2.000	Ea.	.64	30.36	31
Wrought copper Tee for solder joints, ½" diam	2.000	Ea.	1.08	46.92	48
Stop, angle supply, chrome, ½" diam	2.000	Ea.	29.70	28.30	58
TOTAL			652.59	456.31	1,108.90

8.1-434	Service Sink Systems	MAT.	INST.	TOTAL
4260	Service sink w/trim, PE on CI, corner floor, 28" x 28", w/rim guard	655	455	1,110
4300	Wall hung w/rim guard, 22" x 18"	605	530	1,135
4340	24" x 20"	635	530	1,165
4380	Vitreous china, wall hung 22" x 20"	650	530	1,180
C8.1 -401				

254

(Reprinted from Means Plumbing Cost Data 1991.)

Figure 23.14

241

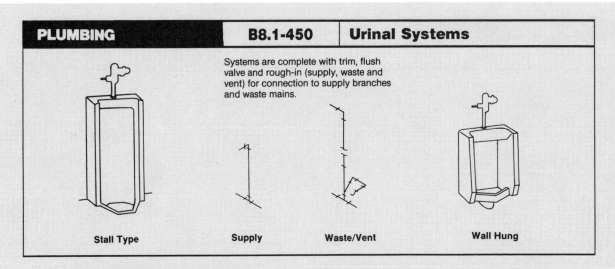

Systems are complete with trim, flush valve and rough-in (supply, waste and vent) for connection to supply branches and waste mains.

Stall Type   **Supply**   **Waste/Vent**   **Wall Hung**

System Components	QUANTITY	UNIT	COST EACH		
			MAT.	INST.	TOTAL
SYSTEM 08.1-450-2000					
URINAL, VITREOUS CHINA, WALL HUNG					
Urinal, wall hung, vitreous china, incl. hanger	1.000	Ea.	368.50	181.50	550
Pipe, steel, galvanized, schedule 40, threaded, 1-½" diam	5.000	L.F.	15.05	34.20	49.25
Copper tubing type DWV, solder joint, hangers 10'OC, 2" diam	3.000	L.F.	14.46	20.64	35.10
Combination Y & ⅛ bend for CI soil pipe, no hub, 3" diam	1.000	Ea.	6.15		6.15
Pipe, CI, no hub, cplg 10' OC, hanger 5' OC, 3" diam	4.000	L.F.	15.64	34.16	49.80
Pipe coupling standard, CI soil, no hub, 3" diam	3.000	Ea.	8.43	43.17	51.60
Copper tubing type L, solder joint, hanger 10' OC ¾" diam	5.000	L.F.	9.75	20	29.75
Wrought copper 90° elbow for solder joints ¾" diam	1.000	Ea.	.54	16.01	16.55
Wrought copper Tee for solder joints, ¾" diam	1.000	Ea.	1.34	25.66	27
TOTAL			439.86	375.34	815.20

8.1-450	Urinal Systems	COST EACH		
		MAT.	INST.	TOTAL
2000	Urinal, vitreous china, wall hung	440	375	815
2040	Stall type	570	435	1,005
C8.1-401				

(Reprinted from Means Plumbing Cost Data 1991.)

Figure 23.15

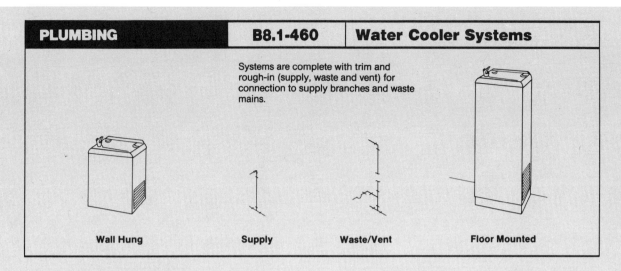

PLUMBING	B8.1-460	Water Cooler Systems

Systems are complete with trim and rough-in (supply, waste and vent) for connection to supply branches and waste mains.

Wall Hung **Supply** **Waste/Vent** **Floor Mounted**

System Components	QUANTITY	UNIT	COST EACH		
			MAT.	INST.	TOTAL
SYSTEM 08.1-460-1840					
WATER COOLER, ELECTRIC, SELF CONTAINED, WALL HUNG, 8.2 GPH					
Water cooler, wall mounted, 8.2 GPH	1.000	Ea.	381.70	138.30	520
Copper tubing type DWV, solder joint, hanger 10'OC 1-¼" diam	4.000	L.F.	11.68	20.32	32
Wrought copper DWV, Tee, sanitary 1-¼" diam	1.000	Ea.	5.23	33.77	39
P trap, copper drainage, 1-¼" diam	1.000	Ea.	26.40	16.60	43
Copper tubing type L, solder joint, hanger 10' OC ⅜" diam	5.000	L.F.	5.45	18.10	23.55
Wrought copper 90° elbow for solder joints ⅜" diam	1.000	Ea.	.75	13.80	14.55
Wrought copper Tee for solder joints, ⅜" diam	1.000	Ea.	1.78	21.22	23
Stop and waste, straightway, bronze, solder, ⅜" diam	1.000	Ea.	4.85	12.65	17.50
TOTAL			437.84	274.76	712.60

8.1-460	Water Cooler Systems	COST EACH		
		MAT.	INST.	TOTAL
1840	Water cooler, electric, wall hung, 8.2 GPH	440	275	715
1880	Dual height, 14.3 GPH	610	280	890
1920	Wheelchair type, 7.5 G.P.H.	1,025	275	1,300
1960	Semi recessed, 8.1 G.P.H.	570	270	840
2000	Full recessed, 8 G.P.H.	900	295	1,195
2040	Floor mounted, 14.3 G.P.H.	465	240	705
2080	Dual height, 14.3 G.P.H.	650	285	935
2120	Refrigerated compartment type, 1.5 G.P.H.	865	240	1,105
2160	Cafeteria type, dual glass fillers, 27 G.P.H.	1,950	360	2,310

257

(Reprinted from Means Plumbing Cost Data 1991.)

Figure 23.16

243

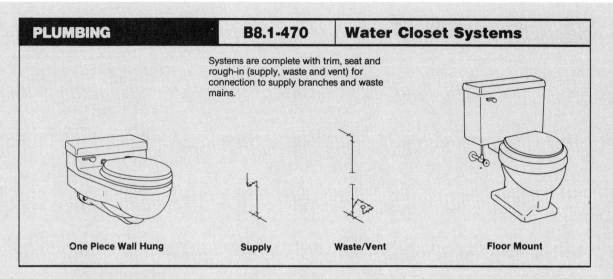

Systems are complete with trim, seat and rough-in (supply, waste and vent) for connection to supply branches and waste mains.

One Piece Wall Hung   **Supply**   **Waste/Vent**   **Floor Mount**

System Components	QUANTITY	UNIT	COST EACH		
			MAT.	INST.	TOTAL
SYSTEM 08.1-470-1840					
WATER CLOSET, VITREOUS CHINA, ELONGATED					
TANK TYPE, WALL HUNG, ONE PIECE					
Wtr closet tank type vit china wall hung 1 pc w/seat supply & stop	1.000	Ea.	544.50	105.50	650
Pipe steel galvanized, schedule 40, threaded, 2″ diam	4.000	L.F.	15.52	34.08	49.60
Pipe, CI soil, no hub, cplg 10′ OC, hanger 5′ OC, 4″ diam	2.000	L.F.	10.06	18.84	28.90
Pipe, coupling, standard coupling, CI soil, no hub, 4″ diam	2.000	Ea.	6.50	33.10	39.60
Copper tubing type L, solder joint, hanger 10′OC, ½″ diam	6.000	L.F.	7.80	22.50	30.30
Wrought copper 90° elbow for solder joints ½″ diam	2.000	Ea.	.64	30.36	31
Wrought copper Tee for solder joints ½″ diam	1.000	Ea.	.54	23.46	24
Support/carrier, for water closet, siphon jet, horiz, single, 4″ waste	1.000	Ea.	121	49	170
TOTAL			706.56	316.84	1,023.40

8.1-470	Water Closet Systems	COST EACH		
		MAT.	INST.	TOTAL
1800	Water closet, vitreous china, elongated			
1840	Tank type, wall hung, one piece	705	315	1,020
1880	Close coupled two piece	535	315	850
1920	Floor mount, one piece	555	345	900
1960	One piece low profile	825	345	1,170
2000	Two piece close coupled	245	345	590
2040	Bowl only with flush valve			
2080	Wall hung	475	320	795
2120	Floor mount	405	345	750

(Rows 1880/1920 left box label: C8.1 -401)

258

(Reprinted from Means Plumbing Cost Data 1991.)

Figure 23.17

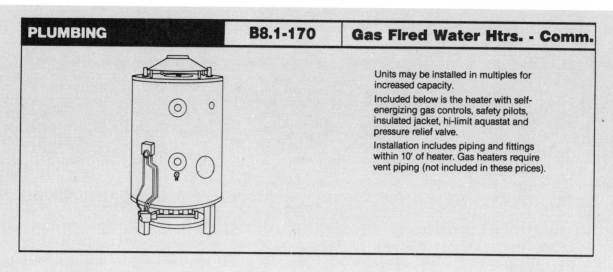

Units may be installed in multiples for increased capacity.

Included below is the heater with self-energizing gas controls, safety pilots, insulated jacket, hi-limit aquastat and pressure relief valve.

Installation includes piping and fittings within 10' of heater. Gas heaters require vent piping (not included in these prices).

System Components	QUANTITY	UNIT	COST EACH		
			MAT.	INST.	TOTAL
SYSTEM 08.1-170-1780					
GAS FIRED WATER HEATER, COMMERCIAL, 100° F RISE					
75.5 MBH INPUT, 63 GPH					
Water heater, commercial, gas, 75.5 MBH, 63 GPH	1.000	Ea.	887.70	212.30	1,100
Copper tubing, type L, solder joint, hanger 10' OC, 1-¼" diam	30.000	L.F.	110.70	156.30	267
Wrought copper 90° elbow for solder joints 1-¼" diam	4.000	Ea.	8.96	79.04	88
Wrought copper Tee for solder joints, 1-¼" diam	2.000	Ea.	14.52	67.48	82
Wrought copper union for soldered joints, 1-¼" diam	2.000	Ea.	15.30	42.70	58
Valve, gate, bronze, 125 lb, NRS, soldered 1-¼" diam	2.000	Ea.	50.60	41.40	92
Relief valve, bronze, press & temp, self-close, ¾" IPS	1.000	Ea.	47.30	10.70	58
Copper tubing, type L, solder joints, ¾" diam	8.000	L.F.	15.60	32	47.60
Wrought copper 90° elbow for solder joints ¾" diam	1.000	Ea.	.54	16.01	16.55
Wrought copper, adapter, CTS to MPT, ¾" IPS	1.000	Ea.	1.24	14.46	15.70
Pipe steel black, schedule 40, threaded, ¾" diam	10.000	L.F.	11.70	49.80	61.50
Pipe, 90° elbow, malleable iron black, 150 lb threaded, ¾" diam	2.000	Ea.	1.40	42.60	44
Pipe, union with brass seat, malleable iron black, ¾" diam	1.000	Ea.	2.53	23.47	26
Valve, gas stop w/o check, brass, ¾" IPS	1.000	Ea.	8.86	14.14	23
TOTAL			1,176.95	802.40	1,979.35

8.1-170	Gas Fired Water Heaters - Commercial Systems		COST EACH		
			MAT.	INST.	TOTAL
1760	Gas fired water heater, commercial, 100° F rise				
1780		75.5 MBH input, 63 GPH	1,175	800	1,975
1860	C8.1-101	100 MBH input, 91 GPH	1,475	835	2,310
1980		155 MBH input, 150 GPH	1,975	975	2,950
2060	C8.1-102	200 MBH input, 192 GPH	2,550	1,175	3,725
2140		300 MBH input, 278 GPH	3,975	1,475	5,450
2180		390 MBH input, 374 GPH	4,475	1,500	5,975
2220		500 MBH input, 480 GPH	5,900	1,600	7,500
2260		600 MBH input, 576 GPH	6,700	1,725	8,425
2300		800 MBH input, 768 GPH	7,850	1,925	9,775
2340		1000 MBH input, 960 GPH	9,225	1,950	11,175
2420		1500 MBH input, 1440 GPH	14,100	2,475	16,575
2460		1800 MBH input, 1730 GPH	15,800	2,725	18,525
2500		2450 MBH input, 2350 GPH	20,700	3,275	23,975
2540		3000 MBH input, 2880 GPH	25,000	3,875	28,875
2580		3750 MBH input, 3600 GPH	31,100	4,075	35,175

(Reprinted from Means Plumbing Cost Data 1991.)

Figure 23.18

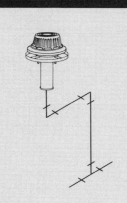

Design Assumptions: Vertical conductor size is based on a maximum rate of rainfall of 4″ per hour. To convert roof area to other rates multiply "Max S.F. Roof Area" shown by four and divide the result by desired local rate. The answer is the local roof area that may be handled by the indicated pipe diameter.

Basic cost is for roof drain, 10′ of vertical leader and 10′ of horizontal, plus connection to the main.

Pipe Dia.	Max. S.F. Roof Area	Gallons per Min.
2″	544	23
3″	1610	67
4″	3460	144
5″	6280	261
6″	10,200	424
8″	22,000	913

System Components	QUANTITY	UNIT	COST EACH		
			MAT.	INST.	TOTAL
SYSTEM 081-310-1880					
ROOF DRAIN, DWV PVC PIPE, 2″ DIAM., 10′ HIGH					
Drain, roof, main, PVC, dome type 2″ pipe size	1.000	Ea.	40.70	39.30	80
Clamp, roof drain, underdeck	1.000	Ea.	11	23	34
Pipe, Tee, PVC DWV, schedule 40, 2″ pipe size	1.000	Ea.	1.53	32.47	34
Pipe, PVC, DWV, schedule 40, 2″ diam.	20.000	L.F.	20.60	185.40	206
Pipe, elbow, PVC schedule 40, 2″ diam.	2.000	Ea.	2.14	39.86	42
TOTAL			75.97	320.03	396

8.1-310	Roof Drain Systems	COST PER SYSTEM		
		MAT.	INST.	TOTAL
1880	Roof drain, DWV PVC, 2″ diam., piping, 10′ high	76	320	396
1920	For each additional foot add	1.03	9.25	10.28
1960	3″ diam., 10′ high	100	385	485
2000	For each additional foot add	1.80	10.30	12.10
2040	4″ diam., 10′ high	130	430	560
2080	For each additional foot add	2.43	11.35	13.78
2120	5″ diam., 10′ high	350	460	810
2160	For each additional foot add	5.85	12.70	18.55
2200	6″ diam., 10′ high	400	550	950
2240	For each additional foot add	4.84	14	18.84
2280	8″ diam., 10′ high	685	865	1,550
2320	For each additional foot add	10.45	17.55	28
3940	C.I., soil, single hub, service wt., 2″ diam. piping, 10′ high	175	330	505
3980	For each additional foot add	2.23	8.65	10.88
4120	3″ diam., 10′ high	210	360	570
4160	For each additional foot add	3.11	9.10	12.21
4200	4″ diam., 10′ high	245	395	640
4240	For each additional foot add	4.04	9.95	13.99
4280	5″ diam., 10′ high	320	435	755
4320	For each additional foot add	5.45	11.20	16.65
4360	6″ diam., 10′ high	425	465	890
4400	For each additional foot add	6.75	11.65	18.40
4440	8″ diam., 10′ high	675	940	1,615
4480	For each additional foot add	10.70	19.30	30
6040	Steel galv. sch 40 threaded, 2″ diam. piping, 10′ high	220	320	540
6080	For each additional foot add	3.88	8.50	12.38

246

(Reprinted from Means Plumbing Cost Data 1991.)

Figure 23.19

Plumbing Estimate Worksheet

Fixture	Table Number	Quantity	Unit Cost	Total Costs	Calculations
Bathtubs		$	$		
Drinking Fount		$	$		
Kitchen Sink		$	$		
Laundry Sink		$	$		
Lavatory	BB.1-433 -1600	21	$ 565	$ 11,865	3 Floors x 7 Lavs.
Service Sink	BB.1-434 -4300	3	$ 1,135	$ 3,405	3 Floors x 1 Sink
Shower	—		$	$	
Urinal	BB.1-450 -2000	6	815	$ 4,890	3 Floors x 2 Urinals
Water Cooler	BB.1-460- -1840 -1920	6 3	715 1,300	4,290 3,900	x 2 Coolers 3 Floors x 1 & Cooler
Water Closet	BB.1-470- -2080	18	795	$ 14,310	3 Floors x W=3 m=2 H=1 } 6 Total
W/C Group		$	$		
Wash Fount. Group		$	$		
Lavatories Group		$	$		
Urinals Group		$	$		
Bathrooms		$	$		
		$	$		
Water Heater	BB.1-170-2060	1	$ 3,725	$ 3,725	.4 Gal. x 142 People x 3 Floors = 170 GPH
			$	$	
			$	$	
Roof Drains	BB.1-310- -4200	6	$ 640	$ 3,840	$\frac{18,900}{3,460}$ = 6 Drains 4"
Add + Length Roof Drains	BB.1-310- -4240	42x6=252	$ 13.99	$ 3,525	14' x 3 Floors + 10' Garage 52' Total - 10' for Basic
SUBTOTAL				$ 53,750	Comments
Water Control	C8.1-031	10 % ST.		$ 5,375	
Pipe & Fittings	C8.1-031	30% ST.		$ 16,125	% of Subtotal
Quality Complexity	C8.1-031	5 % ST.		$ 2,688	
Other		% ST.		$	
TOTAL				$ 77,938	

Figure 23.20

Table 8.1-031 Plumbing Approximations for Quick Estimating

Water Control

Water Meter; Backflow Preventer;
Shock Absorbers; Vacuum Breakers; } . . . . 10 to 15% of Fixtures
Mixer.

Pipe And Fittings: . . . . 30 to 60% of Fixtures

> **Note:** Lower percentage for compact buildings or larger buildings with plumbing in one area.
> Larger percentage for large buildings with plumbing spread out.
> In extreme cases pipe may be more than 100% of fixtures.
> Percentages **do not** include special purpose or process piping.

Plumbing Labor:

1 & 2 Story Residential . . . . Rough-in Labor = 80% of Materials
Apartment Buildings . . . . Rough-in Labor = 90 to 100% of Materials
Labor for handling and placing fixtures is approximately 25 to 30% of fixtures.

Quality/Complexity Multiplier (For all installations)

Economy installation, add . . . . 0 to 5%
Good quality, medium complexity, add . . . . 5 to 15%
Above average quality and complexity, add . . . . 15 to 25%

Table 8.1-032 Pipe Material Consideration

1. Malleable fittings should be used for gas service.
2. Malleable fittings are used where there are stresses/strains due to expansion and vibration.
3. Cast fittings may be broken as an aid to disassembling of heating lines frozen by long use, temperature and minerals.
4. Cast iron pipe is extensively used for underground and submerged service.
5. Type M (light wall) copper tubing is available in hard temper only and is used for low pressure and less severe applications than K and L.
6. Type L (medium wall) copper tubing, available hard or soft for interior service.
7. Type K (heavy wall) copper tubing, available in hard or soft temper for use where conditions are severe. For underground and interior service.
8. Hard drawn tubing requires fewer hangers or supports but should not be bent. Silver brazed fittings are recommended, however soft solder is normally used.
9. Type DWV (very light wall) copper designed for drainage, waste and vent and other non critical pressure services.

Table 8.1-033 Domestic/Imported Pipe and Fittings Cost

The prices shown in this publication for steel/cast iron pipe and steel, cast iron, malleable iron fittings are based on domestic production sold at the normal trade discounts. The above listed items of foreign manufacture may be available at prices of 1/3 to 1/2 those shown. Some imported items after minor machining or finishing operations are being sold as domestic to further complicate the system.

Caution: Most pipe prices in this book also include a coupling and pipe hangers which for the larger sizes can add significantly to the per foot cost and should be taken into account when comparing "book cost" with quoted supplier's cost.

Table 8.1-034 Piping to 10' High

When taking off pipe, it is important to identify the different material types and joining procedures, as well as distances between supports and components required for proper support.

During the takeoff, measure through all fittings. Do not subtract the lengths of the fittings, valves, or strainers, etc. This added length plus the final rounding of the totals will compensate for nipples and waste.

When rounding off totals always increase the actual amount to correspond with manufacturer's shipping lengths.

A. Both red brass and yellow brass pipe are normally furnished in 12' lengths, plain end. Division 151-251 includes in the linear foot costs two field threads and one coupling per 10' length. A carbon steel clevis type hanger assembly every 10' is also prorated into the linear foot costs, including both material and labor.

B. Cast iron soil pipe is furnished in either 5' or 10' lengths. For pricing purposes, Division 151-301 in *Means Plumbing Cost Data* features 10' lengths with a joint and a carbon steel clevis hanger assembly every 5' prorated into the per foot costs of both material and labor.

Three methods of joining are considered in Division 151-301, lead and oakum poured joints, or push-on gasket type joints for the bell and spigot pipe, and a joint clamp for the no-hub soil pipe. The labor and material costs for each of these individual joining procedures are also prorated into the linear costs per foot.

C. Copper tubing in Division 151-401 covers types K, L, M, and DWV which are furnished in 20' lengths. Means pricing data is based on a tubing cut each length and a coupling and two soft soldered joints every 10'. A carbon steel, clevis type hanger assembly every 10' is also prorated into the per foot costs. The prices for refrigeration tubing are for materials only. Labor for full lengths may be based on the type L labor but short cut measures in tight areas can increase the installation man-hours from 20 to 40%.

D. Corrosion-resistant piping in Division 151-451 does not lend itself to one particular standard of hanging or support assembly due to its diversity of application and placement. The several varieties of corrosion-resistant piping shown in this section, therefore, do not include any material or labor costs for hanger assemblies (See Division 151-901 for appropriate selection). Couplings have not been included in this section; the estimator should refer to Division 151-454 in *Means Plumbing Cost Data* for the compatible corrosion-resistant fittings.

E. Glass pipe, as shown in *Means Plumbing Cost Data* Division 151-501, is furnished in standard lengths either 5' or 10' long,

334

(Reprinted from Means Plumbing Cost Data 1991.)

Figure 23.21

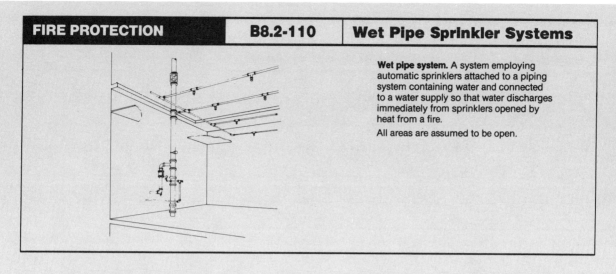

Wet pipe system. A system employing automatic sprinklers attached to a piping system containing water and connected to a water supply so that water discharges immediately from sprinklers opened by heat from a fire.

All areas are assumed to be open.

System Components	QUANTITY	UNIT	COST EACH		
			MAT.	INST.	TOTAL
SYSTEM 08.2-110-0580					
WET PIPE SPRINKLER, STEEL, BLACK, SCH. 40 PIPE					
LIGHT HAZARD, ONE FLOOR, 2000 S.F.					
Valve, gate, iron body, 125 lb, OS&Y, flanged, 4″ diam	1.000	Ea.	177.38	137.63	315.01
Valve, swing check, bronze, 125 lb, regrinding disc, 2-½″ pipe size	1.000	Ea.	61.88	28.13	90.01
Valve, angle, bronze, 150 lb, rising stem, threaded, 2″ diam	1.000	Ea.	82.50	22.50	105
*Alarm valve, 2-½″ pipe size	1.000	Ea.	416.63	142.13	558.76
Alarm, water motor, complete with gong	1.000	Ea.	89.10	60.90	150
Valve, swing check, w/balldrip Cl with brass trim 4″ pipe size	1.000	Ea.	69.71	140.29	210
Pipe, steel, black, schedule 40, 4″ diam	10.000	L.F.	55.88	116.63	172.51
*Flow control valve, trim & gauges, 4″ pipe size	1.000	Set	787.88	318.38	1,106.26
Fire alarm horn, electric	1.000	Ea.	23.10	33.15	56.25
Pipe, steel, black, schedule 40, threaded, cplg & hngr 10'OC, 2-½″ diam	20.000	L.F.	78.90	164.10	243
Pipe, steel, black, schedule 40, threaded, cplg & hngr 10'OC, 2″ diam	12.500	L.F.	30	80.16	110.16
Pipe, steel, black, schedule 40, threaded, cplg & hngr 10'OC, 1-¼″ diam	37.500	L.F.	57.38	173.25	230.63
Pipe steel, black, schedule 40, threaded cplg & hngr 10'OC, 1″ diam	112.000	L.F.	137.76	479.64	617.40
Pipe Tee, malleable iron black, 150 lb threaded, 4″ pipe size	2.000	Ea.	70.95	206.55	277.50
Pipe Tee, malleable iron black, 150 lb threaded, 2-½″ pipe size	2.000	Ea.	20.55	90.45	111
Pipe Tee, malleable iron black, 150 lb threaded, 2″ pipe size	1.000	Ea.	4.42	37.58	42
Pipe Tee, malleable iron black, 150 lb threaded, 1-¼″ pipe size	5.000	Ea.	12.23	145.28	157.51
Pipe Tee, malleable iron black, 150 lb threaded, 1″ pipe size	4.000	Ea.	6	114	120
Pipe 90° elbow, malleable iron black, 150 lb threaded, 1″ pipe size	6.000	Ea.	5.81	106.70	112.51
Sprinkler head, standard spray, brass 135°-286°F ½″ NPT, ⅜″ orifice	12.000	Ea.	43.56	181.44	225
Valve, gate, bronze, NRS, class 150, threaded, 1″ pipe size	1.000	Ea.	12.38	11.63	24.01
*Standpipe connection, wall, single, flush w/plug & chain 2-½″x2-½″	1.000	Ea.	53.63	85.13	138.76
TOTAL			2,297.63	2,875.65	5,173.28
COST PER S.F.			1.15	1.44	2.59

*Not included in systems under 2000 S.F.

8.2-110	**Wet Pipe Sprinkler Systems**	COST PER S.F.		
		MAT.	INST.	TOTAL
0520	Wet pipe sprinkler systems, steel, black, sch. 40 pipe			
0530	Light hazard, one floor, 500 S.F.	.74	1.43	2.17
0560	1000 S.F.	1.08	1.49	2.57
0580	2000 S.F.	1.15	1.44	2.59
0600	5000 S.F.	.57	1.03	1.60
0620	10,000 S.F.	.39	.87	1.26

267

(Reprinted from Means Plumbing Cost Data 1991.)

Figure 23.22a

8.2-110	Wet Pipe Sprinkler Systems	COST PER S.F.		
		MAT.	INST.	TOTAL
0640	50,000 S.F.	.31	.79	1.10
0660	Each additional floor, 500 S.F.	.41	1.19	1.60
0680	1000 S.F.	.36	1.08	1.44
0700	2000 S.F.	.33	.97	1.30
0720	5000 S.F.	.25	.84	1.09
0740	10,000 S.F.	.24	.78	1.02
0760	50,000 S.F.	.27	.74	1.01
1000	Ordinary hazard, one floor, 500 S.F.	.84	1.59	2.43
1020	1000 S.F.	1.06	1.44	2.50
1040	2000 S.F.	1.20	1.54	2.74
1060	5000 S.F.	.63	1.11	1.74
1080	10,000 S.F.	.49	1.15	1.64
1100	50,000 S.F.	.44	1.12	1.56
1140	Each additional floor, 500 S.F.	.52	1.34	1.86
1160	1000 S.F.	.35	1.06	1.41
1180	2000 S.F.	.38	1.07	1.45
1200	5000 S.F.	.39	1.02	1.41
1220	10,000 S.F.	.34	1.05	1.39
1240	50,000 S.F.	.33	.99	1.32
1500	Extra hazard, one floor, 500 S.F.	2.86	2.67	5.53
1520	1000 S.F.	1.74	2.14	3.88
1540	2000 S.F.	1.27	1.99	3.26
1560	5000 S.F.	.88	1.79	2.67
1580	10,000 S.F.	.78	1.63	2.41
1600	50,000 S.F.	.81	1.54	2.35
1660	Each additional floor, 500 S.F.	.59	1.60	2.19
1680	1000 S.F.	.58	1.52	2.10
1700	2000 S.F.	.50	1.52	2.02
1720	5000 S.F.	.43	1.37	1.80
1740	10,000 S.F.	.46	1.22	1.68
1760	50,000 S.F.	.45	1.13	1.58
2020	Grooved steel, black sch. 40 pipe, light hazard, one floor, 2000 S.F.	1.33	1.25	2.58
2060	10,000 S.F.	.58	.78	1.36
2100	Each additional floor, 2000 S.F.	.51	.78	1.29
2150	10,000 S.F.	.35	.66	1.01
2200	Ordinary hazard, one floor, 2000 S.F.	1.37	1.33	2.70
2250	10,000 S.F.	.61	.97	1.58
2300	Each additional floor, 2000 S.F.	.55	.86	1.41
2350	10,000 S.F.	.46	.88	1.34
2400	Extra hazard, one floor, 2000 S.F.	1.52	1.70	3.22
2450	10,000 S.F.	.89	1.25	2.14
2500	Each additional floor, 2000 S.F.	.76	1.24	2
2550	10,000 S.F.	.67	1.10	1.77
3050	Grooved steel black sch. 10 pipe, light hazard, one floor, 2000 S.F.	1.31	1.24	2.55
3100	10,000 S.F.	.48	.74	1.22
3150	Each additional floor, 2000 S.F.	.49	.76	1.25
3200	10,000 S.F.	.33	.64	.97
3250	Ordinary hazard, one floor, 2000 S.F.	1.34	1.32	2.66
3300	10,000 S.F.	.58	.95	1.53
3350	Each additional floor, 2000 S.F.	.52	.85	1.37
3400	10,000 S.F.	.43	.86	1.29
3450	Extra hazard, one floor, 2000 S.F.	1.50	1.69	3.19
3500	10,000 S.F.	.83	1.23	2.06
3550	Each additional floor, 2000 S.F.	.74	1.23	1.97
3600	10,000 S.F.	.64	1.09	1.73
4050	Copper tubing, type M, light hazard, one floor, 2000 S.F.	1.28	1.23	2.51
4100	10,000 S.F.	.54	.75	1.29
4150	Each additional floor, 2000 S.F.	.48	.78	1.26

268

(Reprinted from Means Plumbing Cost Data 1991.)

Figure 23.22b

B8.2-110 | Wet Pipe Sprinkler Systems

8.2-110	Wet Pipe Sprinkler Systems	COST PER S.F.		
		MAT.	INST.	TOTAL
4200	10,000 S.F.	.39	.66	1.05
4250	Ordinary hazard, one floor, 2000 S.F.	1.35	1.39	2.74
4300	10,000 S.F.	.64	.88	1.52
4350	Each additional floor, 2000 S.F.	.55	.87	1.42
4400	10,000 S.F.	.48	.78	1.26
4450	Extra hazard, one floor, 2000 S.F.	1.51	1.72	3.23
4500	10,000 S.F.	1.16	1.34	2.50
4550	Each additional floor, 2000 S.F.	.75	1.26	2.01
4600	10,000 S.F.	.77	1.19	1.96
5050	Copper tubing, type M, T-drill system, light hazard, one floor			
5060	2000 S.F.	1.29	1.15	2.44
5100	10,000 S.F.	.51	.62	1.13
5150	Each additional floor, 2000 S.F.	.48	.70	1.18
5200	10,000 S.F.	.36	.53	.89
5250	Ordinary hazard, one floor, 2000 S.F.	1.30	1.17	2.47
5300	10,000 S.F.	.62	.79	1.41
5350	Each additional floor, 2000 S.F.	.49	.70	1.19
5400	10,000 S.F.	.47	.70	1.17
5450	Extra hazard, one floor, 2000 S.F.	1.37	1.40	2.77
5500	10,000 S.F.	.97	.99	1.96
5550	Each additional floor, 2000 S.F.	.64	.95	1.59
5600	10,000 S.F.	.58	.83	1.41

269

(Reprinted from Means Plumbing Cost Data 1991.)

Figure 23.22c

Mechanical Assemblies

Fire Protection Estimate Worksheet

Type of Component	Hazard or Class	Table Number	Quantity	Unit Cost	Total Costs	Comments
From Table		C8.2 - 102		$	$	
Sprinkler	Light	B8.2-110-0620	18,900	$ 1.26	$ 23,814	Office Floors
		-0740	37,800	$ 1.02	$ 38,556	Office Floors
	Ordinary	B8.2-120-1080	18,900	$ 1.86	$ 35,154	Garage
Standpipe	Class III	B8.2-320- -1540	2	$ 2,250	$ 4,500	2 Standpipes 14'x3 = 42 10'x1 = 10 Roof 2
		- 1560	88	$ 70.00	$ 6,160	54' Total
				$	$	-10'Org. Floor= 44'
Fire Suppress				$	$	
				$	$	
				$	$	
Cabinets & Components		B8.2 - 390- -8400	8	695	$ 5,560	4 Floors × 2 = 8
				$	$	
				$	$	
				$	$	Note: Dry standpipes are used, assuming
				$	$	those in each stairwell at ends
				$	$	of building are subject to freezing
				$	$	
				$	$	
				$	$	
				$	$	
				$	$	
				$	$	
				$	$	
SUBTOTAL					$ 113,744	
Quality Complexity		C8.2- 104	5 %		$ 5,687	
TOTAL					$ 119,431	

Figure 23.23

252

Figures 23.25a and 23.25b, from *Means Plumbing Cost Data*, illustrate a valve standpipe and list its components, material, and labor costs for the three classifications. Line number 8.2-320-1540 (shown in Figure 23.25b) indicates a cost of $2,250 each for the first floor standpipes, including all the system components listed. Note that these components include 20' of pipe, a 10' riser and a 10' allowance for horizontal feed.

For the three additional floors at 14' each, multiplied by two, 84 additional feet of pipe is required 14' above roof. Line number 8.2-320-1560 (Figure 23.25b) indicates $700 for one 10-foot floor, or $70.00 per foot. The 88 additional feet totals $6,160. This additional footage includes a valved hose connection at each floor. The totals for two standpipes should be entered on the fire protection worksheet shown in Figure 23.23.

Six firehose and cabinet assemblies are priced from *Means Plumbing Cost Data*, (shown in Figure 23.26), line number 8.2-390-8400, and also entered on the worksheet.

The sprinkler and standpipe costs are totaled and a quality complexity percentage is added using the parameters from table number 8.2-303, shown in Figure 23.24.

These systems may be modified to match conditions other than those shown in this example. For instance, if these two standpipes were fed from one common siamese inlet connection with interconnecting piping, the system cost could be modified by deducting two 4" x 2-1/2" x 2-1/2" standpipe connections (found in the unit price section). One 6" x 2-1/2" x 2-1/2" standpipe connection must then be added with the appropriate markup percentages. Finally, 6" interconnecting piping for the fire department connection to the two standpipe risers must be added. The pricing information for the additional piping and connections can be found in the unit price section of *Means Plumbing Cost Data*.

Heating and Cooling

The first step in preparing a systems estimate for heating a building is the same as for designing the heating system; determining the heat loss of the structure. In the design stage all of the necessary parameters are known: the types of materials and their coefficient of heat transmission for the building wall areas, and window, door, roof, floor, and slab.

Occupancy, geographic location, building use, and solar orientation all play a part in determining the heating or cooling load for a building. This can be a precise and time-consuming procedure, as these loads are further refined by individual rooms and areas. In the preliminary stages of design most of these details are not known, nor is there sufficient time to complete an accurate heat loss study. Therefore, tables have been prepared to shorten the heat loss determination process. This can be done by combining many of the variables into a single numerical value to be multiplied by the building volume. A heat loss table can be found in *Means Mechanical Cost Data* and is reproduced in Figure 23.27.

This table does not take into consideration today's well insulated building envelope and results in a higher heating load estimation than the actual final design will indicate. However, it provides a quick and easy estimate adequate for the preliminary design stage.

The cubic content of the proposed three-story office building is 793,800 C.F. The geographic location will warrant an outdoor design temperature of 0°F,

Table 8.2-301   Standpipe Systems

The basis for standpipe system design is National Fire Protection Association NFPA 14, however, the authority having jurisdiction should be consulted for special conditions, local requirements and approval.

Standpipe systems, properly designed and maintained, are an effective and valuable time saving aid for extinguishing fires, especially in the upper stories of tall buildings, the interior of large commercial or industrial malls, or other areas where construction features or access make the laying of temporary hose lines time consuming and/or hazardous. Standpipes are frequently installed with automatic sprinkler systems for maximum protection.

There are three general classes of service for standpipe systems:
Class I for use by fire departments and personnel with special training for heavy streams (2-1/2" hose connections).
Class II for use by building occupants until the arrival of the fire department (1-1/2" hose connector with hose).

Class III for use by either fire departments and trained personnel or by the building occupants (both 2-1/2" and 1-1/2" hose connections or one 2-1/2" hose valve with an easily removable 2-1/2" by 1-1/2" adapter).

Standpipe systems are also classified by the way water is supplied to the system. The four basic types are:
Type 1: Wet standpipe system having supply valve open and water pressure maintained at all times.
Type 2: Standpipe system so arranged through the use of approved devices as to admit water to the system automatically by opening a hose valve.
Type 3: Standpipe system arranged to admit water to the system through manual operation of approved remote control devices located at each hose station.
Type 4: Dry standpipe having no permanent water supply.

Table 8.2-302   NFPA 14 Basic Standpipe Design

Class	Design-Use	Pipe Size Minimums	Water Supply Minimums
Class I	2½" hose connection on each floor All areas within 30' of nozzle with 100' of hose Fire Department Trained Personnel	Height to 100', 4" dia. Heights above 100', 6" dia. (275' max. except with pressure regulators 400' max.)	For each standpipe riser 500 GPM flow For common supply pipe allow 500 GPM for first standpipe plus 250 GPM for each additional standpipe (2500 GPM max. total) 30 min. duration 65 PSI at 500 GPM
Class II	1½" hose connection with hose on each floor All areas within 30' of nozzle with 100' of hose. Occupant personnel	Height to 50', 2" dia. Height above 50', 2½" dia.	For each standpipe riser 100 GPM flow For multiple riser common supply pipe 100 GPM 30 min. duration, 65 PSI at 100 GPM
Class III	Both of above. Class I valved connections will meet Class III with addition of 2½" by 1½" adapter and 1½" hose.	Same as Class I	Same as Class I

Combined Systems

Combined systems are systems where the risers supply both automatic sprinklers and 2-1/2" hose connection outlets for fire department use. In such a system the sprinkler spacing pattern shall be in accordance with NFPA 13 while the risers and supply piping will be sized in accordance with NFPA 14. When the building is completely sprinklered the risers may be sized by hydraulic calculation. The minimum size riser for buildings not completely sprinklered is 6".

The minimum water supply of a completely sprinklered, light hazard, high-rise occupancy building will be 500 GPM while the supply required for other types of completely sprinklered high-rise buildings is 1000 GPM.

General System Requirements

1. Approved valves will be provided at the riser for controlling branch lines to hose outlets.
2. A hose valve will be provided at each outlet for attachment of hose.
3. Where pressure at any standpipe outlet exceeds 100 PSI a pressure reducer must be installed to limit the pressure to 100 PSI. Note that the pressure head due to gravity in 100' of riser is 43.4 PSI. This must be overcome by city pressure, fire pumps, or gravity tanks to provide adequate pressure at the top of the riser.
4. Each hose valve on a wet system having linen hose shall have an automatic drip connection to prevent valve leakage from entering the hose.
5. Each riser will have a valve to isolate it from the rest of the system.
6. One or more fire department connections as an auxiliary supply shall be provided for each Class I or Class III standpipe system. In buildings having two or more zones, a connection will be provided for each zone.
7. There will be no shutoff valve in the fire department connection, but a check valve will be located in the line before it joins the system.
8. All hose connections street side will be identified on a cast plate or fitting as to purpose.

Table 8.2-303 Quality/Complexity Adjustment for Sprinkler/Standpipe Systems

Economy installation, add . . . . 0 to 5%
Good quality, medium complexity, add . . . . 5 to 15%
Above average quality and complexity, add . . . . 15 to 25%

(Reprinted from Means Plumbing Cost Data 1991.)

Figure 23.24

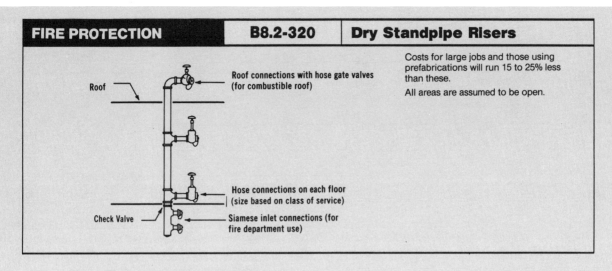

Costs for large jobs and those using prefabrications will run 15 to 25% less than these.

All areas are assumed to be open.

System Components	QUANTITY	UNIT	COST PER FLOOR		
			MAT.	INST.	TOTAL
SYSTEM 08.2-320-0540					
DRY STANDPIPE RISER, CLASS I, PIPE, STEEL, BLACK, SCH 40, 10′ HEIGHT					
4″ DIAMETER PIPE, ONE FLOOR					
Pipe, steel, black, schedule 40, threaded, 4″ diam	20.000	L.F.	214.80	305.20	520
Pipe, Tee, malleable iron, black, 150 lb threaded, 4″ pipe size	2.000	Ea.	94.60	275.40	370
Pipe, 90° elbow, malleable iron, black, 150 lb threaded 4″ pipe size	1.000	Ea.	31.90	93.10	125
Pipe, nipple, steel, black, schedule 40, 2-½″ pipe size x 3″ long	2.000	Ea.	13.42	68.58	82
Fire valve gate NRS 300 lb, brass w/handwheel, 2-½″ pipe size	1.000	Ea.	85.80	44.20	130
Fire valve, pressure restricting, adj, rgh brs, 2-½″ pipe size	1.000	Ea.	104.64	45.36	150
Standpipe conn wall dble flush brs w/plugs & chains 2-½″x2-½″x4″	1.000	Ea.	262.90	112.10	375
Valve swing check w/ball drip Cl w/brs ftngs, 4″pipe size	1.000	Ea.	92.95	187.05	280
Roof manifold, fire, w/valves & caps, horiz/vert brs 2-½″x2-½″x4″	1.000	Ea.	88	117	205
TOTAL			989.01	1,247.99	2,237

8.2-320		Dry Standpipe Risers, Class I	COST PER FLOOR		
			MAT.	INST.	TOTAL
0530		Dry standpipe riser, Class I, steel black sch. 40, 10′ height			
0540		4″ diameter pipe, one floor	990	1,250	2,240
0560	B8.2 -.390	Additional floors	360	450	810
0580		6″ diameter pipe, one floor	1,875	1,900	3,775
0600	C8.2 -.301	Additional floors	605	645	1,250
0620		8″ diameter pipe, one floor	2,675	2,325	5,000
0640	C8.2 -.302	Additional floors	780	790	1,570

8.2-320	Dry Standpipe Risers, Class II	COST PER FLOOR		
		MAT.	INST.	TOTAL
1030	Dry standpipe risers, Class II, steel black sch. 40, 10′ height			
1040	2″ diameter pipe, one floor	500	575	1,075
1060	Additional floor	190	190	380
1080	2-½″ diameter pipe, one floor	580	675	1,255
1100	Additional floors	215	220	435

Figure 23.25a

FIRE PROTECTION	B8.2-320	Dry Standpipe Risers		
8.2-320	**Dry Standpipe Risers, Class III**	COST PER FLOOR		
		MAT.	INST.	TOTAL
1530	Dry standpipe risers, Class III, steel black sch. 40, 10′ height			
1540	4″ diameter pipe, one floor	1,000	1,250	2,250
1560	Additional floors	295	405	700
1580	6″ diameter pipe, one floor	1,900	1,900	3,800
1600	Additional floors	625	645	1,270
1620	8″ diameter pipe, one floor	2,700	2,325	5,025
1640	Additional floor	805	790	1,595

285

(Reprinted from Means Plumbing Cost Data 1991.)

Figure 23.25b

8.2-390	Standpipe Equipment	COST EACH		
		MAT.	INST.	TOTAL
0100	Adapters, reducing, 1 piece, FxM, hexagon, cast brass, 2-½" x 1-½"	23		23
0200	Pin lug, 1-½" x 1"	8.95		8.95
0250	3" x 2-½"	30		30
0300	For polished chrome, add 75% mat.			
0400	Cabinets, D.S. glass in door, recessed, steel box, not equipped			
0500	Single extinguisher, steel door & frame	58	72	130
0550	Stainless steel door & frame	205	72	277
0600	Valve, 2-½" angle, steel door & frame	64	46	110
0650	Aluminum door & frame	110	48	158
0700	Stainless steel door & frame	210	47	257
0750	Hose rack assy, 2-½" x 1-½" valve & 100' hose, steel door & frame	130	92	222
0800	Aluminum door & frame	165	96	261
0850	Stainless steel door & frame	370	94	464
0900	Hose rack assy,& extinguisher,2-½"x1-½" valve & hose,steel door & frame	135	115	250
0950	Aluminum	170	115	285
1000	Stainless steel	390	115	505
1550	Compressor, air, dry pipe system, automatic, 200 gal., ⅓ H.P.	615	245	860
1600	520 gal., 1 H.P.	565	235	800
1650	Alarm, electric pressure switch (circuit closer)	62	12.40	74.40
2500	Couplings, hose, rocker lug, cast brass, 1-½"	18.70		18.70
2550	2-½"	39		39
3000	Escutcheon plate, for angle valves, polished brass, 1-½"	9.35		9.35
3050	2-½"	16.50		16.50
3500	Fire pump, electric, w/controller, fittings, relief valve			
3550	4" pump, 30 H.P., 500 G.P.M.	12,700	1,750	14,450
3600	5" pump, 40 H.P., 1000 G.P.M.	17,300	1,975	19,275
3650	5" pump, 100 H.P., 1000 G.P.M.	19,500	2,275	21,775
3700	For jockey pump system, add	2,150	295	2,445
5000	Hose, per linear foot, synthetic jacket, lined,			
5100	300 lb. test, 1-½" diameter	1.32		1.32
5150	2-½" diameter	2		2
5200	500 lb. test, 1-½" diameter	1.40		1.40
5250	2-½" diameter	2.40		2.40
5500	Nozzle, plain stream, polished brass, 1-½" x 10"	21		21
5550	2-½" x 15" x ¹³⁄₁₆" or 1-½"	64		64
5600	Heavy duty combination adjustable fog and straight stream w/handle 1-½"	225		225
5650	2-½" direct connection	290		290
6000	Rack, for 1-½" diameter hose 100 ft. long, steel	29	28	57
6050	Brass	51	28	79
6500	Reel, steel, for 50 ft. long 1-½" diameter hose	66	39	105
6550	For 75 ft. long 2-½" diameter hose	77	43	120
7050	Siamese, w/plugs & chains, polished brass, sidewalk, 4" x 2-½" x 2-½"	240	230	470
7100	6" x 2-½" x 2-½"	390	285	675
7200	Wall type, flush, 4" x 2-½" x 2-½"	265	110	375
7250	6" x 2-½" x 2-½"	340	125	465
7300	Projecting, 4" x 2-½" x 2-½"	225	115	340
7350	6" x 2-½" x 2-½"	365	120	485
7400	For chrome plate, add 15% mat.			
8000	Valves, angle, wheel handle, 300 Lb., rough brass, 1-½"	26	26	52
8050	2-½"	59	46	105
8100	Combination pressure restricting, 1-½"	36	27	63
8150	2-½"	81	44	125
8200	Pressure restricting, adjustable, satin brass, 1-½"	59	27	86
8250	2-½"	105	45	150
8300	Hydrolator, vent and drain, rough brass, 1-½"	27	27	54
8350	2-½"	79	46	125
8400	Cabinet assy, incls. 2-½" valve, adapter, rack, hose, nozzle & hydrolator	480	215	695

286

(Reprinted from Means Plumbing Cost Data 1991.)

Figure 23.26

and offices need to be heated to 70°. With this information the table in Figure 23.27 can be used to estimate the heat loss.

The heat loss factor for a three-story building to be heated to 70° when the outside temperature is 0° is 4.3. The 4.3 factor is multiplied by the cubic foot area of the building (793,800 C.F.) denoting a heat loss of 3,413,340 British Thermal Units (Btu). Working with figures in the millions can be difficult, therefore, a method to reduce the Btu calculation into thousands has been designed (MBH). Thus, the heat loss for the three-story office building is 3,413 MBH.

Many factors influence the selection of the type of HVAC system for a building. Some of the considerations are initial cost, operating costs, maintenance, available space, fuel cost/availability, geographic location, proposed building use, and quality of construction. In addition to the above, local codes influence the number of air changes per hour for ventilation. At this stage of a preliminary estimate, many of the above factors may be known, in this case, a more precise selection of the systems can be made. For this estimating exercise, however, it will be assumed that building use (office), geographic location (Boston area), and quality of construction (above average) are the only known factors.

The budget estimate will be based on a heating and cooling system, with exhaust and makeup air ventilation. This includes a packaged chiller for cooling and a packaged boiler for heating utilizing one piping system (dual-temperature) feeding air-handling units with ductwork distribution. This system provides either heating or cooling of the entire building. It cannot heat one area while cooling another.

Figure 23.28 is a page from the unit price section of *Means Mechanical Cost Data*, showing costs for cast-iron, jacketed, hot water boilers. A boiler with a minimum output of 3,413 MBH is required. The smallest boiler that will meet this requirement is found on line 155-115-3400. It has an output of 3808 MBH. However, because the heat loss factor is inclined to be conservative, rather than go up 400 MBH to the large boiler, go down 150 MBH to the next smaller boiler of 3264 MBH shown on line 155-115-3380. This unit price cost is entered on the heating and air-conditioning worksheet (shown in Figure 23.29). Unit price costs for the heating boiler are used because a price for the boiler only is needed. The related components required for a complete system will be shared with the cooling system that follows to complete the price for heating and cooling.

The cooling load is determined in a manner similar to the heating load except that it is based on the floor area rather than on building volume. Tables in *Means Mechanical Cost Data* list the cost per square foot for various types of occupancies for several kinds of air-conditioning systems. One such table is shown in Figures 23.30a and 23.30b. Look up the occupancy, in this case for offices, and multiply the cost per square foot by the floor area (tabulated as closely as possible based on the gross floor area of the proposed building). In this example, line 8.4-120-4040 from Figure 23.30b is a cost for 60,000 S.F. offices. This category is chosen rather than the value for 40,000 S.F. offices (or interpolating between them) because the higher number of $7.29 per square foot allows a desirable conservative margin.

If the type of occupancy desired is not priced out in the systems tables, go to Figure 23.31. The lowest requirement for air-conditioning on this chart is for apartment corridors with 550 S.F. per ton; the highest is for bars and taverns at 90 S.F. per ton of air-conditioning. The occupancies priced out are spread

Table 8.3-011 Factor for Determining Heat Loss for Various Types of Buildings

General: While the most accurate estimates of heating requirements would naturally be based on detailed information about the building being considered, it is possible to arrive at a reasonable approximation using the following procedure:

1. Calculate the cubic volume of the room or building.
2. Select the appropriate factor from Table 8.3-011. Note that the factors apply only to inside temperatures listed in the first column and to 0°F outside temperature.

3. If the building has bad north and west exposures, multiply the heat loss factor by 1.1
4. If the outside design temperature is other than 0°F, multiply the factor from Table 8.3-011 by the factor from Table 8.3-012
5. Multiply the cubic volume by the factor selected from Table 8.3-011. This will give the estimated BTUH heat loss which must be made up to maintain inside temperature.

Building Type	Conditions	Qualifications	Loss Factor*
Factories & Industrial Plants General Office Areas 70°F	One Story	Skylight in Roof No Skylight in Roof	6.2 5.7
	Multiple Story	Two Story Three Story Four Story Five Story Six Story	4.6 4.3 4.1 3.9 3.6
	All Walls Exposed	Flat Roof Heated Space Above	6.9 5.2
	One Long Warm Common Wall	Flat Roof Heated Space Above	6.3 4.7
	Warm Common Walls on Both Long Sides	Flat Roof Heated Space Above	5.8 4.1
Warehouses 60°F	All Walls Exposed	Skylights in Roof No Skylights in Roof Heated Space Above	5.5 5.1 4.0
	One Long Warm Common Wall	Skylight in Roof No Skylight in Roof Heated Space Above	5.0 4.9 3.4
	Warm Commom Walls on Both Long Sides	Skylight in Roof No Skylight in Roof Heated Space Above	4.7 4.4 3.0

*Note: This table tends to be conservative particularly for new buildings designed for minimum energy consumption.

Table 8.3-012 Outside Design Temperature Correction Factor (for Degrees Fahrenheit)

Outside Design Temperature	50	40	30	20	10	0	-10	-20	-30
Correction Factor	0.29	0.43	0.57	0.72	0.86	1.00	1.14	1.28	1.43

(Reprinted from Means Mechanical Cost Data 1991.)

Figure 23.27

155 100 | Boilers

			CREW	DAILY OUTPUT	MAN-HOURS	UNIT	BARE COSTS				TOTAL INCL O&P	
							MAT.	LABOR	EQUIP.	TOTAL		
110	2820	3600 KW, 12,283 MBH	Q-21	.16	200	Ea.	51,850	4,825		56,675	64,000	110
115	0010	**BOILERS, GAS FIRED** Natural or propane, standard controls										115
	1000	Cast iron, with insulated jacket										
	2000	Steam, gross output, 81 MBH	Q-7	1.40	22.860	Ea.	1,055	555		1,610	2,000	
	2020	102 MBH		1.30	24.620		1,220	600		1,820	2,225	
	2040	122 MBH		1	32		1,335	780		2,115	2,625	
	2060	163 MBH		.90	35.560		1,615	865		2,480	3,075	
	2080	203 MBH		.90	35.560		1,840	865		2,705	3,325	
	2100	240 MBH		.85	37.650		2,065	915		2,980	3,650	
	2120	280 MBH		.80	40		2,420	975		3,395	4,125	
	2140	320 MBH		.70	45.710		2,640	1,125		3,765	4,575	
	2160	360 MBH		.65	49.230		2,980	1,200		4,180	5,075	
	2180	400 MBH		.60	53.330		3,180	1,300		4,480	5,425	
	2200	440 MBH		.55	58.180		3,400	1,425		4,825	5,850	
	2220	544 MBH		.50	64		4,510	1,550		6,060	7,275	
	2240	765 MBH		.45	71.110		5,740	1,725		7,465	8,900	
	2260	892 MBH		.40	80		6,970	1,950		8,920	10,600	
	2280	1275 MBH		.36	88.890		8,600	2,175		10,775	12,700	
	2300	1530 MBH		.31	103		10,550	2,525		13,075	15,400	
	2320	1875 MBH		.28	114		11,900	2,775		14,675	17,200	
	2340	2170 MBH		.26	123		13,500	3,000		16,500	19,300	
	2360	2675 MBH		.24	133		15,600	3,250		18,850	22,000	
	2380	3060 MBH		.23	139		16,900	3,400		20,300	23,600	
	2400	3570 MBH		.18	178		18,950	4,325		23,275	27,300	
	2420	4207 MBH		.16	200		20,950	4,875		25,825	30,300	
	2440	4720 MBH		.14	229		23,050	5,575		28,625	33,700	
	2460	5660 MBH		.13	246		36,450	6,000		42,450	49,000	
	2480	6100 MBH		.12	267		41,650	6,500		48,150	55,500	
	2500	6390 MBH		.10	320		44,000	7,800		51,800	60,000	
	2520	6680 MBH		.09	356		45,150	8,650		53,800	62,500	
	2540	6970 MBH		.08	400		46,750	9,750		56,500	66,000	
	3000	Hot water, gross output, 80 MBH		1.46	21.920		935	535		1,470	1,825	
	3020	100 MBH		1.35	23.700		1,100	575		1,675	2,075	
	3040	122 MBH		1.10	29.090		1,215	710		1,925	2,400	
	3060	163 MBH		1	32		1,495	780		2,275	2,800	
	3080	203 MBH		1	32		1,720	780		2,500	3,050	
	3100	240 MBH		.95	33.680		1,945	820		2,765	3,375	
	3120	280 MBH		.90	35.560		2,245	865		3,110	3,750	
	3140	320 MBH		.80	40		2,465	975		3,440	4,175	
	3160	360 MBH		.75	42.670		2,805	1,050		3,855	4,625	
	3180	400 MBH		.70	45.710		3,000	1,125		4,125	4,950	
	3200	440 MBH		.65	49.230		3,220	1,200		4,420	5,325	
	3220	544 MBH		.60	53.330		4,310	1,300		5,610	6,675	
	3240	765 MBH		.55	58.180		5,540	1,425		6,965	8,200	
	3260	1088 MBH		.50	64		7,410	1,550		8,960	10,500	
	3280	1275 MBH		.46	69.570		8,400	1,700		10,100	11,800	
	3300	1530 MBH		.42	76.190		10,340	1,850		12,190	14,100	
	3320	2000 MBH		.38	84.210		12,920	2,050		14,970	17,300	
	3340	2312 MBH		.36	88.890		14,860	2,175		17,035	19,600	
	3360	2856 MBH		.33	96.970		16,900	2,350		19,250	22,100	
	3380	3264 MBH		.30	107		18,000	2,600		20,600	23,700	
	3400	3808 MBH		.26	123		19,900	3,000		22,900	26,400	
	3420	4488 MBH		.22	145		22,100	3,550		25,650	29,600	
	3440	4720 MBH		.18	178		32,150	4,325		36,475	41,800	
	3460	5520 MBH		.14	229		36,450	5,575		42,025	48,400	
	3480	6100 MBH		.12	267		41,650	6,500		48,150	55,500	
	3500	6390 MBH		.10	320		44,000	7,800		51,800	60,000	

151

(Reprinted from Means Mechanical Cost Data 1991.)

Figure 23.28

Heating & Air Conditioning Estimate Worksheet

Equipment	Type	Table Number	Quantity	Unit Cost	Total Costs	Comments
Heat Source	Gas	15.5- 115 -3380	1	$ 23,700	$ 23,700	Volume 18,900 x 14 = 264,600 x 3 — 793,800
				$	$	
				$	$	
				$	$	x 4.3 — 3,413,340
						BTUH
Pipe	with A/c			$	$	
Duct	with A/c			$	$	
Terminals	with A/c			$	$	
Other				$	$	
				$	$	
				$	$	
				$	$	
Cold Source		B8.4 -120 -4040	56,700	$ 7.29	$ 413,343	18,900 x 3 = 56,700 SF — By taking the higher number rather than interpolating gives an allowance for connecting Heating Pipe.
				$	$	
				$	$	
				$	$	
Pipe	included			$	$	
Duct	included			$	$	See mechanical book Assembly B 84-120
Terminals	included			$	$	for discussion of terminals vs. ducts
Other				$	$	
SUBTOTAL					$ 437,043	
Quality Complexity		C8.4 - 008	10 %		$ 43,704	
TOTAL					$ 480,747	

Figure 23.29

between these two extremes. Pick the one desired that has been priced as an assembly. The result is entered on the heating and cooling worksheet shown in Figure 23.29.

The graphics and components listed for this chilled water system do not indicate a ductwork distribution system, but are based on many individual fan-coil units. Using fewer units per floor and increasing their size or capacity, and relying on ductwork distribution would be approximately the same cost per square foot. The system table 8.4-120 does not include any provision for heating. As discussed earlier, the heating boiler selected from the unit price section must be added to the worksheet to round out a complete HVAC system. This price can then be compared to the prices for other systems from *Means Mechanical Cost Data* to select the most cost effective system.

The Estimate Summary

To summarize the mechanical systems prices from separate subcontracting firms, or from one total mechanical contractor, add the three subtotals shown in Figure 23.32 together and round off to a manageable figure. In the event that the owner or designer has not informed the estimator of the type or quality of the system to be priced, the estimator should, at this time, qualify his proposal by stating the quality and the type of systems being priced.

These rounded off prices all include subcontractors' overhead and profit. At this time, a location factor adjustment, as discussed in Chapter 22, "Using *Means Mechanical Cost Data* and *Means Plumbing Cost Data*" (Figure 22.12), should be added.

Any sales taxes and other adjustments (extra overhead, contingencies, etc.) that are required must also be applied. When these factors have been included, a budget price for the mechanical work is complete.

General: Water cooled chillers are available in the same sizes as air cooled units. They are also available in larger capacities.

Design Assumptions: The chilled water systems with water cooled condenser, include reciprocating hermetic compressors, water cooling tower, pumps, piping and expansion tanks and are based on a two pipe system. Chilled water piping is insulated. No ducts are included and fan-coil units are cooling only. Area distribution is through use of multiple fan coil units. Fewer but larger fan coil units with duct distribution would be approximately the same S.F. Cost (See C8.4-001 for methods of adding heat). Water treatment and balancing are not included.

System Components	QUANTITY	UNIT	COST EACH		
			MAT.	INST.	TOTAL
SYSTEM 08.4-120-1320					
PACKAGED CHILLER, WATER COOLED, WITH FAN COIL UNIT					
APARTMENT CORRIDORS, 4,000 S.F., 7.33 TON					
Fan coil air conditioner unit, cabinet mounted & filters, chilled water	2.000	Ea.	3,399.43	326.14	3,725.57
Water chiller, reciprocating, water cooled, 1 compressor semihermetic	1.000	Ea.	7,174.11	1,848.49	9,022.60
Cooling tower, draw thru single flow, belt drive	1.000	Ea.	548.28	67.44	615.72
Cooling tower pumps & piping	1.000	System	274.14	165.66	439.80
Chilled water unit coil connections	2.000	Ea.	768.66	1,381.34	2,150
Chilled water distribution piping	520.000	L.F.	4,524	11,076	15,600
TOTAL			16,688.62	14,865.07	31,553.69
COST PER S.F.			4.17	3.72	7.89

*Cooling requirements would lead to choosing multiple chillers

8.4-120	Chilled Water, Cooling Tower Systems	COST PER S.F.		
		MAT.	INST.	TOTAL
1300	Packaged chiller, water cooled, with fan coil unit			
1320	Apartment corridors, 4,000 S.F., 7.33 ton	4.17	3.72	7.89
1360	6,000 S.F., 11.00 ton	3.51	3.15	6.66
1400	10,000 S.F., 18.33 ton	2.70	2.38	5.08
1440	20,000 S.F., 26.66 ton	2.13	1.78	3.91
1480	40,000 S.F., 73.33 ton	2.66	1.88	4.54
1520	60,000 S.F., 110.00 ton	2.58	1.94	4.52
1600	Banks and libraries, 4,000 S.F., 16.66 ton	5.90	4.15	10.05
1640	6,000 S.F., 25.00 ton	5.05	3.62	8.67
1680	10,000 S.F., 41.66 ton	4.22	2.74	6.96
1720	20,000 S.F., 83.33 ton	4.92	2.73	7.65
1760	40,000 S.F., 166.66 ton	4.79	3.29	8.08
1800	60,000 S.F., 250.00 ton	4.69	3.57	8.26
1880	Bars and taverns, 4,000 S.F., 44.33 ton	10.25	5.35	15.60
1920	6,000 S.F., 66.50 ton	10.70	5.55	16.25
1960	10,000 S.F., 110.83 ton	10.75	4.76	15.51
2000	20,000 S.F., 221.66 ton	10.05	5.15	15.20
2040	40,000 S.F., 440 ton*			
2080	60,000 S.F., 660 ton*			
2160	Bowling alleys, 4,000 S.F., 22.66 ton	7.05	4.53	11.58
2200	6,000 S.F., 34.00 ton	6	4.01	10.01
2240	10,000 S.F., 56.66 ton	5.30	3.01	8.31

324

(Reprinted from Means Mechanical Cost Data 1991.)

Figure 23.30a

8.4-120	Chilled Water, Cooling Tower Systems	COST PER S.F.		
		MAT.	INST.	TOTAL
2280	20,000 S.F., 113.33 ton	5.80	2.97	8.77
2320	40,000 S.F., 226.66 ton	5.65	3.52	9.17
2360	60,000 S.F., 340 ton*			
2440	Department stores, 4,000 S.F., 11.66 ton	4.98	3.96	8.94
2480	6,000 S.F., 17.50 ton	4.31	3.37	7.68
2520	10,000 S.F., 29.17 ton	3.23	2.50	5.73
2560	20,000 S.F., 58.33 ton	2.74	1.90	4.64
2600	40,000 S.F., 116.66 ton	3.34	2.04	5.38
2640	60,000 S.F., 175.00 ton	3.82	3.20	7.02
2720	Drug stores, 4,000 S.F., 26.66 ton	7.55	4.67	12.22
2760	6,000 S.F., 40.00 ton	6.75	4.06	10.81
2800	10,000 S.F., 66.66 ton	6.75	3.71	10.46
2840	20,000 S.F., 133.33 ton	6.65	3.21	9.86
2880	40,000 S.F., 266.67 ton	6.55	4.01	10.56
2920	60,000 S.F., 400 ton*			
3000	Factories, 4,000 S.F., 13.33 ton	5	3.94	8.94
3040	6,000 S.F., 20.00 ton	4.46	3.37	7.83
3080	10,000 S.F., 33.33 ton	3.56	2.59	6.15
3120	20,000 S.F., 66.66 ton	3.67	2.35	6.02
3160	40,000 S.F., 133.33 ton	3.70	2.16	5.86
3200	60,000 S.F., 200.00 ton	4.14	3.35	7.49
3280	Food supermarkets, 4,000 S.F., 11.33 ton	4.88	3.94	8.82
3320	6,000 S.F., 17.00 ton	3.89	3.28	7.17
3360	10,000 S.F., 28.33 ton	3.16	2.48	5.64
3400	20,000 S.F., 56.66 ton	2.77	1.91	4.68
3440	40,000 S.F., 113.33 ton	3.30	2.04	5.34
3480	60,000 S.F., 170.00 ton	3.79	3.19	6.98
3560	Medical centers, 4,000 S.F., 9.33 ton	4.18	3.61	7.79
3600	6,000 S.F., 14.00 ton	3.69	3.22	6.91
3640	10,000 S.F., 23.33 ton	2.98	2.39	5.37
3680	20,000 S.F., 46.66 ton	2.45	1.84	4.29
3720	40,000 S.F., 93.33 ton	2.95	1.95	4.90
3760	60,000 S.F., 140.00 ton	3.45	3.13	6.58
3840	Offices, 4,000 S.F., 12.66 ton	4.84	3.89	8.73
3880	6,000 S.F., 19.00 ton	4.37	3.44	7.81
3920	10,000 S.F., 31.66 ton	3.52	2.61	6.13
3960	20,000 S.F., 63.33 ton	3.60	2.35	5.95
4000	40,000 S.F., 126.66 ton	4	3.09	7.09
4040	60,000 S.F., 190.00 ton	3.99	3.30	7.29
4120	Restaurants, 4,000 S.F., 20.00 ton	6.35	4.20	10.55
4160	6,000 S.F., 30.00 ton	5.40	3.75	9.15
4200	10,000 S.F., 50.00 ton	4.85	2.88	7.73
4240	20,000 S.F., 100.00 ton	5.50	2.89	8.39
4280	40,000 S.F., 200.00 ton	4.99	3.32	8.31
4320	60,000 S.F., 300.00 ton	5.30	3.76	9.06
4400	Schools and colleges, 4,000 S.F., 15.33 ton	5.55	4.06	9.61
4440	6,000 S.F., 23.00 ton	4.74	3.55	8.29
4480	10,000 S.F., 38.33 ton	3.97	2.68	6.65
4520	20,000 S.F., 76.66 ton	4.69	2.68	7.37
4560	40,000 S.F., 153.33 ton	4.49	3.20	7.69
4600	60,000 S.F., 230.00 ton	4.32	3.42	7.74

323

(Reprinted from Means Mechanical Cost Data 1991.)

Figure 23.30b

Table 8.4-001

General: The purpose of air conditioning is to control the environment of a space so that comfort is provided for the occupants and/or conditions are suitable for the processes or equipment contained therein. The several items which should be evaluated to define system objectives are:

Temperature Control
Humidity Control
Cleanliness
Odor, smoke and fumes
Ventilation

Efforts to control the above parameters must also include consideration of the degree or tolerance of variation, the noise level introduced, the velocity of air motion and the energy requirements to accomplish the desired results.

The variation in **temperature** and **humidity** is a function of the sensor and the controller. The controller reacts to a signal from the sensor and produces the appropriate suitable response in either the terminal unit, the conductor of the transporting medium (air, steam, chilled water, etc.), or the source (boiler, evaporating coils, etc.).

The **noise level** is a by-product of the energy supplied to moving components of the system. Those items which usually contribute the most noise are pumps, blowers, fans, compressors and diffusers. The level of noise can be partially controlled through use of vibration pads, isolators, proper sizing, shields, baffles and sound absorbing liners.

Some **air motion** is necessary to prevent stagnation and stratification. The maximum acceptable velocity varies with the degree of heating or cooling which is taking place. Most people feel air moving past them at velocities in excess of 25 FPM as an annoying draft, however, velocities up to 45 FPM may be acceptable in certain cases. Ventilation, expressed as air changes per hour and percentage of fresh air, is usually an item regulated by local codes.

Selection of the system to be used for a particular application is usually a trade-off. In some cases the building size, style, or room available for mechanical use limits the range of possibilities. Prime factors influencing the decision are first cost and total life (operating, maintenance and replacement costs). The accuracy with which each parameter is determined will be an important measure of the reliability of the decision and subsequent satisfactory operation of the installed system.

Heat delivery may be desired from an air conditioning system. Heating capability usually is added as follows: A gas fired burner or hot water/steam/electric coils may be added to the air handling unit directly and heat all air equally. For limited or localized heat requirements the water/steam/electric coils may be inserted into the duct branch supplying the cold areas. Gas fired duct furnaces are also available. **Note:** when water or steam coils are used the cost of piping and boiler must also be added. For a rough estimate use the cost per square foot of the appropriate sized hydronic system with unit heater. This will give the boiler, piping, and the unit heaters which would approximate the cost of the heating coils. The installed cost of electric and gas heaters, boilers and other heat related items on a unit basis may be located in Section 15.5 of *Means Mechanical Cost Data*.

Table 8.4-002 Air Conditioning Requirements
BTU's Per Hour Per S.F. of Floor Area and S.F. Per Ton of Air Conditioning

Type Building	BTU per S.F.	S.F. per Ton	Type Building	BTU per S.F.	S.F. per Ton	Type Building	BTU per S.F.	S.F. per Ton
Apartments, Individual	26	450	Dormitory, Rooms	40	300	Libraries	50	240
Corridors	22	550	Corridors	30	400	Low Rise Office, Exterior	38	320
Auditoriums & Theaters	40	300/18*	Dress Shops	43	280	Interior	33	360
Banks	50	240	Drug Stores	80	150	Medical Centers	28	425
Barber Shops	48	250	Factories	40	300	Motels	28	425
Bars & Taverns	133	90	High Rise Office—Ext. Rms.	46	263	Office (small suite)	43	280
Beauty Parlors	66	180	Interior Rooms	37	325	Post Office, Individual Office	42	285
Bowling Alleys	68	175	Hospitals, core	43	280	Central Area	46	260
Churches	36	330/20*	Perimeter	46	260	Residences	20	600
Cocktail Lounges	68	175	Hotel, Guest Rooms	44	275	Restaurants	60	200
Computer Rooms	141	85	Corridors	30	400	Schools & Colleges	46	260
Dental Offices	52	230	Public Spaces	55	220	Shoe Stores	55	220
Dept. Stores, Basement	34	350	Industrial Plants, Offices	38	320	Shop'g. Ctrs., Super Markets	34	350
Main Floor	40	300	General Offices	34	350	Retail Stores	48	250
Upper Floor	30	400	Plant Areas	40	300	Specialty	60	200

*Persons per ton

12,000 BTU = 1 ton of air conditioning

(Reprinted from Means Mechanical Cost Data 1991.)

Figure 23.31

Means' Forms

CONDENSED ESTIMATE SUMMARY

SHEET NO.

PROJECT Office Building

ESTIMATE NO.

LOCATION Boston Area

TOTAL AREA/VOLUME 56,700 S.F.

DATE 2/6/91

ARCHITECT

COST PER S.F./C.F. 10.55

NO. OF STORIES 3

PRICES BY: JJM

EXTENSIONS BY: RG

CHECKED BY: RG

	Systems Estimate	Total		Cost/S.F.
	Plumbing	78000		1.38
	Fire Protection	120000		2.12
	HVAC	481000		8.48
	Total mechanical	679000		11.98

Figure 23.32

Chapter 24

UNIT PRICE ESTIMATING EXAMPLE

This chapter contains a sample mechanical estimate for a three-story office building. To prepare the Unit Price Estimate, procedures and forms are utilized, and calculations made, as previously described in Part I of this book. The estimate that follows is prepared from the perspective of a prime mechanical contractor who has been invited by a familiar general contractor to bid on a regional office building for a nationwide firm. All costs and reference tables are from the annual *Means Mechanical Cost Data* or *Means Plumbing Cost Data*. All forms are from *Means Forms for Building Construction Professionals* or *Means Forms for Contractors*.

Project Description

The sample project is a three-story office building with an open parking garage beneath the building. The structure is steel frame with curtain wall enclosure. Plans for the mechanical work are shown in Figures 24.1 through 24.5. These drawings have been reduced in size and are not to be scaled. The plans are for illustration only, and, while fairly representative, do not show all of the detail one would expect from a project of this magnitude. A full set of specifications would normally accompany a set of plans, spelling out the scope of work and level of quality of materials and equipment.

The quantities given on the following takeoff, summary, and cost analysis forms are representative of an actual takeoff of the various systems made from the original plans.

Some liberties have been taken in material selection, etc., to provide the reader with a variety of estimating procedures. For example, in the piping estimates, steel pipe, cast iron, copper tube, and polyvinyl chloride (PVC) have all been used. The estimate also includes a wide variety of joining methods, including threaded, butt weld, flanged, mechanical joint, lead and oakum, no hub, solvent weld, and (no lead) soft solder. This practice has been continued throughout the entire mechanical estimate to provide a diversified estimate while staying within the confines of practicality.

In keeping with accepted practices of the trade in many areas of this country, all building service piping, gas, water, storm, and sanitary systems either stop or begin at the appropriate meter or 5' outside the building foundation.

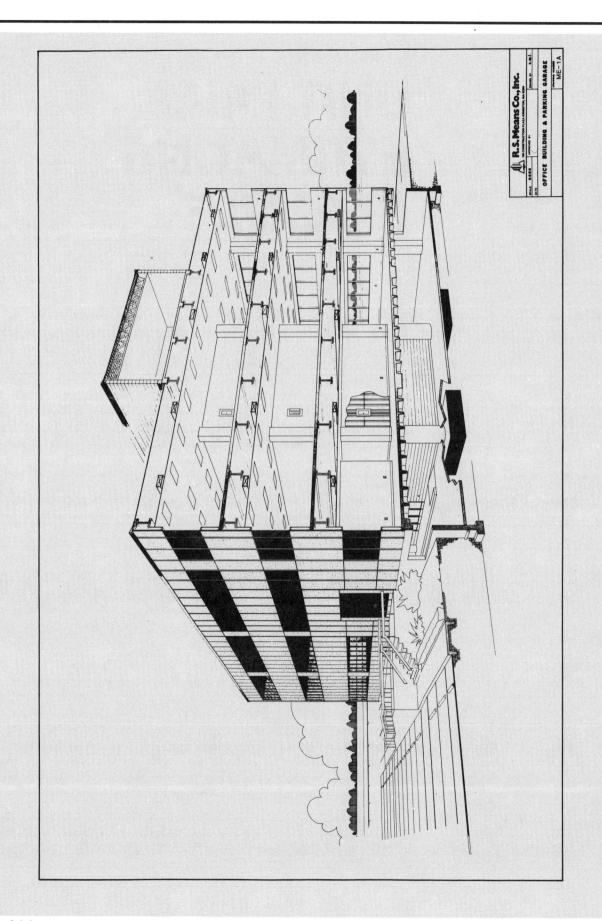

Figure 24.1

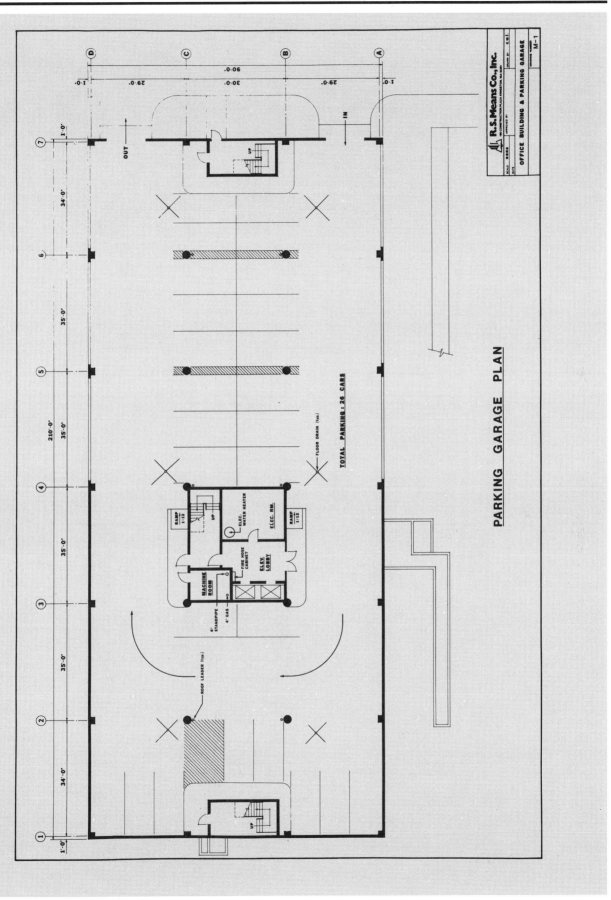

PARKING GARAGE PLAN

Figure 24.2

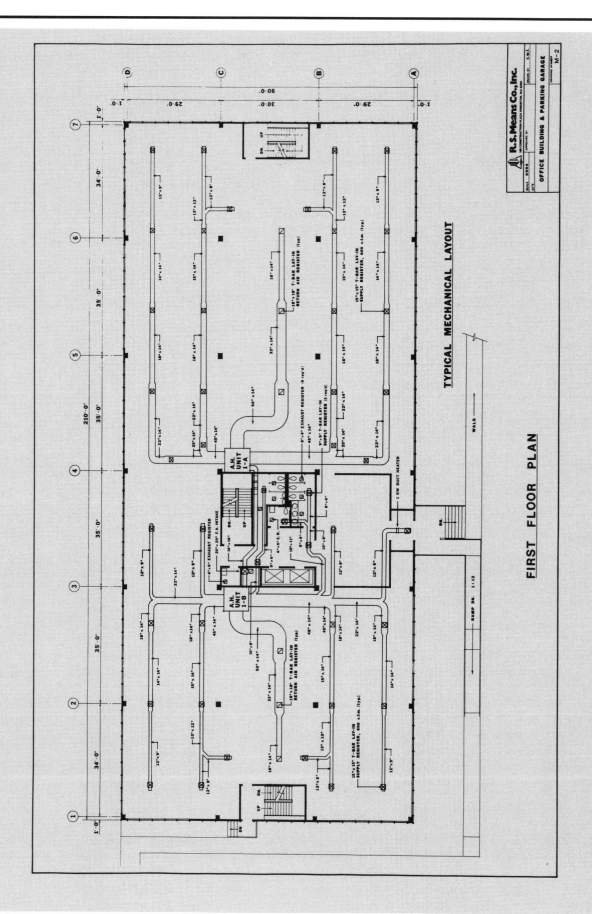

TYPICAL MECHANICAL LAYOUT

FIRST FLOOR PLAN

Figure 24.3

Getting Started

A good estimator must be able to visualize the proposed mechanical installation from source (i.e., water heaters or chillers) to end use (i.e., fixture). This process helps identify each component of the mechanical portion of a building. Before starting the takeoff, the estimator should follow some basic steps:

- Read the mechanical plans and specifications carefully (with special attention given to the General and Special Conditions).
- Examine the drawings for general content; type of structure, number of floors, etc.
- Clarify any unclear areas with the designer, making sure that the scope of work is understood. Addenda to the contract documents may be necessary to clarify certain items of work so that the responsibility for the performance of all work is defined.
- Immediately contact suppliers, manufacturers, and subsystem specialty contractors in order to get their quotations, and sub-bids.

Certain information should be taken from the complete architectural plans and specifications if the mechanical work is to be properly estimated. The following items should be given special attention:

- Temporary heat
- Temporary water or sanitary facilities
- Job completion dates (partial occupancy)
- Cleaning

Forms are a necessary tool in the preparation of estimates. This is especially true for the mechanical trades, which include a wide variety of items and materials. Careful measurements and a count of components cannot compensate for an oversight such as forgetting to include pipe insulation. A well designed form acts as a checklist, a guide for standardization, and a permanent record. A typical preprinted form and checklist is shown in Figure 24.6.

Plumbing

As in any estimate, the first step in preparing the plumbing estimate is to visualize the scope of the job. This can be done by scanning the drawings and specifications. The next step is to make up a material "takeoff" sheet. Having read through the plumbing section of the specifications, list each of the item headings on the summary sheet. This will serve as a quantity checklist. Also include labor that will not be specified in the plans, such as cleaning, adjusting, testing, and balancing.

The easiest way to make a material "quantity count" or takeoff is by system, as most of the pipe components of a system will tend to be of the same material, class weight, and grade. A waste system, for example, would consist of pipe varying from 2" to 15" in diameter, but it would all be of service weight cast iron or DWV copper up to a specified size, and extra heavy cast iron for the larger sizes.

Fixture takeoff usually involves nothing more than counting the various types, sizes, and styles, and entering them on a fixture form. It is important, however, that each fixture be fully identified. A common source of error is to overlook what does or does not come with the fixture (i.e., trim, carrier, flush valve). Equipment such as pumps, water heaters, water softeners, and all items not previously counted are also listed at this time. The order of proceeding is not as important as the development of a consistent method which will assist the estimator to improve in speed while minimizing the chances of overlooking any item or

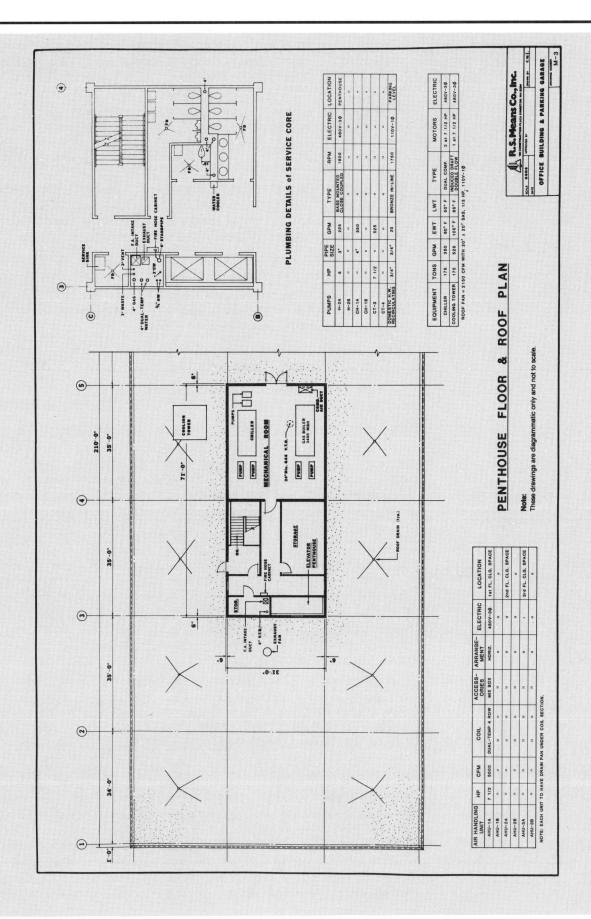

Figure 24.4

272

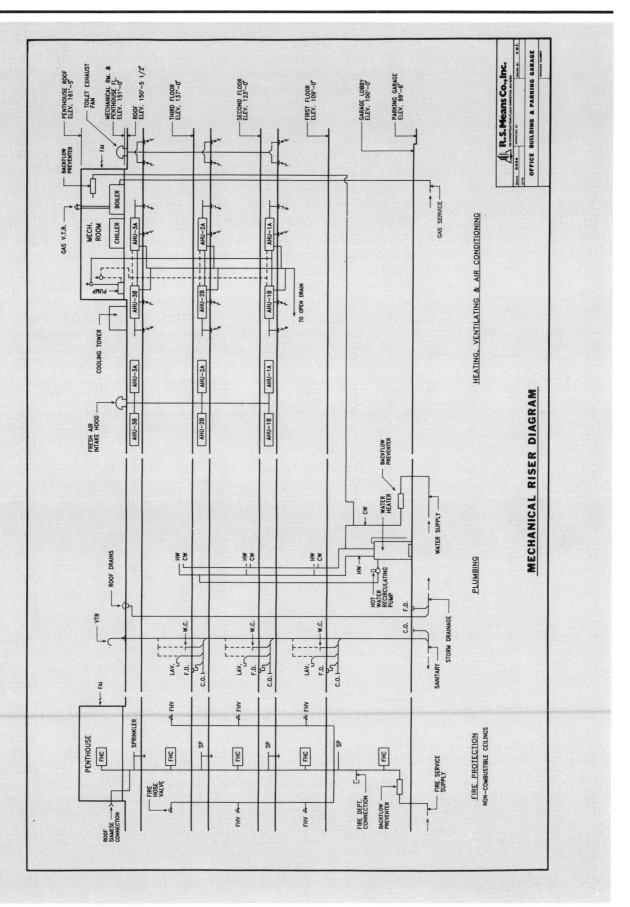

Figure 24.5

class of items. The various counts should then be totaled and transferred to the summary quantity sheet (Figure 24.7). Miscellaneous items which add to the plumbing contract should also be listed. Many of these would have been identified during the preliminary review of the plans and specifications. It is desirable to take off the fixtures and equipment before the piping for several reasons. The fixture and equipment lists can be given to suppliers for pricing while the estimator is performing the more arduous and time-consuming piping takeoff. Taking off the fixtures and equipment first also gives the estimator a good perspective of the building and its systems. Coloring the fixture and equipment as the takeoff is made aids the estimator when he returns for the piping takeoff by acting as targets for piping networks.

Pipe runs for any type of system consist of straight section and fittings of various shapes, styles, and purposes. Depending on job size, separate forms may be used for pipe and fittings, or they may be combined on one sheet. The estimator should start at one end of each system and using a tape, ruler, or wheeled indicator set at the corresponding scale, should measure and record the straight lengths of each size of pipe. When a fitting is encountered, it should be recorded with a check in the appropriate column. The use of colored pencils will aid in marking runs that have been completed, or the termination points on the main line where measurements are stopped, so that a branch may be taken off. Any changes in material should be noted. This would only occur at a joint; therefore, the estimator should always verify that the piping material going into a joint is the same as that leaving the joint. The plumbing takeoff is shown in Figures 24.8a through 24.8e. Subcontractors (i.e., insulation, controls, etc.) should be notified of the availability of your plans after you have completed your takeoff.

Rounding off piping quantities to the nearest five or ten feet is good practice. In larger projects, rounding might be carried to the nearest hundred or thousand feet. This makes pricing and extending less prone to error. Rounding off pipe quantities is also practical because pipe is purchased in uniform lengths (i.e., rigid copper tubing and steel pipe is rounded off to 20' lengths, cast-iron soil pipe to 5' or 10' lengths, etc.).

The plumbing totals are transferred to cost analysis forms for pricing and extending, as shown in Figures 24.9a through 24.9e.

Units of cost are assigned to each item. These costs may be taken from historical data, or from a reliable reference, such as *Means Plumbing Cost Data*. Be sure to match the description of items to be priced with the historical data as closely as possible when obtaining unit costs. These costs are then extended (multiplied by the total quantity of each item) in order to obtain a total cost per item.

On the final sheet of the plumbing estimate, (Figure 24.9e) page subtotals are recorded and totaled to arrive at the "bare" cost for plumbing. These values must then be adjusted for sales tax (if applicable), overhead, and profit on material, labor, and equipment as discussed in Chapter 22, "Using *Means Mechanical Cost Data* and *Means Plumbing Cost Data*." Finally, the marked up totals are adjusted by the appropriate location factor (Boston is used in the example) also as discussed in Chapter 22. This completes the plumbing estimate for the office building. The resultant price is what a general contractor would receive as a complete bid for the plumbing work.

		1	2

NAME OF JOB:
LOCATION:
ARCHITECT:

DUE DATE:
BID SENT TO:
ENGINEER:

	1	2
Labor w/Payroll Taxes, Welfare		
Travel		
Fixtures		
Drains, Carriers, etc.		
Sanitary - Interior		
Storm - Interior		
Water - Interior		
Valves - Interior		
Water Exterior - Street Connection		
Water Exterior - Extension to Building		
Sanitary & Storm Exterior - Street Connection		
Sanitary & Storm Exterior - Extension to Building		
Water Meter		
Gas Service		
Gas Interior		
Accessories		
Flashings for: Vents, Floor & Roof Drains, Pans		
Hot Water Tanks: Automatic w/trim		
Hot Water Tanks: Storage, w/stand and trim		
Hot Water Circ. Pumps, Aquastats, Mag. Starters		
Sump Pumps and/or Ejectors; Controls, Mag. Starters		
Flue Piping		
Rigging, Painting, Excavation and Backfill		
Therm. Gauges, P.R.V. Controls, Specialties, etc.		
Record Dwgs., Tags and Charts		
Fire Extinguishers, Equipment, and/or Standpipes		
Permits		
Sleeves, Inserts and Hangers		
Staging		
Insulation		
Acid Waste System		
Sales Tax		
Trucking and Cartage		

Figure 24.6

QUANTITY SHEET

SHEET NO. P6 of 6

PROJECT Office Building

ESTIMATE NO.

~~LOCATION~~ Plumbing

ARCHITECT

DATE 1/4/91

TAKE OFF BY MJM

EXTENSIONS BY: MJM

CHECKED BY:

DESCRIPTION	NO.	DIMENSIONS			Eq.	UNIT		UNIT		UNIT		UNIT
Plumbing Fixtures + Appliances												
Water Heater 36 KW, 148 GPH		80	Gal.		1	Eq.						
Recirculating Pump		3/4"			1	Eq.						
Water Closet, white, Wm w/ Flush Valve					15	Eq.						
Water Closet, white, Fm w/ Flush (Paraplegic)					3	Eq.						
Urinal, Stall Type					6	Eq.						
Lav. 19" x 16" Oval Vanity mount					9	Eq.						
Lav. 19" x 17" Wall Hung					9	Eq.						
Lav. 28 x 21 Wall Hung (Paraplegic)					3	Eq.						
Service Sinks					3	Eq.						
Electric Water Cooler Wall Hung					6	Eq.						

Figure 24.7

PIPING SCHEDULE

JOB Office Building

SYSTEM Plumbing – Storm System / Gas Piping

PAGE P1 of 6

DATE 1/4/91

BY JJM

			PIPE DIAMETER IN INCHES															
			12	10	8	6	5	4	3 1/2	3	2 1/2	2	1 1/2	1 1/4	1	3/4	1/2	
Storm System Roof Drains PVC Sch 40										12								
PVC Pipe DWV										1100								
DWV 1/4 Bend (90° Ells)										24								
Floor Drains Galv.								6										
Cleanout Tee C.I.								12										
San Y (Tee) C.I.						1		6										
C.I. Pipe						220		120										
Gas B.S. Pipe	Sch 40							160							20 (vent)			
90° Ells Blk mall								7							4			
Tees Blk mall								3										
Caps								3										
Cocks								2										

Figure 24.8a

PIPING SCHEDULE

JOB _Office Building_

SYSTEM _Plumbing — Water Pipe, Cold Water_

PIPE DIAMETER IN INCHES

	1/2	3/4	1	1 1/4	1 1/2	2	2 1/2	3	3 1/2	4	5	6	8	10	12
Cold Water															
"L" Copper		460	60			160									
90° Ells Copper		26	18			6									
Tees Copper						10									
Gate Valve Copper		12				3									
Water Meter						1									
Hose Bibbs		6													
Backflow Preventer			1 Boiler Room			1 Bldg.									
Shock Absorber		4													
Vacuum Breaker		4													

Figure 24.8b

PIPING SCHEDULE

PAGE P3 of 6

JOB Office Building

DATE 1/4/91

SYSTEM Plumbing — Water Pipe, Hot Water

BY JJM

	PIPE DIAMETER IN INCHES																
			12	10	8	6	5	4	3 1/2	3	2 1/2	2	1 1/2	1 1/4	1	3/4	1/2	
Hot Water "L" Copper																220		
90° Ells Copper																12		
Tees Copper																18		
Check Valve Copper																1		
Relief Valve																1		
Unions																2		
Temp Reg./mixing Valve																1		
Gate Valve Copper																6		

Figure 24.8c

PIPING SCHEDULE

JOB Office Building

SYSTEM Plumbing — Sanitary & Vent

PAGE P4 of 6

DATE 1/4/91

BY JJM

			PIPE DIAMETER IN INCHES														
			12	10	8	6	5	4	3 1/2	3	2 1/2	2	1 1/2	1 1/4	1	3/4	1/2	
San/vent STD WT Hub C.I. Pipe								90		40								
Comb Y & 1/8 Bend								5		2								
Floor cleanout (FCO)								3		1								
1/4 Bend (90° Ells)										2								
No Hub std wt C.I. Pipe								170		210								
Y (San T)								30		20								
Floor cleanout (FCO)								3										
Floor Drains										12								
Vent Flashing Thru Roof								2		2								
Wall cleanout (WCO)								6										
1/4 Bend (90° Ells)								6										

Figure 24.8d

PIPING SCHEDULE

JOB Office Building

SYSTEM Plumbing — Sanitary & Vent

PAGE P5 of 6

DATE 1/4/91

BY JJM

			PIPE DIAMETER IN INCHES															
			12	10	8	6	5	4	3 1/2	3	2 1/2	2	1 1/2	1 1/4	1	3/4	1/2	
1/8 Bend (45° Ells)								11		4								
Stainless Steel Coupling								90		72								
Copper DWV Tube												320						
San Tee DWV Copper												48						
90° Ells DWV Copper												24						

Figure 24.8e

Means' Forms
COST ANALYSIS

PROJECT Office Building
LOCATION
TAKE OFF BY JJM | QUANTITIES BY JJM | PRICES BY MJM | CLASSIFICATION Div. 15 - Plumbing | EXTENSIONS BY MJM
ARCHITECT | CHECKED BY JJM
ESTIMATE NO. 91-1
DATE 1-7-91

DESCRIPTION	SOURCE/DIMENSIONS			QUANTITY	UNIT	MATERIAL		LABOR		EQUIPMENT		SUBCONTRACT		TOTAL	
						UNIT COST	TOTAL	UNIT COST	TOTAL	UNIT COST	TOTAL	UNIT COST	TOTAL	UNIT COST	TOTAL
Water Heater 36 Kw 80Gal. (148 GPH)	153	110	4200	1	Ea.	2690	2690	135	135						
Circulating Pump 3/4"	152	410	0640	1	Ea.	94.60	95	23	23						
Water Closet, White, WmFlush	152	180	3100	15	Ea.	275	4125	63	945						
w/c Rough In			3200	15		156.94	2354	180	2700						
w/c White, Fm/Flush Parapl			3380	3		310	930	73	219						
w/c Rough In		↓	3400	3	↓	95.28	286	205	615						
Urinal- Stall Type	152	168	5000	6	Ea.	410	2460	145	870						
Urinal- Rough In	152	168	6980	6	Ea.	107.85	647	185	1110						
Lav.19x16 Oval Vanity, Mtd	152	136	0680	9	Ea.	128	1152	57	513						
Lav. v/m Rough In			3580	9		88.75	799	160	1440						
Lav. 9 x17 Wall Mtd.			4120	9		195	1755	46	414						
Lav. w/m Rough In			6960	9		130.10	1171	220	1980						
Lav. 28 x 21 Wall Mtd. Para			6210	3		233	699	52	156						
Lav. w/m Rough In		↓	6460	3	↓	130.10	390	220	660						
Service Sink 24 x 20 w/m	152	152	7100	3	Ea.	342	1026	92	276						
Service Sink w/m Rough In	152	152	8980	3	Ea.	215.72	647	280	840						
Water Cooler Wall Hung	153	105	0220	6	Ea.	367	2202	92	552						
Water Cooler w/H Rough In			9800	6		51.01	306	92	552						
Water Cooler-Stainless Cabinet		↓	0640	6	↓	41.50	249								
Page P1 Subtotal							23983		14000						

Figure 24.9a

282

COST ANALYSIS

PROJECT: Office Building
LOCATION:
CLASSIFICATION: Div. 15 - Plumbing
ARCHITECT:

TAKE OFF BY: JJM | QUANTITIES BY: JJM | PRICES BY: MJM | EXTENSIONS BY: MJM | CHECKED BY: JJM

Description	Size		Source/Dimensions	Quantity	Unit	Material Unit Cost	Material Total	Labor Unit Cost	Labor Total
PVC									
Roof Drain	3"	151	125 4780	12	Ea.	37	444	26	312
Pipe	3"		551 4470	1100	LF	1.64	1804	1.45	1590
90° Ells	3"	⌄	558 5080	24	Ea.	2.15	52	22	528
Cast Iron B&S Type									
Floor Drain Galv.	4"	151	125 0490 0480	6	Ea.	316	1896	46	276
Tee Clean Out	4"		115 0240	12		32	384	62	744
San Y (Tee)	4"		320 0620	6		11.15	67	46	276
	6"		320 0800	1		25	25	52	52
Floor Clean Outs	3"		110 0120	1		49	49	25	25
	4"		110 0140	3		65	195	34	102
Comb. Y + 1/8 Bend	2"		320 1420	2		7.05	14	37	74
	4"		320 1520	5		15	75	46	230
1/4 Bend (90° Ell)	3"		320 0120	2	→	5.35	11	26	52
Pipe, Cast Iron B&S	3"		301 2140	40	LF	2.83	113	6.10	244
	4"		301 2160	210		3.67	771	6.65	1397
	6"	⌄	301 2200	220	→	6.12	1346	7.80	1716
Cast Iron No Hub									
Pipe, Cast Iron No Hub	3"	151	301 4140	210	LF	3.55	746	5.75	1208
"	4"		301 4160	170	LF	4.57	777	6.30	1071
San Y (Tee)	3"		320 6520	20	Ea.	4.20	84		
	4"		320 6600	30	→	6.55	197		
Floor Clean Out	4"		110 0140	3		65	195	34	201
Floor Drains	3"		125 2220	12		51	612	31	273
Wall Cleanout	4"	⌄	110 4100	6	→	54	324	20	120
				Page P2 Subtotal			10181		10491

Figure 24.9b

Means' Forms
COST ANALYSIS

PROJECT Office Building CLASSIFICATION Div. 15 – Plumbing SHEET NO. **3** of 5

LOCATION ARCHITECT ESTIMATE NO. 91-1

TAKE OFF BY JJM QUANTITIES BY JJM PRICES BY MJM EXTENSIONS BY MJM CHECKED BY JJM DATE 1-7-91

Description	Source/Dimensions	Quantity	Unit	Material Unit Cost	Material Total	Labor Unit Cost	Labor Total	Equipment Unit Cost	Equipment Total	Subcontract Unit Cost	Subcontract Total	Total Unit Cost	Total
Cast Iron No Hub (cont'd)													
90° Ell) ¼ Bend 4"	151 320 6120	6	Ea.	4.95	30								
(45° Ell) ⅛ Bend 3"	6260	4	→	2.70	11								
4"	6280	11		3.60	40								
St. St. Couplings 3"	8650	72		9.15	659	9.65	695						
4"	8660	90	↓	9.75	878	11.10	999						
Vent thru Roof Flash 3"	195 1450	2		17.40	35	12	24						
4"	195 1460	2	↓	19.20	38	12.75	26						
Black Steel													
Pipe 1"	151 701 0580	20	LF	1.49	30	3.84	77						
4"	0650	160	LF	9.76	1562	10.20	1632						
90° Ell Blk. mall. 1"	716 5100	4	Ea.	1.17	5	15.65	63						
4"	5170	7	→	29	203	61	427						
Tee Blk. Mall. 4"	5580	3	↓	43	129	92	276						
Caps 4"	5780	3	↓	22	66	19	57 (note: This is ½ coupling labor)						
Gas Cocks 4"	151 990 7030	2	Ea.	350	700	120	240						
Copper (Sweat Joints)													
Type L Tube ¾"	151 401 2180	680	LF	1.77	1204	2.68	1822						
1"	2200	60	↓	2.45	147	2.99	179						
2"	2260	160	↓	6.33	1013	4.85	**776**						
90° Ells ¾"	430 0120	38	Ea.	.49	19	10.70	**407**						
1"	0130	18	→	.99	18	12.75	**230**						
2"	0160	6	↓	6.90	41	18.50	**111**						
Page P3 Subtotal					6828		8041						

Figure 24.9c

284

COST ANALYSIS

PROJECT Office Building

LOCATION

TAKE OFF BY JJM QUANTITIES BY JJM PRICES BY MJM EXTENSIONS BY MJM CHECKED BY JJM

CLASSIFICATION Div. 15 - Plumbing

ARCHITECT

DESCRIPTION		SOURCE/DIMENSIONS		QUANTITY	UNIT	MATERIAL		LABOR		EQUIPMENT		SUBCONTRACT		TOTAL	
						UNIT COST	TOTAL	UNIT COST	TOTAL	UNIT COST	TOTAL	UNIT COST	TOTAL	UNIT COST	TOTAL
Copper (Sweat Joints, cont'd)															
Tee	3/4"	151	430 0500	18	Ea.	1.22	22	16.95	305						
	2"		430 0540	10		12.20	122	29	290						
Shock Absorber	3/4"	165	165 0500	4		37	148	16.95	68						
Vacuum Breaker	3/4"	185	185 1080	4		16.15	67	10.20	41						
Unions	3/4"	430	430 0900	2		2.78	6	11.30	23						
Gate Valve	3/4"	955	955 2940	18		14.40	259	10.20	184						
	2"		2980	3		40	120	18.50	56						
Check Valve	3/4"		1860	1		11	11	10.20	10						
Relief Valve	3/4"		5640	1		43	43	7.25	7						
Temp. Mix Valve	3/4"		8440	1		31	31	10.20	10						
DWV Copper (Sweat)															
DWV Tube	2"	151	401 4140	320	LF	4.38	1402	4.63	1482						
San Tee	2"		430 2290	48	Ea.	8.10	389	29	1392						
90° Ell	2"		430 2070	24	Ea.	6.50	156	20	480						
Water meter	2"	153	160 2360	1	Ea.	325	325	34	34						
Hose Bibbs	3/4"	151	141 5000	6		4.50	27	8.50	51						
Back Flow Preventer	1"		105 1120	1		155	155	14.55	15						
	2"		105 1160	1		270	270	29	29						
							3553		4477						

Figure 24.9d

Means' Forms
COST ANALYSIS

PROJECT Office Building
LOCATION
TAKE OFF BY JJM QUANTITIES BY JJM PRICES BY MJM EXTENSIONS BY MJM
CLASSIFICATION Div. 15 - Plumbing
ARCHITECT
CHECKED BY JJM

DESCRIPTION	SOURCE/DIMENSIONS		QUANTITY	UNIT	MATERIAL		LABOR		EQUIPMENT		SUBCONTRACT		TOTAL	
					UNIT COST	TOTAL	UNIT COST	TOTAL	UNIT COST	TOTAL	UNIT COST	TOTAL	UNIT COST	TOTAL
Insulation for Hot/Cold Water Copper Tube														
Fiberglass w/ASJ 1"Wall 3/4"	155	651 6840	680	LF	Subcontract						3.37	2292		
1"		6860	60		Subcontract						3.63	218		
2"		6890	160		Subcontract						4.16	666		
Miscellaneous														
Demonstration Tests			4	Hrs			25.45	102						
Sleeving			16				25.45	407						
Warranty			6				25.45	153						
Valves, Tags & Charts			4				25.45	102						
Plumbing Page P5 Subtotal								764				3176		
Plumbing Subtotals P1						23983		14000						
P2						10181		16491						
P3						6828		8041						
P4						3553		4477						
P5								764						
Subtotal Total						44545		43773				3176		3176
Material Sales Tax					5%	2227								
Handling, Overhead + Profit Markup					10%	4455	44.25%	21536			10%	318		
Total						51227		65309				3494		120030
City modifier for Boston					X 1.040		X 4.185				X 4.112			
Boston Total						53276		77391				3885		134552

Figure 24.9e

286

Fire Protection

After having scanned the plans and specifications to visualize the scope and type of fire protection proposed for this project, estimate sheets are prepared, as shown in Figure 24.10a through 24.10c. When properly used, such forms will serve as a summary of the specifications.

The specifications should be studied, and all components of fire protection systems should be entered on the takeoff sheets (Figures 24.10a through 24.10c). Material pricing and labor considerations should be listed.

The General Conditions and the Special Conditions should indicate the fire protection contractor's responsibilities regarding excavation and backfill, cutting and patching, masonry and concrete, painting, electrical work, temporary services, and any other items that may affect the estimate. Note on the estimate sheet the services that will be provided "by others." Indicate "labor only" items, such as cleaning, testing, flushing, application of decals, distribution of equipment to be stored in the cabinet, etc.

The material or quantity takeoff should be executed systematically, floor by floor. The fire protection estimator should be familiar with the architectural plans as this may be the only place that equipment such as extinguishers or other specialized equipment will be shown.

Hazardous areas requiring special considerations which may have been overlooked by the designer may be found on the architectural plans. The reflected ceiling plans and ductwork layout should also be studied to identify the degree of coordination required for the sprinkler head layout. The piping takeoff for standpipes should be kept separate from the sprinkler distribution piping. As in other piping takeoffs, pipe should be listed by material and joining method. The sprinkler heads should be taken off first, floor by floor, by type and configuration.

Fire protection materials after the takeoff should be totaled and transferred to a cost analysis form for pricing and extending, as shown in Figures 24.11a through 24.11c. As with the plumbing estimate, bare costs are summarized, totaled, and marked up on the final page (Figure 24.11c) to arrive at the "bid" price for fire protection.

Heating, Ventilating, and Air Conditioning

Visualizing the air-conditioning system and its varied and numerous components requires more than just scanning the plans and specifications. An in-depth perusal is required to understand the operation, distribution, and control of what appears to be, and can be, a complex installation. A detailed form should be utilized for estimating HVAC systems, which may be three or more pages long.

The General and Special Conditions define the scope of the project and should delegate financial responsibility for such pertinent items as excavation and backfill, masonry and concrete, cutting and patching, temporary heat (and operation), power and control wiring, hoisting, rigging, etc. When these essential items have been noted on the estimate either as "by others" or "to be priced," the equipment takeoff can begin. The estimator should begin a plan by plan takeoff of the equipment, starting either at the roof or at the basement. Each floor should be listed separately. Starting at the roof provides a takeoff point at a less cluttered location. The complexity increases as the estimator moves down gradually floor by floor. The starting point is a matter of personal preference; however, it should be consistent throughout the takeoff. Marking off each system on the plans with a different colored pencil will help assure that every item has been accounted for. Consistency is the key to accuracy.

After the equipment is taken off and grouped together for pricing, price requests can be mailed out to equipment suppliers (see Figure 24.12). This may be

PIPING SCHEDULE

JOB Office Building

SYSTEM Fire Protection Standpipe / Sprinkler

PAGE FP 1 of 3

DATE 1/7/91

BY JJm

	PIPE DIAMETER IN INCHES																	
			12	10	8	6	5	4	3 1/2	3	2 1/2	2	1 1/2	1 1/4	1	3/4	1/2	
Standpipe B.S. Pipe Sch.40					60	60		300			80							
90° Ells C.I. Screwed					3	1		4			16							
Tees C.I. Screwed					3	5-6x2½ 3-6x5		8										
F.D. Valves											13							
Reducers C.I.					2													
Roof Siamese						6x2½x2½ 1												
OS & Y Gate Valve					1	3												
Swing Check Valve w/Ball Drip						1												
Wall Siamese						1												
Fire Hose													500					
Hose Adapter											5							
Hose Rack											5							
Hose nozzle											5							

Figure 24.10a

PIPING SCHEDULE

JOB Office Building

SYSTEM Fire Protection Standpipe / Sprinkler

PAGE FP 2 of 3

DATE 1/7/91

BY JJm

			PIPE DIAMETER IN INCHES															
			12	10	8	6	5	4	3 1/2	3	2 1/2	2	1 1/2	1 1/4	1	3/4	1/2	
Standpipe (con't) Hydrolator											5							
Fire Hose Cabinet											5							
Sprink Distribution Alarm Valve						1												
Alarm water mtr/Gong																		Size std. 1
Check Valve						1												
Flow Control Valve						1												
B.S. Pipe Grooved Sch. 10								50			300	960						
B.S. Pipe Grooved Sch. 40													1150	570	1150			
Tee Grooved								10			30							

Figure 24.10b

Means® Forms

PIPING SCHEDULE

PAGE FP 3 of 3

JOB Office Building

DATE 1/7/91

SYSTEM Fire Protection Standpipe/Sprinkler

BY JJM

									PIPE DIAMETER IN INCHES									
			12	10	8	6	5	4	3 1/2	3	2 1/2	2	1 1/2	1 1/4	1	3/4	1/2	
Sprink Distribution (Con't) mech. Joint Tees												2x1 80	1½x1 80	1¼x¾ 40				
90° Ells C.I. Screwed															50			
Coupling Grooved Jt.								20			80	20						
90° Ells Groove Jt.												10						
Sprinkler Heads																	300	
Tee C.I. Screwed															40			

Figure 24.10c

Means' Forms
COST ANALYSIS

PROJECT: Office Building
LOCATION:
TAKE OFF BY: JJM QUANTITIES BY: JJM PRICES BY: MJM EXTENSIONS BY: MJM CHECKED BY: JJM

CLASSIFICATION: Div. 15 - Fire Protection
ARCHITECT:

DESCRIPTION		SOURCE/DIMENSIONS		QUANTITY	UNIT	MATERIAL UNIT COST	MATERIAL TOTAL	LABOR UNIT COST	LABOR TOTAL	EQUIPMENT UNIT COST	EQUIPMENT TOTAL	SUBCONTRACT UNIT COST	SUBCONTRACT TOTAL	TOTAL UNIT COST	TOTAL
Black Steel (Screwed)															
Pipe Sch. 40	2½"	151	701 0620	80	LF							16.20	1296		
4"			0650	300								26	7800		
6"			0670	60								52	3120		
8"			0680	60	↓							65	3900		
90° Ells C.I.	2½"	716	0150	16	Ea.							48	768		
4"			0180	4								120	480		
6"			0200	1								180	180		
8"			0210	3								270	810		
Tees C.I.	4"		0620	8								175	1400		
8"			0650	3								460	1380		
6" x 2½"			0672	5								310	1550		
6" x 5"			0672	3								310	930		
Reducers C.I.	8"		0699	2	↓							260	520		
F.D. Bronze Valves	2½"	154	160 0090	13	Ea.							105	1365		
Roof Siamese 6" x 2½" x 2½"		154	160 6060	1								255	255		
OS+Y Gate Valve	6"	151	960 3700	3								675	2025		
8"		151	960 3720	1								1025	1025		
Swing Check Valve w/Ball 6"		154	170 6540	1								495	495		
Wall Siamese	6"	154	135 7370	1	↓							640	640		
Fire Hose	1½"	154	135 7260	500	LF							1.32	660		
" " Adapter	2½"		1600	5	Ea.							9.35	47		
" " Rack	1½"		2640	5								57	285		
" " Nozzle	1½"		6700	5								21	105		
" " Coupling	1½"		1410	5								18.70	94		
" " Rack Nipple	1½"		2820	5	↓							18.70	94		
												Page FP1 Subtotal	31,224		

Figure 24.11a

Means' Forms

COST ANALYSIS

PROJECT: Office Building

LOCATION

TAKE OFF BY: JJM QUANTITIES BY: JJM PRICES BY: MJM EXTENSIONS BY: MJM CHECKED BY: JJM

CLASSIFICATION: Div. 15 - Fire Protection

ARCHITECT

DESCRIPTION	SOURCE/DIMENSIONS			QUANTITY	UNIT	MATERIAL UNIT COST	MATERIAL TOTAL	LABOR UNIT COST	LABOR TOTAL	EQUIPMENT UNIT COST	EQUIPMENT TOTAL	SUBCONTRACT UNIT COST	SUBCONTRACT TOTAL	TOTAL UNIT COST	TOTAL
Hydrolator 1½"	154	160	4200	5	Ea.							54	270		
Fire Hose Cabinet	154	115	4200	5	Ea.							220	1100		
Alarm Valve 6"	154	170	6300	1	Ea.							925	925		
Water Alarm Mtr/Gong			1220	1								200	200		
Check Valve 6"			6840	1								585	585		
Flow Control Valve 6"			8860	1								1950	1950		
Blk. Stl. Grooved															
Pipe Sch. 10 2"	151	801	0550	960	LF							10.35	9936		
2½"			0560	300								13.10	3930		
4"			0590	60								17.60	1056		
Pipe Sch. 40 1"			1050	1160								7.25	8410		
1¼"			1060	580								8.20	4756		
1½"			1070	1160								9.25	10730		
Tee Grvd. 2½"			4780	30	Ea.							42	1260		
4"			4800	10								78	780		
90° Ells Grvd. 2"			4070	10								22	220		
Coupling Grvd. 2"			4990	20								22	440		
2½"			5000	80								26	2080		
4"			5030	20								41	820		
Tee Mech Jt 1¼" x ½"	720		9510	40								41	1640		
1½" x 1"			9560	80								43	3440		
2" x 1"			9590	80								50	4000		
Page FP2 Subtotal															58528

Figure 24.11b

Means® Forms

COST ANALYSIS

SHEET NO. FP3 of 3

PROJECT Office Building　　CLASSIFICATION Div. 15 - Fire Protection　　ESTIMATE NO. 91-1

LOCATION　　ARCHITECT　　DATE 1-7-91

TAKE OFF BY JJm　　QUANTITIES BY JJm　　PRICES BY mJm　　EXTENSIONS BY mJm　　CHECKED BY JJm

DESCRIPTION	SOURCE/DIMENSIONS			QUANTITY	UNIT	MATERIAL		LABOR		EQUIPMENT		SUBCONTRACT		TOTAL	
						UNIT COST	TOTAL	UNIT COST	TOTAL	UNIT COST	TOTAL	UNIT COST	TOTAL	UNIT COST	TOTAL
90° Ell SCR C.I. 1"	151	716	0110	50	Ea.							25	1250		
Tee SCR C.I. 1"	151	716	0550	40	Ea.							40	1600		
Sprinkler Heads	154	170	4830	300	Ea.							38	11400		
							Page FP3 Subtotal						14250		
Fire Protection Subtotals															
FP 1													31224		
FP 2													58528		
FP 3													14250		
									Total				104002		104002
									Total				104002		
City modifier for Boston													x 1.112		
									Boston Total				115650		115650

Figure 24.11c

done by postcard. This point is too soon, however, to notify subcontractors, as they will tie up the plans while the estimator is trying to complete the ductwork and piping takeoffs.

Equipment schedules are often included on the drawings rather than listed in the specifications. Typical details, such as trap assemblies, pump connections, coil connections, and duct assemblies, are also a great asset when they are provided with the plans. From the equipment totals many of these details will give the material quantities instantly, thereby avoiding an item by item takeoff while going through the floor plans.

The piping takeoff for the HVAC estimate is similar to those previously covered in plumbing and fire protection. Welding fittings for the larger pipe sizes will be more predominant in the heating and cooling systems. The HVAC takeoff is shown in Figures 24.13a through 24.13i).

Ductwork is scaled off the plans much like the piping. It is then converted to pounds of metal for takeoff and pricing. The takeoff for ductwork is shown in Figures 24.14a through 24.14d.

The tables for calculating the weight of duct from the linear measure and for obtaining the surface area of ductwork are shown in Figures 18.3 and 18.4.

Insulation follows the same takeoff procedures as for piping and ductwork, converting the fittings to linear feet for pricing as previously discussed. Figures 18.3 and 18.4 indicate the method of obtaining the duct surface area in square feet for insulation purposes.

The pipe, valve, fitting, ductwork, and equipment takeoffs are combined and the totals transferred to cost analysis sheets for pricing (see Figures 24.15a through 24.15g). Any subcontractor totals are also included on the cost analysis sheets when appropriate. All page totals are summarized on the last sheet (Figure 24.15g). Miscellaneous and subcontractor items are added and totaled to apply the final markups. Appropriate sales taxes, markups, and modifiers (as discussed earlier), are then added to complete the HVAC estimate.

Alternate Pricing Method

An alternate pricing method for the HVAC takeoff of the proposed project is illustrated in Figures 24.16a through 24.16d. The alternate method involves giving a list of material to the suppliers for lump sum quotations. This method is used when the estimator has the luxury of time and the advantage of reliable suppliers with pricing capabilities. It can save countless hours of pricing and extending, and is a free service provided by suppliers.

The estimator can break these lump sums down into pricing units for future reference. For example, bevel end pipe and threaded and coupled (T&C) pipe are priced separately. From the T&C total, the estimator can derive a certain percentage for screwed fittings and nipples. The estimator should retain costs from previous jobs, as they are good references for these percentages. Often, a pattern will emerge for valve or pipe hanger percentages based on pipe costs and job type. Labor is estimated from the lump sums instead of by a line item basis (shown in Figure 24.16c). Figure 24.16d is the estimate summary sheet.

This alternative method is based on the office building estimated earlier in this chapter, using only the heating and cooling system as an example. This pricing method could also be used for plumbing and/or fire protection depending on the estimator's preference.

Means' Forms

QUANTITY SHEET

PROJECT *Office Building*

ESTIMATE NO.

~~LOCATION~~ *HVAC*   ARCHITECT   DATE *1/4/91*

TAKE OFF BY *MJM*   EXTENSIONS BY: *MJM*   CHECKED BY: *JJM*

DESCRIPTION	NO.	DIMENSIONS			Ea.	UNIT	LF	UNIT		UNIT		UNIT
Heating												
Boiler Flue												
Dbl. Wall Galv. Chimney		24"	∅				12	LF				
Roof Flashing Collar		24"	∅		1	Ea.						
Dbl. Wall Tee		24"	∅		1	Ea.						
Tee Cap		24"	∅		1	Ea.						
Rain Cap & Screen		24"	∅		1	Ea.						
Boiler 3500 MBH Gas C.I.		Water			1	Ea.						
Electric Duct Heater		4 KW			1	Ea.						
Expansion Tank ASME		79 Gal.			1	Ea.						
Hot Water Pump		5 HP			2	Ea.						
Roof Exhaust Fan *		2200 CFM			1	Ea.						
Air-Conditioning												
Chiller		175 Tons			1	Ea.						
Fan Coil Unit		30 Tons			6	Ea.						
Cooling Tower		175 Tons			1	Ea.						
Condenser Water Pump		7½ HP			2	Ea.						
Chilled Water Pump		5 HP			2	Ea.						
Expansion Tank ASME		30 Gal.			1	Ea.						
* Fan supplied by HVAC contractor												
Installed by sheet metal contractor												

Figure 24.12

Means® Forms

PIPING SCHEDULE

PAGE HVAC 1 of 10

JOB Office Building

DATE 1/4/91

SYSTEM HVAC- Heating

BY JJM

	PIPE DIAMETER IN INCHES																
		12	10	8	6	5	4	3 1/2	3	2 1/2	2	1 1/2	1 1/4	1	3/4	1/2	
Heating B.S. Pipe Sch. 40 B.E.							40										
90° Ells Welded							6										
Tees Welded							3										
Weld Neck Flange							6										
Air Control Valve							1										
OS+Y Gate Valve Flanged							2										
Iron Body Check Valve Flanged							1										
Thread-O-Lets															2		
Thermometers and Wells															1		
Bolts/Nuts/Gaskets							6										
Automatic Air Vent															1		
Bronze Gate Valve Boiler Drain Screwed													2				

Figure 24.13*a*

Means® Forms

PIPING SCHEDULE

PAGE HVAC 2 of 10

JOB Office Building

DATE 1/4/91

SYSTEM HVAC- Chiller

BY JJm

	PIPE DIAMETER IN INCHES																	
			12	10	8	6	5	4	3 1/2	3	2 1/2	2	1 1/2	1 1/4	1	3/4	1/2	
Chiller B.S. Pipe B.E.	Sch.40							20										
90° Ells Welded								4										
Tees Welded								2										
Weld Neck Flange								4										
OS+Y Gate Valve Flanged								2										
Reducer Welded								2										
Thread-O-Lets																4		
Drain Valve/ Hose Bibb																2		
Thermometers and Wells																2		
Bolts/nuts Gaskets								4										

Figure 24.13b

Means® Forms

PIPING SCHEDULE

JOB Office Building

SYSTEM HVAC - Pumps

PAGE HVAC 3 of 10

DATE 1/4/91

BY JJM

			PIPE DIAMETER IN INCHES															
			12	10	8	6	5	4	3 1/2	3	2 1/2	2	1 1/2	1 1/4	1	3/4	1/2	
Pumps B.S. Pipe B.E.	Sch.40							40										
90° Ells Welded								8										
Tees Welded								6										
Weld Neck Flange								48										
Strainer								4										
OS + Y Gate Valve Flanged								4										
Gauges																8		
Flex Conn								8										
Lube Plug Shut off and Balance Valve								4										
Thread-O-Lets																16		
Bolts/nuts Gaskets								48										
Drain Valve/ Hose Bibb																8		
Check Valve Iron Body								4										

Figure 24.13c

Means® Forms

PIPING SCHEDULE

JOB Office Building

SYSTEM HVAC- Mech. Room Loop Piping/Risers

PAGE HVAC 4 of 10

DATE 1/4/91

BY JJM

	PIPE DIAMETER IN INCHES																	
			12	10	8	6	5	4	3 1/2	3	2 1/2	2	1 1/2	1 1/4	1	3/4	1/2	
Mech. Room Loop										Bev.	⟷		T&C					
B.S. Pipe Sch. 40 B.E.								120										
90° Ells Welded								8										
Tees Welded								4										
Risers B.E/T&C B.S. Pipe Sch. 40								20		40	40	40						
Tees Welded/THD								2		2	2	C.I. 3						
Reducers Welded								2		2	2							
Drain Valve/ Hose Bibb																2		
Thread-O-Lets																2		
Air Vent																1		

Figure 24.13d

Means® Forms

PIPING SCHEDULE

JOB Office Building

SYSTEM HVAC – Condensate Drain

PAGE HVAC 5 of 10

DATE 1/4/91

BY JJM

	PIPE DIAMETER IN INCHES																	
			12	10	8	6	5	4	3 1/2	3	2 1/2	2	1 1/2	1 1/4	1	3/4	1/2	
Cond. Drain PVC													200					
San Tee PVC													24					
90° Ell PVC													2					
Adapt PVC to THD													6					
P Traps PVC													6					

Figure 24.13e

Means' Forms
PIPING SCHEDULE

JOB Office Building

SYSTEM HVAC – Condenser Water Piping (Cooling Tower)

PIPE DIAMETER IN INCHES

Item	1/2	3/4	1	1 1/4	1 1/2	2	2 1/2	3	3 1/2	4	5	6	8	10	12
Cond. Water Piping B.S. Pipe Sch.40 B.E.											80				
90° Ells welded											11				
Tees welded											5				
Strainers											2				
Thermometers and wells		2													
Flex Conn											4				
Iron Body Check											2				
Shut off and balancing valve											2				
OS&Y Gate Valve											2				
Weld neck Flange											20				
Nuts/Bolts Gaskets											20				
Drain/Valve Hose Bibb		2													
Thread-O-Lets		8													

Gauges 4

Figure 24.13f

Means® Forms

PIPING SCHEDULE

PAGE HVAC 7 of 10

JOB Office Building

DATE 1/4/91

SYSTEM HVAC – Cooling Tower Drain System

BY JJM

	PIPE DIAMETER IN INCHES																	
			12	10	8	6	5	4	3 1/2	3	2 1/2	2	1 1/2	1 1/4	1	3/4	1/2	
Cool Tower Drain Galv. Stl. Sch 40 T&C												20						
90° Ells Cast												5						
Tees Cast												1						
Union												1						
Gate Valve												1						

Figure 24.13*g*

Means® Forms

PIPING SCHEDULE

PAGE HVAC 8 of 10

JOB Office Building

DATE 1/4/91

SYSTEM HVAC — Boiler/Chiller/Cooling Tower Cold Water Make Up

BY JJM

	PIPE DIAMETER IN INCHES																
			12	10	8	6	5	4	3 1/2	3	2 1/2	2	1 1/2	1 1/4	1	3/4	1/2
Make Up																	
L Copper															40		
90° Ells Copper															12		
Tees Copper															4		
Gate Valve Copper															4		
Globe Valve Copper															1		
Check Valve Copper															3		
Reduce Pressure Valve-Copper															1		
Expansion Tank																	
B.S. Pipe Sch 40		T+C													20		
90° Ells C.I.															4		
Gate Valve															1		
Drain Valve/ Hose Bibb																1	
Union															1		

Figure 24.13*h*

Means' Forms

PIPING SCHEDULE

JOB Office Building

SYSTEM HVAC — Air Handling Units (AHU)

PAGE HVAC 9 of 10

DATE 1/4/91

BY JJM

			PIPE DIAMETER IN INCHES															
			12	10	8	6	5	4	3 1/2	3	2 1/2	2	1 1/2	1 1/4	1	3/4	1/2	
AHU B.S. Pipe Sch40		T+C										180						
90° Ells C.I.												36						
Tees C.I.												42						
Unions												21						
Gate Valve												12						
Air vent																6		
Drain Valve/ Hose Bibb																6		
Thermo-meters																12		
Gauges																12		
3-way Temp. Cont. Valve												6						

Figure 24.13*i*

DUCTWORK SCHEDULE

JOB *Office Building*

SYSTEM *Noted: Duct* HVAC

MATERIAL *Galv. Steel*

BY *MJM*

DRAWING NO. *ME-1*

DUCT SIZE		LINING	INSUL	GAUGE	LENGTH	TOTAL LENGTH	LBS/ FOOT	TOTAL POUNDS
X					SUPPLY DUCT AHU #1-A (AHU #2 & 3A IDENTICAL)			
	X							
65 X 14		NO	2"	20	4	4	22.9	92
45 X 14				22	6, 34	40	14.8	592
25 X 14				24	20, 8, 18	46	8.4	386
22 X 14					50, 16, 16, 50	132	7.8	1030
18 X 14					24, 24, 24, 24	96	6.9	662
15 X 14					23, 23	46	6.2	285
14 X 14					23, 23	46	6.0	276
13 X 12					4, 4	8	5.4	43
12 X 9					24, 24	48	4.5	216
12 X 8		↓	↓	↓	20, 10, 10, 20	60	4.3	258
X								
	X						TOTAL	3840#
X								
	X							
X					SUPPLY DUCT AHU #1-B (AHU #2 & 3B IDENTICAL)			
	X							
65 X 14		No	2°	20	4	4	22.9	92
48 X 14				22	16	16	15.5	248
45 X 14				22	12, 10	22	14.8	326
22 X 14				24	16, 16	32	7.8	250
18 X 14					12, 12, 12, 12	48	6.9	331
14 X 14					24, 24, 24, 24	96	6.0	576
13 X 12					12, 12	24	5.4	130
12 X 9					20, 20, 24, 24, 20, 20	128	4.5	576
12 X 8					16, 16, 8, 8	48	4.3	206
11 X 10					12	12	4.5	54
10 X 8					8, 12	20	3.9	78
8 X 6		↓	↓	↓	16, 18	34	3.0	102
X								
	X						TOTAL	2969
X						1 FLOOR	TOTAL	6809
	X							
X								
	X				SURFACE AREA FOR INSUL. (TOP FLOOR ONLY)			
X								
	X			20	92 + 92 = 184/1.656 =	111		
X				22	592 + 248 + 326 = 1166/1.406 =	829		
	X			24	5459 / 1.156	4722		
X								
	X				1 FLOOR TOTAL	5662	SF	
X								
	X							
X								

Figure 24.14a

Means' Forms

DUCTWORK SCHEDULE

JOB Office Building

DATE 1-8-91

SYSTEM noted: Duct   HVAC   BY mJm

MATERIAL Galv. Steel   DRAWING NO. ME-1

DUCT SIZE			LINING	INSUL.	GAUGE	LENGTH	TOTAL LENGTH	LBS./FOOT	TOTAL POUNDS
	X								
	X			Return Air		AHU #1-A (AHU #2+3A Same)			
	X								
30	X	30	no	no	24	8	8	12.9	103
50	X	14			22	32	32	16.0	512
33	X	14			22	24	24	11.8	283
18	X	14	↓	↓	24	22	22	6.9	152
	X								
	X							Total	1050#
	X								
	X			Return Air		AHU #2-B (AHU #2+3B Same)			
	X								
30	X	30	no	no	24	6	16	12.9	77
50	X	14			22	16,7	23	16.0	368
33	X	14			22	16	16	11.8	189
18	X	14	↓	↓	24	18,1	19	6.9	131
	X								
	X							Total	765#
	X								
	X						1 Floor	Total	1815#
	X								
	X			Fresh Air Intake (Insulated)					
	X								
42	X	20	no	2"	22	8	8	15.5	124
32	X	20			22	15	15	13.0	195
20	X	20			24	15	15	8.6	129
16	X	16	↓	↓	24	108	108	6.9	745
	X								
	X							Total	1193#
	X								
	X			Insulation (FAI)					
	X				22	124+195 = 319/1.406	227SF		
	X				24	129+745 = 874/1.156	756SF		
	X								
	X					Total	983SF		
	X								
	X								
	X								
	X								
	X								
	X								
	X								
	X								
	X								

Figure 24.14b

DUCTWORK
SCHEDULE

PAGE Sm 3 OF 4
DATE 1 / 8 /91
BY MUM
DRAWING NO. ME-1

JOB *Office Building*

SYSTEM *noted: Duct*   HVAC

MATERIAL *Galv. Steel*

DUCT SIZE		LINING	INSUL.	GAUGE	LENGTH	TOTAL LENGTH	LBS./ FOOT	TOTAL POUNDS
	X							
	X				TOILET EXHAUST			
18 X 18		NO	NO	24	8	8	7.8	62
15 X 15					15	15	6.5	98
12 X 12					15	15	5.2	78
8 X 8					30	30	3.4	102
8 X 6					30	30	3.0	90
6 X 6					36	36	2.6	94
6 X 4		↓	↓	↓	78	78	2.2	172
	X							
	X						TOTAL	696 #
	X							
	X							
	X							
	X				MECHANICAL ROOM (FAI)			
	X							
48 X 24		NO	NO	22	5	5	18	90
48 X 12		''	''	''	6	6	15	90
	X							
	X						TOTAL	180 #
	X							
	X							
	X							
	X							
	X							
	X							
	X							
	X							
	X							
	X							
	X							
	X							
	X							
	X							
	X							
	X							
	X							
	X							
	X							
	X							

Figure 24.14c

Means' Forms

QUANTITY SHEET

PROJECT Office Building

ESTIMATE NO.

LOCATION   ARCHITECT   DATE 1/8/91

TAKE OFF BY MJM   EXTENSIONS BY: MJM   CHECKED BY: JJM

DESCRIPTION	NO.	DIMENSIONS		Galv. Steel	# UNIT	Duct Insul.	SF UNIT	Ea.	UNIT	LF	UNIT
Duct											
Supply Top Floor				6809	Lbs.	5662	SF				
" Floor 1 + 2				13618	Lbs.						
Return Top Floor				1815	Lbs.						
" Floor 1 + 2				3628	Lbs.						
Fresh Air Intake				1193	Lbs.	983	SF				
Toilet Exhaust				696	Lbs.						
Mechanical Room (FAI)				180	Lbs.						
Total				27939	Lbs.	6645	SF				
Diffusers											
T-Bar Lay-In		15	15					120	Ea.		
" " "		9	9					15	Ea.		
Return Air Reg.											
T-Bar Lay-In		18	18					18	Ea.		
Flex Connector										180	LF
Fresh Air Intake Hood/Curb		20	42					1	Ea.		
" " " Damper								6	SF		
Volume Dampers		16	16					6	Ea.		
Toilet Exhaust Reg.		6	6					6	Ea.		
" " "		8	8					18	Ea.		
Fire Damper		12	12					1	Ea.		
" "		15	15					1	Ea.		
" " Access Door								2	Ea.		
Roof Fan (Labor Only- Fan furnished by HVAC)								1	Ea.		
Elevator Relief Wall Louver		18	12					1	Ea.		
" " Trans Grille		18	12					1	Ea.		
Mech. Room Wall Louver		48	48					1	Ea.		
" " Damper		48	24					1	Ea.		
" " "		48	12					1	Ea.		
(Boiler Flue By Others)											

Figure 24.14d

308

Means' Forms
COST ANALYSIS

PROJECT: Office Workbook
LOCATION
TAKE OFF BY MJM
QUANTITIES BY MJM
PRICES BY MJM
CLASSIFICATION Div. 15 - HVAC
ARCHITECT
EXTENSIONS BY MJM

DESCRIPTION	SOURCE/DIMENSIONS		QUANTITY	UNIT	MATERIAL		LABOR		EQUIPMENT		SUBCONTRACT		TOTAL	
					UNIT COST	TOTAL	UNIT COST	TOTAL	UNIT COST	TOTAL	UNIT COST	TOTAL	UNIT COST	TOTAL
Sheet Metal														
Total Duct	157	250 0580	28,000	Lbs.	.45	12600	1.96	54880						
Diffusers														
T-Bar 15"×15"	157	450 2060	120	Ea.	92.40	11088	18.20	2184						
9"×9"	157	450 2020	15	Ea.	56.10	842	14.30	215						
Ret. Air Reg.														
T-Bar 18"×18"	157	450 2080	18	Ea.	93.40	1681	20	360						
Flex Connector	157	480 2000	180	LF	1.39	250	2.00	360						
FAI Hood/Curb 20"×42"		490 5680	1	Ea.	875	875	90	90						
FAI Damper 20"×42"	↓	490 7600	6	SF	46.90	281	3.60	22						
Volume Dampers 16"×16"	157	480 8180	6	Ea.	34.80	209	13.35	80						
Toilet Exh. Reg. 6"×6"	157	470 5040	6	Ea.	12.10	73	8.35	50						
8"×8"	157	470 5100	18	Ea.	14.85	267	10.55	190						
Fire Damper 12"×12"	157	480 3180	1	Ea.	20.75	21	10	10						
" 15"×15"		3240	1	Ea.	24.75	25	11.10	11						
" Access Door	↓	1020	2	Ea.	11.55	23	18.20	36						
Elev. Relf. Wall Lvr. 18"×12"	157	482 3200	1.5	SF	26.30	54	7.15	11						
" Trans Grille 18"×12"		460 5160	1	Ea.	28.50	29	10	10						
Toilet Exhaust Fan	↓	290 7180	1	Ea.	-		115	115						
Page HVAC 1						283318		586624						

(Labor Only Material Supplied by HVAC) For Sheet Metal

Figure 24.15a

Means' Forms
COST ANALYSIS

PROJECT Office Building

LOCATION

TAKE OFF BY MJM QUANTITIES BY MJM PRICES BY MJM EXTENSIONS BY MJM

CLASSIFICATION Div.15- HVAC

ARCHITECT

CHECKED BY JJM

DESCRIPTION	SOURCE/DIMENSIONS		QUANTITY	UNIT	MATERIAL		LABOR		EQUIPMENT		SUBCONTRACT		TOTAL	
					UNIT COST	TOTAL	UNIT COST	TOTAL	UNIT COST	TOTAL	UNIT COST	TOTAL	UNIT COST	TOTAL
Sheet metal (Cont'd)														
Mech. Rm. Wall Lvr. 48"x48"	157	482 3200	16	SF	26.30	4211	7.15	1114						
" " Relief Damp 48x24"		3500	8	↓	13.15	105	2.15	17						
" " Manual 48"x12"	↓	3500	4	↓	13.15	53	2.15	9						
Page HVAC 2 Subtotal						579		140						
Sheet metal Subtotals														
HVAC 1						28318		58624						
HVAC 2						579		140						
Subtotals Total						28897		58764						
Material Sales Tax					5%	1445								
Handling, Overhead + Profit Markup					10%	2890	51.9%	30499						
Total						33232		89263						122495
City Modifier for Boston					x 1.040		x 1.185							
Boston Total						34561		105777						140338

Figure 24.15*b*

Means' Forms
COST ANALYSIS

PROJECT: Office Building
LOCATION
TAKE OFF BY: MJM
QUANTITIES BY: MJM
PRICES BY: MJM
CLASSIFICATION: Div. 15 - HVAC
ARCHITECT
EXTENSIONS BY: MJM
DATE: 1-9-91
CHECKED BY: JJM

DESCRIPTION	SOURCE/DIMENSIONS	QUANTITY	UNIT	MATERIAL UNIT COST	MATERIAL TOTAL	LABOR UNIT COST	LABOR TOTAL	EQUIPMENT UNIT COST	EQUIPMENT TOTAL	SUBCONTRACT UNIT COST	SUBCONTRACT TOTAL	TOTAL UNIT COST	TOTAL
Piping													
B.S. Pipe Sch. 40 T&C 1"	151 701 0580	20	LF	1.49	30	3.84	77						
2"	0610	220		2.91	640	5.75	1265						
BE 2½"	2082	40		3.74	150	7.80	312	.83	33				
3"	2090	40		4.19	188	8.50	340	.91	36				
4"	2110	240		6.77	1625	9.90	2376	1.05	252				
5"	2120	80		11.33	906	11.45	916	1.22	98				
90° Ell Cast 1"	151 716 0110	4	Ea.	1.30	5	15.65	63						
2"	0140	39		3.92	153	20	780						
90° Ell weld 4"	720 3130	26		16.80	437	73	1898	7.80	203				
5"	3140	11		41	451	115	1265	7.80	86				
Tee Cast 2"	151 716 0580	42	Ea.	5.55	233	33	1386						
weld 2½"	720 3420	2		20	40	73	146	7.80	16				
3"	3430	2		23	46	92	184	9.75	20				
4"	3440	17		31	527	120	2040	13	221				
5"	3450	5		57	285	190	950	13	65				
Union 1"	151 716 7050	1	Ea.	3.37	3	16.95	17						
2"	7080	12		7.15	86	22	264						
Gate Valve 1"	151 955 3450	1	Ea.	15	15	10.70	11						
2"	3480	12		35	420	18.50	222						
Page HVAC 3 Subtotal					6304		14710		1030				

Figure 24.15c

Means' Forms
COST ANALYSIS

PROJECT Office Building   CLASSIFICATION Div. 15 - HVAC

LOCATION   ARCHITECT

TAKE OFF BY MJM   QUANTITIES BY MJM   PRICES BY MJM   EXTENSIONS BY MJM

DESCRIPTION	SOURCE/DIMENSIONS			QUANTITY	UNIT	MATERIAL		LABOR		EQUIPMENT		SUBCONTRACT		TOTAL	
						UNIT COST	TOTAL	UNIT COST	TOTAL	UNIT COST	TOTAL	UNIT COST	TOTAL	UNIT COST	TOTAL
Piping (Cont'd)															
Reducer Welded 2½"	151	720	3443	2	Ea.	16.80	34	41	82	4.33	9				
3"	151	720	3444	2		17.90	36	46	92	4.88	10				
4"	151	720	3495	4		21	84	61	244	6.50	26				
Weld Neck Flange 4"	151	720	6500	58	Ea.	21	1218	37	2146	3.90	226				
5"	151	720	6510	20		27	540	46	920	4.88	98				
Flanged Fittings + Valves															
Water Check Valve 4"	151	960	6680	2	Ea.	210	420	95	190						
OS+Y Gate Valve 4"			3680	8		215	1720	120	960						
5"			3690	2		355	710	170	340						
Strainer 4"	156	612	1060	4		541	780	120	480						
5"	156	612	1080	2		305	610	170	340						
Iron Body Check 4"	151	960	6060	5		190	950	120	600						
Flex Connectors 4"	156	225	0540	8		298	2384	46	368						
5"	156	225	0560	4		329	1316	52	208						
Lube Plug Valve 4"	151	990	7030	4		350	1400	120	480						
5"		990	7040	2		585	1170	190	380						
Bolt+Gasket Set 4"		720	0670	58		8.85	513	7.55	438						
5"		720	0680	20		12.95	259	7.85	157						
Air Control Valve 4"	156	205	0120	1		950	950	120	120						
3-way Temp. Cont. Valve 2"	157	420	7400	6		460	2760	22	132						
Thread-O-Let 3/4"	151	720	5233	32		580	1860	17.45	558	1.86	60				
Auto. Air Vent 3/4"	156	201	0220	8		174	1392	20	160						
Thermometers + Wells 3/4"	157	420	4620	17		43	731	7.30	124						
Page HVAC 4 Subtotal							2011163		9519		429				

Figure 24.15d

Means® Forms

COST ANALYSIS

SHEET NO. HVAC 5 of 7

PROJECT Office Building | CLASSIFICATION Div. 15- HVAC | ESTIMATE NO. 91-1

LOCATION | ARCHITECT | DATE 1-9-91

TAKE OFF BY mjm | QUANTITIES BY mjm | PRICES BY mjm | EXTENSIONS BY mjm | CHECKED BY JJM

DESCRIPTION		SOURCE/DIMENSIONS			QUANTITY	UNIT	MATERIAL		LABOR		EQUIPMENT		SUBCONTRACT		TOTAL	
							UNIT COST	TOTAL	UNIT COST	TOTAL	UNIT COST	TOTAL	UNIT COST	TOTAL	UNIT COST	TOTAL
Piping (Cont'd)																
Bronze Gate Valve Scr	1½"	151	955	3470	2	Ea.	24	48	15.65	31						
Gauges	¾"	157	420	2300	24		12	288	6.40	154						
Drain Valve (Hose Bibb)	¾"	151	141	5000	21		4.50	95	8.50	179						
PVC Piping DWV																
Pipe Sch. 40	1½"	151	551	4420	200	LF	.83	166	5.65	1130						
90° Ell	1½"		558	5170	2	Ea.	.70	1	12.75	26						
San Tee	1½"		558	5270	24		.96	23	20	480						
P-Traps	1½"		181	6910	6		4.11	25	11.30	68						
Adapt PVC/THD	1½"		558	8520	6		.53	3	11.30	68						
Galv. Steel Piping																
Pipe Sch. 40	2"	151	701	1360	20	LF	3.53	71	5.75	115						
90° Ell C.I.	2"		716	0800	5	Ea.	8.25	41	20	100						
Tee C.I.	2"			1180	1		9.45	9	33	33						
Union	2"			7080 7250	1		8.22	8	22	22						
Gate Valve	2"		960	1700	1		140	140	18.50	19						
Copper (Sweat)																
"L" Tube	1"	151	401	2200	40	LF	2.45	98	2.99	120						
Tees	1"		430	0510	4	Ea.	3.55	14	20	80						
Gate Valve	1"		955	2950	4		17.50	70	10.70	43						
Globe Valve	1"		955	4970	1		27	27	10.70	11						
90° Ells	1"		430	0130	12		.99	12	12.75	153						
		Page HVAC 5 Subtotal						1139		2832						

Figure 24.15e

Means' Forms
COST ANALYSIS

PROJECT Office Building
LOCATION
TAKE OFF BY MJM   QUANTITIES BY MJM   PRICES BY MJM
CLASSIFICATION Div. 15 - HVAC
ARCHITECT
EXTENSIONS BY MJM   CHECKED BY JJM

DESCRIPTION	SOURCE/DIMENSIONS	QUANTITY	UNIT	MATERIAL UNIT COST	MATERIAL TOTAL	LABOR UNIT COST	LABOR TOTAL	EQUIPMENT UNIT COST	EQUIPMENT TOTAL	SUBCONTRACT UNIT COST	SUBCONTRACT TOTAL	TOTAL UNIT COST	TOTAL
Copper (Sweat) Cont'd													
Check Valve 1"	151 955 1870	3	Ea.	14.50	44	10.70	32						
Reduce Press. Valve 1"	151 955 6960	1	Ea.	100	100	10.70	11						
Heating Equipment													
Boiler Flue													
Dbl Wall Straight 24Ø	155 680 0340	12	LF	102	1224	17.50	210						
Roof Flashing Collar 24Ø	1170	1	Ea.	172	172	47	47						
Dbl Wall Tee 24Ø	1330	1		506	506	47	47						
Tee Cap 24Ø	1590	1		33.20	33	27	27						
Rain Cap + Screen 24Ø	1880	1	↓	470	470	28	28						
Toilet Exhaust Fan	157 290 7180	1		1400	1400	*	—						
Boiler 3500 MBH Gas													
C.I. Water	155 115 3400	1	Ea.	19900	19900	3000	3000		650				650
Electric Duct Htr. 4kw	408 0100	1		277	277	29	29						
Expansion Tank ASME 79 G	671 3060	1		1590	1590	73	73						
Hot Water Pump 5 HP	152 410 4300	2	↓	1115	2230	205	410						
Air Conditioning													
Chiller 175 Ton	157 190 1600	1	Ea.	7180	71800	13000	13000						
Fan Coil Units 30 Ton	150 0240	6	Ea.	4830	28980	950	5700						
Cooling Tower 175 Ton	240 1900	175	Ton	55	9625	4.53	793						
Conders. Water Pump 7.5 HP	152 410 4420	2	Ea.	1505	3010	230	460						
Chilled Water Pump 5 HP	410 4410	2		1235	2470	230	460						
Expansion Tank ASME 30 G	155 671 3020	1	↓	1050	1050	46	46						
Page HVAC 6 Subtotal					114488 11		243373		650				

* Labor is by Sheetmetal

Figure 24.15f

Means' Forms — COST ANALYSIS

PROJECT Office Building
LOCATION
TAKE OFF BY MJM QUANTITIES BY MJM PRICES BY MJM
CLASSIFICATION Div. 15 – HVAC
ARCHITECT
EXTENSIONS BY MJM CHECKED BY JJM

DESCRIPTION	SOURCE/DIMENSIONS	QUANTITY	UNIT	MATERIAL UNIT COST	MATERIAL TOTAL	LABOR UNIT COST	LABOR TOTAL	EQUIPMENT UNIT COST	EQUIPMENT TOTAL	SUBCONTRACT UNIT COST	SUBCONTRACT TOTAL	TOTAL UNIT COST	TOTAL
HVAC Subtotals													
HVAC 1				sheet metal	6304		14710		1030				
2				sheet metal									
3					20163		9519		429				
4					1139		2832						
5													
6					14881		24373		650				
Miscellaneous													
Record Drawings		8	Hrs.			25.50	204						
Operating Instructions		8	→			→	204						
Maintenance Manuals, Tags & Charts							102						
Cleaning System		16	↓			↓	408						
Subcontracts													
Insulation s/m+Pipe (HVAC)	155 / 651			–	Quoted –						181192		
Balancing Air	157 / 601			–	Quoted –						5781		
Water	157 / 604			–	Quoted –						1600		
Temp. Control Electronic	157 / 430			–	Quoted –						357732		
Crane Service	016 / 460			–	Quoted –						1275		
Piping Subtotal					172487		523153		21109		62580		
Sales Tax				5%	8624								
Handling, Overhead + Profit Markup				10%	17249	49.2%	257758	10%	2111	10%	6258		
Piping Total					198360		781111		23320		68838		
Sheet Metal Total					33232		89263						
Combined Total					231592		167374		23320		688838		470112
City Modifier for Boston				x 1.040		x 1.185		x 1.040		x 1.1112			
Boston Total					240856		198338		24413		765548		518815

Figure 24.15g

**COST
ANALYSIS**

SHEET NO. 1 of 4

PROJECT Office Building

ESTIMATE NO.

ARCHITECT HVAC Piping

DATE 3/17/91

TAKE OFF BY: JJM   QUANTITIES BY: JJM   PRICES BY: CDS   EXTENSIONS BY: CHECKED BY:

80'-5" Blk Stl Pipe, B.E.
240'-4"
40'-3"
40'-2½" $

220'-2" Blk Stl Pipe T+C
20'-1"   "   " $

11-5" Weld Elbows
26-4"   "
5-5" Weld Tees
17-4"
2-3" $
2-2½"
4-4"x3" Weld Reducers
2-3"x 2½"   "
2-2½" x 2"   "
20-5" Weld Neck Flanges
40-4"   "

2-5" L.R., C.I. Flanged Ells
4-4"   "   "   "   "
2-5" C.I. Flanged Base Ells $
4-4" "   "   "   "

20 Sets 5" Bolts, Nuts, Gaskets $
40 "   4"   "

2-5" OS+Y Flanged Gates
8-4"
2-5" Flanged Check Valves
4-4" $
12-2" Bronze Gate Valves
1-2" Ball Valve
21-3/4" Hose End Drain Cocks

2-5" I.B Flanged Strainers $
4-4"   "   "   "
2-5" Flanged Plug Cocks
4-4"   "   "   " $

Figure 24.16a

▲ Means· Forms

TELEPHONE QUOTATION

PROJECT Office Building

TIME 1:15 p.m.

FIRM QUOTING C.D.S. Piping Supply

PHONE (617) 747-1270

ADDRESS

BY D.B.

ITEM QUOTED Duel Temp. System

RECEIVED BY

WORK INCLUDED		AMOUNT OF QUOTATION
Bevel End Steel Pipe	$ 2,869	
T+C Steel Pipe	$ 670	
Weld Fittings + Flanges	$ 7,398	
Bolts, Nuts and Gaskets	$ 772	
Valves	$ 7,424	
Strainers	$ 1,390	
Lubricated Plug Cocks	$ 2,570	

DELIVERY TIME Stock Items   **TOTAL BID**

DOES QUOTATION INCLUDE THE FOLLOWING: If ☐ NO is checked, determine the following:

STATE & LOCAL SALES TAXES	☐ YES	☑ NO	MATERIAL VALUE	
DELIVERY TO THE JOB SITE	☑ YES	☐ NO	WEIGHT	
COMPLETE INSTALLATION	☐ YES	☐ NO	QUANTITY	
COMPLETE SECTION AS PER PLANS & SPECIFICATIONS	☐ YES	☐ NO	DESCRIBE BELOW	

EXCLUSIONS AND QUALIFICATIONS

note: Substitute butterfly valves for
     lubricated plugs   Deduct $1,960

ADDENDA ACKNOWLEDGEMENT   **TOTAL ADJUSTMENTS**

                                                 ADJUSTED TOTAL BID

ALTERNATES

ALTERNATE NO.
ALTERNATE NO.
ALTERNATE NO.
ALTERNATE NO.
ALTERNATE NO.
ALTERNATE NO.
ALTERNATE NO.

Figure 24.16b

Means® Forms

CONDENSED ESTIMATE SUMMARY

			SHEET NO. 3 of 4
PROJECT Office Building			ESTIMATE NO.
LOCATION Boston		TOTAL AREA/VOLUME	DATE 3-25-91
ARCHITECT means		COST PER S.F./C.F.	NO. OF STORIES
PRICES BY: JJm		EXTENSIONS BY: JJm	CHECKED BY: mJm

	LABOR	DAYS (Pair)	
1	Cast Iron Boiler Packaged Work with Riggers		4
2	Expansion Tanks and miscellaneous Trim		3
1	Packaged Chiller Work with Riggers		5
6	Air Handling Units Work with Riggers		6
1	Cooling Tower Work with Riggers		5
6	Base mounted Pumps		8
	Piping		35
	welding		25
	Sleeves, Hangers and Inserts		3
	Supports, Anchors and Guides		3
	Testing		5
	Supervision		6
	Record Drawings		3
	Operating Instructions + manuals		2
	Valve Tags, Charts, Arrows		2
	Clean Systems		2
	Gauges, Thermometers, Air Vents		2
	Call Backs		2
			121
	Lost Time 5%		6
	Total		127
	127 Days × 410 = $52,070		

Figure 24.16c

318

Means' Forms

**CONDENSED
ESTIMATE SUMMARY**

PROJECT	Office Building		ESTIMATE NO.	
LOCATION	Boston, ma	TOTAL AREA/VOLUME	DATE	3/27/91
ARCHITECT	R.S. means	COST PER S.F./C.F.	NO. OF STORIES	3
PRICES BY: JJm		EXTENSIONS BY: JJm	CHECKED BY: mJm	

1	Boiler C.I. Gas 3500 MBH	19900
2	Expansion Tanks and misc. Specialties	3820
	Boiler Flue – Prefab	2405
1	Chiller 175 Ton	71800
6	Air Handling Units	28980
1	Duct Coil Electric	280
1	Roof Exhaust Fan	1400
1	Cooling Tower	9625
6	Pumps	6530
28,000#	Ductwork	122495
	Black Steel Pipe 670 T+C 2869 B.E.	3539
	Fittings Cast Iron 100% 670 Weld 7398	8068
	Galv. Pipe	71
	Galv. Fitting	60
	Bolts, Nuts + Gaskets	772
	PVC Pipe	166
	PVC Fittings	52
	Copper Tube Type L	98
	Copper Fittings	26
	Pipe Hangers + Supports	180
	Miscellaneous Steel, Supports, Anchors, Guides	500
	Valves	7424
	Air Vents	1392
	Gauges and Thermometers, 17	731
12	Flexible Pump Connectors	3700
	Strainers	1390
	Trucking and Cartage	200
	Subcontractors	
	Balancing	7381
	Rigging	1275
	Insulation	18192
	Temperature Control	35732
	Labor 127 days/pair @ 410/day	52070
	Cost	410254

Figure 24.16d

Subcontractors: The prime mechanical contractor usually subcontracts a portion of the mechanical work. These subcontractors will employ estimators who specialize in their individual fields. Typical mechanical subcontractors are Air and Water Balancing, Hoisting and Rigging, Insulation, Sheet Metal, and Temperature Control. In this example, the prime mechanical contractor will accomplish the sheet-metal work, and the other four trades will be subcontracted. Competitive bids will be solicited for the four subcontracted trades. Figures 24.17a through 24.17d are representative subcontractors' estimates for the proposed project, which are either mailed or telephoned to the prime mechanical contractor.

In a competitive situation, the subcontractors will usually submit their bids by telephone at the last possible minute to prevent any "bid shopping." Figures 24.18a through 24.18d are typical subcontractor quotes received by telephone and recorded by the prime mechanical contractor.

Overhead and Profit: Since fire protection, plumbing, and sheet-metal contractors do not usually have to purchase large quantities of expensive equipment, their overhead markups are based primarily on labor. In this case the markup for labor makes up a relatively large percentage of the total cost, and, therefore, the overhead markup percentage will likely be low in relation to total project cost. A contractor whose work involves large equipment purchases and several subcontractors, on the other hand, must include overhead percentages for these items in addition to the labor percentage. Published industry guides such as *Means Mechanical Cost Data* or *Means Plumbing Cost Data* provide useful overhead percentages. However, such books should be used primarily as a reference. Every mechanical task has unique labor and material requirements, and markups for overhead and profit should ultimately be based on the contractor's individual situation.

The Recap Sheet: It is generally good practice to list the cost totals from each segment of the work onto one sheet of paper, often called a "Recap" (shown in Figure 24.19). This procedure will allow the estimator to review all of the segments of work and to add them into a total contractor's cost.

The Project Schedule: When the work for all sections is priced, the estimator should complete the project schedule so that time-related costs in the Project Overhead Summary can be determined. When preparing the schedule, the estimator must visualize the entire construction process in order to determine the correct sequence of work. Certain tasks must be completed before others are begun. Different trades will work simultaneously. Material deliveries will also affect scheduling. All such variables must be incorporated into the Project Schedule. An example of a Project Schedule is shown in Figure 24.20. The man-hour figures, which have been calculated for each section, are used to assist with scheduling. The estimator must be careful not only to use the man-hours for each section independently, but to coordinate each section with related work.

The schedule shows that the project will last approximately one year. The duration of the mechanical work will be about six months. Time-dependent items, such as equipment rental and superintendent costs, can be analyzed on a Project Overhead Summary form (Figures 24.21a and 24.21b). Some items, such as permits and insurance, are dependent on total job costs. The total direct costs for the project from the Estimate Summary can be used to estimate these costs. These items should be analyzed individually if possible. Alternatively, percentages can be added to material and labor to cover these

Means' Forms

COST ANALYSIS

PROJECT Office Building

Quote Preparation
By Subcontractor
CLASSIFICATION Div.15- Balancing

LOCATION

ARCHITECT

DATE 1/9/91

TAKE OFF BY MJM   QUANTITIES BY MJM   PRICES BY MJM   EXTENSIONS BY MJM   CHECKED BY JJM

DESCRIPTION	SOURCE/DIMENSIONS			QUANTITY	UNIT	MATERIAL		LABOR		EQUIPMENT		SUBCONTRACT		TOTAL	
						UNIT COST	TOTAL	UNIT COST	TOTAL	UNIT COST	TOTAL	UNIT COST	TOTAL	UNIT COST	TOTAL
Balancing Air Flow															
Roof Exhaust Fan	157	601	1400	1	Ea.							120	120		120
Air Supply Diff.			3000	135								37	4995		4995
Air Supply Reg.			3000	18								37	666		666
Air Total													5781		5781
Balancing Water Flow															
Packaged Chiller	157	604	0200	1	Ea.							260	260		260
Cooling Tower			0500	1								200	200		200
Fan Coil Units			0600	6								55	330		330
Pumps			1000	6								135	810		810
Water Total													1600		1600
Total															7381

Figure 24.17a

Means® Forms

TELEPHONE QUOTATION

Q 2-A

DATE 1/11/91

PROJECT Office Building

TIME 10:45 a.m.

FIRM QUOTING Wind + Water Balance Co., Inc.

PHONE (543) 548-5980

ADDRESS

BY Jack Frost

ITEM QUOTED Independent HVAC testing, balance, adjustment

RECEIVED BY Marcia

WORK INCLUDED TBA	AMOUNT OF QUOTATION
Air balancing of roof exhaust fan air handling units, registers, diffusers and tower fan.	$ 5781
Water balance of pumps, hot and chilled water systems including the flow through the chiller evaporator and condenser	$ 1600
Price is based on sending the original balancing report to the architect plus a marked up set of the contract drawings will be left with the owner.	

	TOTAL BID	$ 7381

DOES QUOTATION INCLUDE THE FOLLOWING:

If ☐ NO is checked, determine the following:

STATE & LOCAL SALES TAXES	☑ YES ☐ NO	MATERIAL VALUE
DELIVERY TO THE JOB SITE	☐ YES ☐ NO	WEIGHT
COMPLETE INSTALLATION	☐ YES ☐ NO	QUANTITY
COMPLETE SECTION AS PER PLANS & SPECIFICATIONS	☑ YES ☐ NO	DESCRIBE BELOW

EXCLUSIONS AND QUALIFICATIONS

"Entire system must be in and operating before we are able to do this work. All areas must be available to our technicians. HVAC contractor is to supply us with a complete set of plans."

ADDENDA ACKNOWLEDGEMENT

TOTAL ADJUSTMENTS

ADJUSTED TOTAL BID

Figure 24.18a

🔺 Means' Forms

COST ANALYSIS

PROJECT **Office Building**

LOCATION

TAKE OFF BY **MJM** QUANTITIES BY **MJM** PRICES BY **MJM** EXTENSIONS BY **MJM** CHECKED BY **JJM**

Quote Preparation By Subcontractor

CLASSIFICATION **Div. 15 - Crane**

ARCHITECT

SHEET NO. **Q4 of 4**

ESTIMATE NO. **91-1**

DATE **1-7-91**

DESCRIPTION	SOURCE/DIMENSIONS	QUANTITY	UNIT	MATERIAL UNIT COST	MATERIAL TOTAL	LABOR UNIT COST	LABOR TOTAL	EQUIPMENT UNIT COST	EQUIPMENT TOTAL	SUBCONTRACT UNIT COST	SUBCONTRACT TOTAL	TOTAL UNIT COST	TOTAL TOTAL
Crane Service													
Truck Crane - 1 Day	016 460 1800	1	Day							575	575		
Equip. Operating Cost	016 460 1800	8	Hrs.							23.70	190		
Equipment Operator	Inside Back Cover	1	Day							280	280		
Equipment Oiler	Inside Back Cover	1	Day							230	230		
Total											1275		1275

Figure 24.17b

Means® Forms

**TELEPHONE
QUOTATION**

PROJECT _Office Building_

FIRM QUOTING _Sky High Co., Inc._

ADDRESS

ITEM QUOTED _Truck Crane - 1 Day_

DATE _1/11/91_

TIME _2:30 P.m._

PHONE (_543_) _789-1234_

BY _Hy Jinks_

RECEIVED BY _Marcia_

WORK INCLUDED	AMOUNT OF QUOTATION
Supply 30 ton crane for 1 day with oiler and operator	
To place max. weight 4 tons a max. distance 45' in from edge of 52' high building	¢ 1275

	TOTAL BID	$ 1275

DELIVERY TIME

DOES QUOTATION INCLUDE THE FOLLOWING:

If ☐ NO is checked, determine the following:

STATE & LOCAL SALES TAXES	☐ YES	☐ NO	MATERIAL VALUE
DELIVERY TO THE JOB SITE	☐ YES	☐ NO	WEIGHT
COMPLETE INSTALLATION	☐ YES	☐ NO	QUANTITY
COMPLETE SECTION AS PER PLANS & SPECIFICATIONS	☐ YES	☐ NO	DESCRIBE BELOW

EXCLUSIONS AND QUALIFICATIONS

ADDENDA ACKNOWLEDGEMENT

TOTAL ADJUSTMENTS

ADJUSTED TOTAL BID

Figure 24.18b

Means' Forms
COST ANALYSIS

PROJECT: Office Building
LOCATION:
TAKE OFF BY: MJM QUANTITIES BY: MJM PRICES BY: MJM EXTENSIONS BY: MJM CHECKED BY: JJM

CLASSIFICATION: Div. 15 - Insulation
ARCHITECT:
SHEET NO. Q1 of 4
ESTIMATE NO. 91-1
DATE 1/9/91

DESCRIPTION	SOURCE/DIMENSIONS			QUANTITY	UNIT	MATERIAL UNIT COST	MATERIAL TOTAL	LABOR UNIT COST	LABOR TOTAL	EQUIP. UNIT COST	EQUIP. TOTAL	SUBCON. UNIT COST	SUBCON. TOTAL	TOTAL UNIT COST	TOTAL
Insulation															
Plumbing															
Copper Tube															
Fiberglass w/ASJ 1" Wall 3/4"	155	651	6840	680	LF	.97	660	1.48	1006						
1"			6860	60		1.12	67	1.55	93						
2"		⌄	6890	160	⌄	1.39	222	1.69	270						
Plumbing Total							949		1369						
Sales Tax						5%	47								
Handling, Overhead + Profit Markup						10%	95	55.2%	756						
Total							1091		2125						3216
Sheet metal															
Blanket 1½"	155	651	3170	6645	SF	.30	1994	1.11	7376						
HVAC Piping															
Fiberglass w/ASJ 1" Wall 2"	155	651	6900	210	LF	1.51	317	1.78	374						
2½"			6910	30		1.68	50	1.87	56						
3"			6920	30		1.88	56	1.98	59						
4"		⌄	6940	240	⌄	2.42	581	2.37	569						
S/m + HVAC Subtotal							2998		8434						
Sales Tax						5%	150								
Handling, Overhead + Profit markup						10%	300	55.2%	4656						
Subtotal							3448		13090						
Total of material + Labor									16538						
+10% for Valves + Flanges									1654						
Total									18192						18192

Figure 24.17c

Means° Forms

TELEPHONE QUOTATION

Q 1-A

DATE 1/11/91

PROJECT Office Building

TIME 10:30 a.m.

FIRM QUOTING Thermal Insulation, Inc.

PHONE (466) 747-4321

ADDRESS

BY Sven Olsen

ITEM QUOTED Insulation

RECEIVED BY Marcia

WORK INCLUDED		AMOUNT OF QUOTATION
Plumbing Insulation	$	3 2 1 6
Sheet Metal + HVAC Insulation		1 8 1 9 2

DELIVERY TIME				**TOTAL BID**	2 1 4 0 8

DOES QUOTATION INCLUDE THE FOLLOWING:

If ☐ NO is checked, determine the following:

STATE & LOCAL SALES TAXES	☑ YES	☐ NO	MATERIAL VALUE	
DELIVERY TO THE JOB SITE	☑ YES	☐ NO	WEIGHT	
COMPLETE INSTALLATION	☑ YES	☐ NO	QUANTITY	
COMPLETE SECTION AS PER PLANS & SPECIFICATIONS	☑ YES	☐ NO	DESCRIBE BELOW	

EXCLUSIONS AND QUALIFICATIONS

Price includes valves and flanges

ADDENDA ACKNOWLEDGEMENT none	**TOTAL ADJUSTMENTS**	
	ADJUSTED TOTAL BID	

ALTERNATES

ALTERNATE NO.
ALTERNATE NO.
ALTERNATE NO.
ALTERNATE NO.
ALTERNATE NO.
ALTERNATE NO.
ALTERNATE NO.

Figure 24.18c

Means' Forms
COST ANALYSIS

PROJECT **Office Building**

LOCATION

TAKE OFF BY MJM QUANTITIES BY MJM PRICES BY MJM EXTENSIONS BY MJM CHECKED BY JJM

Quote Preparation
By Subcontractor
CLASSIFICATION Div. 15 - Temperature Control

ARCHITECT

DESCRIPTION	SOURCE/DIMENSIONS	QUANTITY	UNIT	MATERIAL UNIT COST	MATERIAL TOTAL	LABOR UNIT COST	LABOR TOTAL	EQUIPMENT UNIT COST	EQUIPMENT TOTAL	SUBCONTRACT UNIT COST	SUBCONTRACT TOTAL	TOTAL UNIT COST	TOTAL
Control Systems													
Basic Control of Units	157 425 0220	1	Ea.							3650	3650		
Add for over 20 tons	0260	1								17%	621		
Heating/Cooling Coils	0320	6								2625	15750		
Cooling Tower	0620	1								4575	4575		
Boiler Room Air	3000	1								2050	2050		
Program Optimizer	4040	1								4425	4425		
Subtotal											31671		
Electronic Control	157 430 0020	1								15%	4661		
Total											35732	35732	35732

Figure 24.17d

327

△ Means® Forms

TELEPHONE QUOTATION

Q3-A

DATE 1/11/91

PROJECT Office Building

TIME 11:15 A.m.

FIRM QUOTING Barber Honeyshaw Controls, Inc.

PHONE (543) 762-5861

ADDRESS

BY Bob Jones

ITEM QUOTED Electronic controls

RECEIVED BY Marcia

WORK INCLUDED	AMOUNT OF QUOTATION
Install electronic control system for HVAC equipment.	
System to include controls tie in for the chiller, boiler, boiler room air, heating/cooling coils with specified program optimizer.	$ 35732

DELIVERY TIME				TOTAL BID	35732

DOES QUOTATION INCLUDE THE FOLLOWING: If ☐ NO is checked, determine the following:

STATE & LOCAL SALES TAXES	☑ YES ☐ NO	MATERIAL VALUE	
DELIVERY TO THE JOB SITE	☑ YES ☐ NO	WEIGHT	
COMPLETE INSTALLATION	☑ YES ☐ NO	QUANTITY	
COMPLETE SECTION AS PER PLANS & SPECIFICATIONS	☑ YES ☐ NO	DESCRIBE BELOW	

EXCLUSIONS AND QUALIFICATIONS

Job installed per plans + specs complete.

ADDENDA ACKNOWLEDGEMENT

TOTAL ADJUSTMENTS

ADJUSTED TOTAL BID

Figure 24.18d

328

⚓ Means® Forms

**CONDENSED
ESTIMATE SUMMARY**

SHEET NO. _____

PROJECT **Office Building**   ESTIMATE NO. _____

LOCATION **Boston, Ma**   TOTAL AREA/VOLUME **56,700 S.F.**   DATE **3/27/91**

ARCHITECT **R.S. means**   COST PER S.F./C.F. _____   NO. OF STORIES **3**

PRICES BY: **JJM**   EXTENSIONS BY: **JJM**   CHECKED BY: **RG**

Unit Price		Round Off	
Plumbing	134552	134550	
Fire Protection	127215	127200	
HVAC	51855	51850	

Figure 24.19

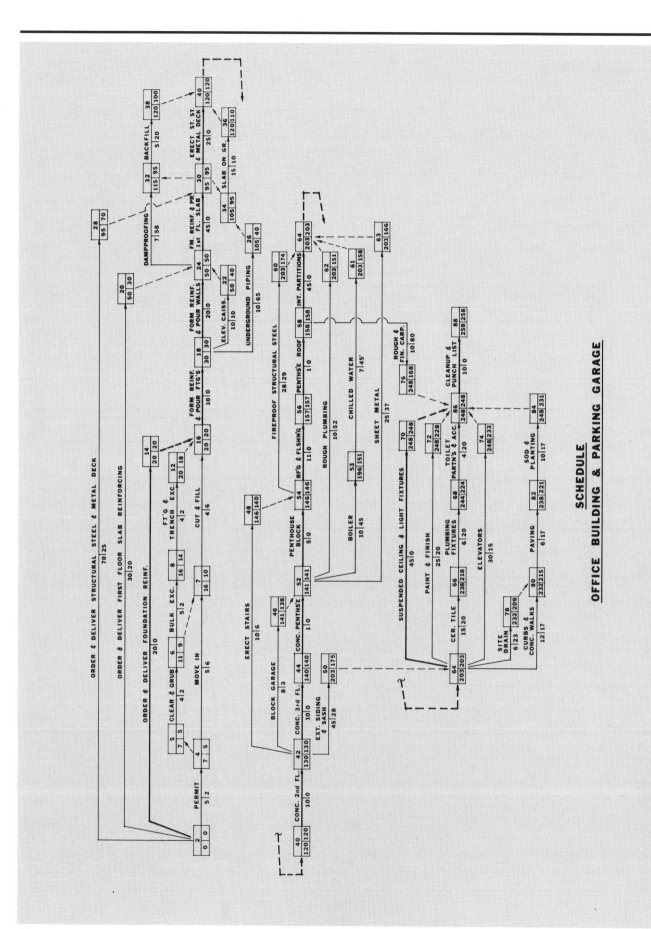

SCHEDULE

OFFICE BUILDING & PARKING GARAGE

Figure 24.20

330

items. Since the example estimate has been prepared using *Means Mechanical Cost Data* or *Means Plumbing Cost Data*, the percentage method has been used.

The Bottom Line: The estimator is now able to complete the Estimate Summary form as shown in Figure 24.19. Appropriate contingency, sales tax, and overhead and profit costs must be added to the direct costs of the project. The overhead and profit percentage for labor obtained from Figure 22.3 (see Chapter 22) has already been added to all labor costs. Wherever possible, contractors should determine appropriate markups for their own companies as discussed earlier in this section and in Part I. Finally, Means prices represent national averages, and should be adjusted with the City Cost Index for your locality as shown in Chapter 22, Figure 22.12.

Since all totals for our estimate of the office building project have already been marked up to cover these items, the Estimate Summary shown here lists only the previously calculated totals for each portion of the project. Either applying markups individually or at the final summary is acceptable, depending on the estimator's preference.

The prime mechanical contractor may also wish to or need to add a final markup for profit if the cost of administering subcontracted items has not been covered previously. Market conditions and project overhead will dictate the amount of final markup, if any, to be applied. The complete mechanical estimate is now ready for submittal to the general contractor.

Means Forms

**PROJECT
OVERHEAD SUMMARY**

PROJECT _____

SHEET NO. _____

ESTIMATE NO. _____

LOCATION _____ ARCHITECT _____ DATE _____

QUANTITIES BY: _____ PRICES BY: _____ EXTENSIONS BY: _____ CHECKED BY: _____

DESCRIPTION	QUANTITY	UNIT	MATERIAL/EQUIPMENT		LABOR		TOTAL COST	
			UNIT	TOTAL	UNIT	TOTAL	UNIT	TOTAL
Job Organization: Superintendent								
Project Manager								
Timekeeper & Material Clerk								
Clerical								
Safety, Watchman & First Aid								
Travel Expense: Superintendent								
Project Manager								
Engineering: Layout								
Inspection/Quantities								
Drawings								
CPM Schedule								
Testing: Soil								
Materials								
Structural								
Equipment: Cranes								
Concrete Pump, Conveyor, Etc.								
Elevators, Hoists								
Freight & Hauling								
Loading, Unloading, Erecting, Etc.								
Maintenance								
Pumping								
Scaffolding								
Small Power Equipment/Tools								
Field Offices: Job Office								
Architect/Owner's Office								
Temporary Telephones								
Utilities								
Temporary Toilets								
Storage Areas & Sheds								
Temporary Utilities: Heat								
Light & Power								
Water								
PAGE TOTALS								

Page 1 of 2

Figure 24.21a

◢◣ Means Forms

DESCRIPTION	QUANTITY	UNIT	MATERIAL/EQUIPMENT		LABOR		TOTAL COST	
			UNIT	TOTAL	UNIT	TOTAL	UNIT	TOTAL
Totals Brought Forward								
Winter Protection: Temp. Heat/Protection								
Snow Plowing								
Thawing Materials								
Temporary Roads								
Signs & Barricades: Site Sign								
Temporary Fences								
Temporary Stairs, Ladders & Floors								
Photographs								
Clean Up								
Dumpster								
Final Clean Up								
Punch List								
Permits: Building								
Misc.								
Insurance: Builders Risk								
Owner's Protective Liability								
Umbrella								
Unemployment Ins. & Social Security								
Taxes								
City Sales Tax								
State Sales Tax								
Bonds								
Performance								
Material & Equipment								
Main Office Expense								
Special Items								
TOTALS:								

Figure 24.21b

APPENDIX

Table of
Contents

⚜ Means Forms

SPEC-AID

DATE _____

DIVISION 15: MECHANICAL

PROJECT _____ LOCATION _____

Building Drainage: Design Rainfall _____ □ Roof Drains _____ □ Court Drains _____
   □ Floor Drains _____ □ Yard Drains _____ □ Lawn Drains _____ □ Balcony Drains _____
   □ Area Drains _____ □ Sump Drains _____ Shower Drains _____ □ _____
   □ Drain Piping: Size _____ Describe _____
   □ Drain Gates _____ □ Clean Outs _____ □ Grease Traps _____
Sanitary System: □ No □ Yes □ Site Main _____ □ Manholes _____
   □ Sump Pumps _____ □ Bilge Pumps _____ □ Ejectors _____
   □ Soils, Stacks _____ □ Wastes, Vents _____ □ _____
Domestic Cold Water: □ No □ Water Meters _____ □ Law Sprinkler Connection _____
   □ Water Softening _____ □ Water Filtering _____
   □ Boiler Feed Water _____ □ Conditioning Apparatus _____
   □ Standpipe System _____ □ Hose Bibbs _____
   □ Pressure Tank _____ □ Booster Pumps _____
   □ Reducing Valves _____ □ _____
Domestic Hot Water: □ No □ Electric □ Gas □ Oil □ Solar _____
   □ Boiler _____ □ Conditioner _____ □ Fixture Connections _____
   □ Storage Tanks _____ Capacity _____
   □ Pumps _____
Piping: □ No □ Yes Material _____
   □ Air Chambers _____ □ Escutcheons _____ □ Expansion Joints _____
   □ Shock Absorbers _____ □ Hangers _____
   □ Valves _____ □ Paint _____
Special Piping: □ No □ Compressed Air _____ □ Vacuum _____
   □ Oxygen _____ □ Nitrous Oxygen _____
   □ Carbon Dioxide _____ □ Process Piping _____
Insulation Cold: □ No □ Yes Material _____ Jacket _____
   Hot: □ No □ Yes Material _____ Jacket _____
Fixtures Bathtub: □ No □ C.I. □ Steel □ Fiberglass □ _____ Color _____
   □ Curtain □ Rod □ Enclosure □ Wall Shower _____
   Drinking Fountain: □ No □ Yes □ Wall Hung □ Pedestal _____
   Hose Bibb: □ No □ Yes Describe _____
   Lavatory: □ No □ China □ C.I. □ Steel □ _____ Color _____
   □ Wall Hung □ Legs □ Acid Resisting _____
   Shower: □ No □ Individual □ Group □ Heads □ _____ Size _____
   Compartment: □ No □ Metal □ Stone □ Fiberglass □ _____ □ Door □ Curtain
   Receptor: □ No □ Plastic □ Metal □ Terrazzo □ _____
   Sinks: □ No □ Kitchen _____ □ Janitor _____
   □ Laundry _____ □ Pantry _____
   □ _____
   Urinals: □ No □ Floor Mounted □ Wall Hung _____
   Screens: □ No □ Floor Mounted □ Wall Hung _____
   Wash Centers: □ No □ Yes Describe _____
   Wash Fountains: □ No □ Floor Mounted □ Wall Hung _____ Size _____
   Describe _____
   Water Closets: □ No □ Floor Mounted □ Wall Hung Color _____
   Describe _____
   Water Coolers: □ No □ Floor Mounted □ Wall Hung _____ Capacity _____ gph.
   □ Water Supply □ Bottle □ Hot □ Compartment _____
   Other Fixtures: _____

Means Forms

SPEC-AID

DIVISION 15: MECHANICAL

Fire Protection: ☐ Carbon Dioxide System _____ ☐ Standpipe _____
☐ Sprinkler System ☐ Wet ☐ Dry _____ Spacing _____
☐ Fire Department Connection _____ ☐ Building Alarm _____
☐ Hose Cabinets _____ ☐ Hose Racks _____
☐ Roof Manifold _____ ☐ Compressed Air Supply _____
☐ Hydrants _____ ☐ _____
Special Plumbing _____

Gas Supply System: ☐ No ☐ Natural Gas ☐ Manufactured Gas _____
Pipe: Schedule _____ Fittings _____
Shutoffs: _____ Master Control Valve: _____
Insulation: _____ Paint: _____
Oil Supply System: ☐ No ☐ Tanks ☐ Above Ground ☐ Below Ground _____
☐ Steel ☐ Plastic ☐ _____ Capacity _____
Heating Plant: ☐ No ☐ Electric ☐ Gas ☐ Oil ☐ Solar _____
☐ Boilers _____ ☐ Pumps _____
☐ PRV Stations _____ ☐ Piping _____
☐ Heat Pumps _____
Cooling Plant: ☐ No ☐ Yes _____ Tons _____
Chillers: ☐ Steam ☐ Water ☐ Air _____
Condenser—Compressor ☐ Air ☐ Water _____
Pumps _____ Cooling Towers _____
System Type: _____
☐ Single Zone _____ ☐ Multi-Zone _____
☐ All Air _____ ☐ Terminal Reheat _____
☐ Double Duct _____ ☐ Radiant Panels _____
☐ Fan Coil _____ ☐ Unit Ventilators _____
☐ Perimeter Radiation _____ ☐ _____
Air Handling Units: Area Served _____ Number _____
Total CFM _____ % Outside Air _____
Cooling, Tons _____ Heating, MBH _____
Filtration _____ Supply Fans _____
Economizer _____
Fans: ☐ No ☐ Return ☐ Exhaust ☐ _____
Describe _____
Distribution: Ductwork _____ Material _____
Terminals: ☐ Diffusers _____ ☐ Registers _____
☐ Grilles _____ ☐ Hoods _____
Volume Dampers: _____
Terminal Boxes: ☐ High Velocity _____ ☐ With Coil _____
☐ Double Duct _____ ☐ _____
Coils: _____
☐ Preheat _____ ☐ Reheat _____
☐ Cooling _____ ☐ _____
Piping: See Previous Page _____
Insulation: Cold: ☐ No ☐ Yes Material _____ Jacket _____
☐ Hot: ☐ No ☐ Yes Material _____ Jacket _____
Automatic Temperature Controls: _____
Air & Hydronic Balancing: _____
Special HVAC: _____

Useful Formulas for the Mechanical Trades

Rectangle

$A = W \times L$

Parallelogram

$A = H \times L$

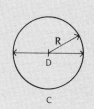

Trapezoid

$A = H \times \dfrac{L_1 + L_2}{2}$

Triangle

$A = \dfrac{W \times H}{2}$

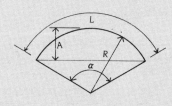

Circle

$A = 3.142 \times R \times R$

$C = 3.142 \times D$

$R = \dfrac{D}{2}$

$D = 2 \times R$

Sector of Circle

$A = \dfrac{3.142 \times R \times R \times \alpha}{360}$

$L = .01745 \times R \times \alpha$

$\alpha = \dfrac{L}{.01745 \times R}$

$R = \dfrac{L}{.01745 \times \alpha}$

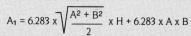

Ellipse

$A = 3.142 \times A \times B$

$C = 6.283 \times \sqrt{\dfrac{A^2 + B^2}{2}}$

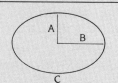

Rectangular Solid

$A_1 = 2[W \times L + L \times H + H \times W]$

$V = W \times L \times H$

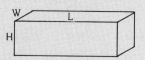

Cone

$A_1 = 3.142 \times R \times S + 3.142 \times R \times R$

$V = 1.047 \times R \times R \times H$

Cylinder

$A_1 = 6.283 \times R \times H + 6.283 \times R \times R$

$V = 3.142 \times R \times R \times H$

Elliptical Tanks

$V = 3.142 \times A \times B \times H$

$A_1 = 6.283 \times \sqrt{\dfrac{A^2 + B^2}{2}} \times H + 6.283 \times A \times B$

Sphere

$A_1 = 12.56 \times R \times R$

$V = 4.188 \times R \times R \times R$

For above containers:

Capacity in gallons = $\dfrac{V}{231}$ when V is in cubic inches.

Capacity in gallons = $7.48 \times V$ when V is in cubic feet.

A = Area   C = Circumference   V = Volume

Conversion Tables

Heat and Power		Volume and Weight	
1 Btu =	778.3 ft.-lb. 0.2930 Int. whr. 252.0 I.T. calorie	1 cu. ft. =	7.481 gal. 1728 cu. in.
1000 LT. calories 1 LT. Kilocalorie =	3.968 Btu 3088 ft.-lb. 1.1628 Int. whr.	1 gallon (U.S.A.) =	231 cu. in. 0.1337 cu. ft.
		1 gal. (British or Imperial) =	277.42 cu. in.
1 horsepower =	0.7455 Int. kw 42.40 Btu per minute 33,000 ft.-lb. per minute 550 ft.-lb. per second	1 cu. ft. water at 60 F =	62.37 lb.
		1 cu. ft. water at 212 F =	59.83 lb.
		1 gal. water at 60 F =	8.338 lb.
		1 gal. water at 212 F =	7.998 lb.
1 boiler horsepower =	33,475 Btu per hour 9.809 Int. kw		
1 Int. watthour =	2656 ft.-lb. 3.413 Btu 3600 Int. joules 860 LT. calories		
1 Int. kilowatt (1000 watts) =	1.341 hp 56.88 Btu per minute 44,267 ft.-lb. per minute		
1 Int. kilowatthour =	3,413 Btu 3,517 lb. water evaporated from and at 212 F		
1 ton refrigeration =	12,000 Btu per hour 200 Btu per minute		
Latent heat of ice (fusion) =	143.4 Btu per pound		

Conversion Tables

Pressure		Metric	
1 lb. per square inch =	144 lb. per square foot 2.0360 in. mercury at 32 F 2.0422 in. mercury at 62 F 2.309 ft. water at 62 F 27.71 in. water at 62 F	1 cm. =	0.3937 in. = 0.0328 ft.
		1 in. =	2.540 cm.
		1 m =	3.281 ft.
		1 ft. =	0.3048 m.
		1 sq. cm. =	0.1550 sq. in.
1 oz. per square inch =	0.1276 in. mercury at 62 F 1.732 in. water at 62 F	1 sq. in. =	6.452 sq. cm.
		1 sq. m. =	10.76 sq. ft.
		1 sq. ft. =	0.09290 sq. m.
1 atmosphere =	14.696 lb. per square inch 2116 lb. per square foot 33.94 ft. water at 62 F 30.01 in. mercury at 62 F 29.921 in. mercury at 32 F	1 cu. cm. =	0.06102 cu. in.
		1 cu. in. =	16.39 cu. cm.
		1 cu. m. =	35.31 cu. ft.
		1 cu. ft. =	0.02832 cu. m.
		1 liter =	1000 cu. cm. = 0.2642 gal.
1 in. water at 62 F =	0.03609 lb. per squre inch 0.5774 oz. per square inch 5.197 lb. per square foot	1 kg. =	2.205 lb. (avdp)
		1 lb. =	0.4536 kg.
		1 metric ton =	2205 lb. (avdp)
1 ft. water at 62 F =	0.4330 lb. per square inch 62.37 lb. per square foot		
1 in. mercury at 62 F =	0.4897 lb. per square inch 7.835 oz. per square inch 1.131 ft. water at 62 F 13.57 in. water at 62 F		
1 in. mercury at 32 F =	0.49115 lb. per square inch		

Glossary of Piping Terms

ANSI
American National Standards Institute

API
American Petroleum Institute

ASHRAE
American Society of Heating, Refrigerating, and Air-Conditioning Engineers

ASME
American Society of Mechanical Engineers

ASTM
American Society for Testing Materials

AWWA
American Water Works Association

Backing Ring
A metal ring used during the welding process. Its purpose is to prevent melted metal from entering a pipe when making a butt-weld joint, and to provide a uniform gap for welding. Often referred to as "chill rings."

BE
Bevelled end. (The end of a pipe or fitting prepared for welding.)

Bell and Spigot Joint
The most commonly used joint in cast-iron soil pipe. Each length is made with an enlarged bell at one end into which the spigot end of another piece is inserted. The joint is then made tight by lead and oakum or a rubber ring caulked into the bell around the spigot.

Black Steel Pipe
Steel pipe that has not been galvanized.

Blank Flange
A flange in which the bolt holes are not drilled.

Blind Flange
A flange used to close off the end of a pipe.

Branch
The outlet or inlet of a fitting that is not in line with the run, and takes off at an angle to the run (e.g., tees, wyes, crosses, laterals, etc.).

Building Sewer
The pipe running from the outside wall of the building drain to the public sewer.

Building Storm Drain
A drain for carrying rain, surface water, condensate, etc., to the building drain or sewer.

Bull Head T ee
A tee with the branch larger than the run.

Butt-Weld Joint
A welded pipe joint made with the ends of the two pipes butting each other.

Butt Weld Pipe
Pipe welded along the seam and not lapped.

Carbon Steel Pipe
Steel pipe which owes its properties chiefly to the carbon which it contains.

Companion Flange
A pipe flange which connects with another flange or with a flanged valve from fitting. This flange differs from a flange that is an integral part of a fitting or valve.

Cooling Tower
A device for cooling water by evaporation. A natural draft cooling tower is one where the airflow through the tower is due to its natural chimney effect. A mechanical draft tower employs fans to force or induce a draft.

Couplings
Fittings for joining two pieces of pipe.

Drainage System
The piping system in a building up to where it discharges to the sewer system.

Elbow
A fitting that makes an angle in a pipe run. The angle is 90° unless another angle is specified.

Equivalent Length
The resistance of a duct or pipe elbow, valve, damper, orifice, bend, fitting, or other obstruction to flow expressed in the number of feet of straight duct or pipe of the same diameter which would have the same resistance.

ERW
Electric resistance weld. (A method of welding pipe in the manufacturing process.)

Expansion Joint
A joint the primary purpose of which is to absorb the longitudinal expansion and contraction in the line due to temperature changes.

Expansion Loop
A large radius loop in a pipe line which absorbs the longitudinal expansion and contraction in the line due to temperature changes.

Flange
A ring-shaped plate at right angles to the end of the pipe. It is provided with holes for bolts to allow fastening of the pipe to a similar flange.

Flat Face
Pipe flanges which have the entire face of the flange faced straight across and which use a full face gasket. These are commonly employed for pressures less than 125 pounds.

Galvanizing
Coating iron or steel surfaces with a protective layer of zinc.

Gate Valve
A valve utilizing a gate, usually wedge-shaped, which allows fluid flow when the gate is lifted from the seat. Gate valves have less resistance to flow than globe valves, and should always be used fully open or fully closed.

Globe Valve
A valve with a rounded body utilizing a manually raised or lowered disc which, when closed, seats so as to prevent fluid flow. Globe valves are ideal for throttling in a semi-closed position.

Header
A large pipe or drum into which each of a group of boilers, chillers, or pumps are connected. (see Manifold)

Horizontal Branch
In plumbing, the horizontal line from the fixture drain to the waste stack.

Hot water Heating System
One in which hot water is the heating medium. Flow is either gravity or forced circulation.

ID
Inside diameter

IDHA
International District Heating Association

Insulation
Thermal insulation is a material used for covering pipes, ducts, vessels, etc., to effect a reduction of heat loss or gain.

Lapped Joint
A lapped joint, like a van stone joint, is a type of pipe joint made using loose flanges on lengths of pipe. The ends of this pipe are lapped over to give a bearing surface for a gasket or metal to metal joint.

Lap Weld Pipe
Pipe made by welding along a scarfed longitudinal seam in which one part is overlapped by another.

LCL
Less carload lot. (Pipe is either ordered from the mill by the carload, or LCL.)

Lead Joint
A joint made by pouring molten lead into the space between a bell and spigot, and making the joint tight by caulking.

Malleable Iron
Cast iron which has been heat treated to reduce its brittleness.

Manifold
A fitting with several branch outlets.

Mill Length
Also known as "random length." The usual run-of-the-mill pipe is 16'-20' in length. Line pipe for power plant or oil field use is often made in double random lengths of 30'-35'

Nipple
A piece of pipe less than 12'' long and threaded on both ends. Pipe over 12'' long is regarded as a cut measure.

Nominal
Name given to standard pipe size designations through 12'' nominal O.D. For example, 2'' nominal is 2-3/8'' O.D.

OD.
Outside diameter.

OS&Y
Outside screw and yoke. A valve configuration in which the valve stem, having exposed external threads supported by a yoke, indicates the open or closed position of the valve.

PE
Plain end. (Used to describe the ends of pipe which are shipped from the mill with unfinished ends. These ends may eventually be threaded, beveled, or grooved in the field.)

Pipe
A hollow cylinder or tube for conveyance of a fluid.

Plug Valve
A valve containing a tapered plug through which a hole is drilled so that fluid can flow through when the holes line up with the inlet and the outlet, but when the plug is rotated 90° the flow is stopped.

Plumbing Fixtures
Devices which receive water, liquid, or waterborne wastes, and discharge the wastes into a drainage system.

Plumbing System
Arrangements of pipes, fixtures, fittings, valves, and traps, in a building which supply water and remove liquid-borne wastes, including storm water.

Potable Water
Water suitable for human consumption.

PSI
Pounds per square inch.

Reducer
A pipe coupling with a larger size at one end than the other. The larger size is designated first. Reducers are threaded, flanged, welded, etc. Reducing couplings are available in either eccentric or concentric configurations.

Riser
A vertical pipe extending one or more floors.

Roof Drain
A fitting which collects water on the roof surface and discharges it into the leader.

Schedule Number
Schedule numbers are American Standards Association designations for classifying the strength of pipe. Schedule 40 is the most common form of steel pipe used in the mechanical trades.

Screwed Joint
A pipe joint consisting of threaded male and female parts joined together.

Seamless Pipe
Pipe or tube formed by piercing a billet of steel and then rolling, rather than having welded seams.

Service Pipe
A pipe connecting water or gas mains into a building from the street.

Slip-on Flange
A flange slipped onto the end of a pipe and then welded in place.

Socket Weld
A pipe joint made by use of a socket weld fitting which has a

female end or socket for insertion of the pipe to be welded.

Stainless Steel
An alloy steel having unusual corrosion-resisting properties, usually imparted by nickel and chromium.

Storm Sewer
A sewer carrying surface or storm water from roofs or exterior surfaces of a building.

Street Elbow
An elbow with a male thread or one end, and a female thread on the other end.

Swing Joint
An arrangement of screwed fittings to allow for movement in a pipe line.

Swivel Joint
A special pipe fitting designed to be pressure tight under continuous or intermittent movement of the equipment to which it is connected.

TBE
Thread both ends. Term used when specifying or ordering cut measures of pipe.

T&C
Threaded and coupled; an ordering designation for threaded pipe.

Tee
A pipe fitting that has a side port at right angles to the run.

TOE
Thread one end. Term used when specifying or ordering cut measures of pipe.

Union
A fitting used to join pipes. It commonly consists of three pieces. Unions are extensively used, because they allow dismantling and reassembling of piping assemblies with ease and without distorting the assembly.

Van Stone
A type of joint made by using loose flanges on lengths of pipe the ends of which are lapped over to give a bearing surface for the flange.

Vents
Vents are used to permit air to escape from hydronic systems, condensate receivers, fuel oil storage tanks, as a breather line for gas regulators, etc.

Vent Stack
A vertical vent pipe which provides air circulation to and from the drainage system.

Vent System
Piping which provides a flow of air to or from a drainage system to protect trap seals from siphonage or back pressure.

Welding Fittings
Wrought steel elbows, tees, reducers, saddles, and the like, beveled for butt welding to pipe. Forged fittings with hubs or with ends counter-bored for fillet welding to pipe are used for small pipe sizes and high pressures.

Welding Neck Flange
A flange with a long neck, beveled for butt welding to pipe.

Wye
A pipe fitting with a side outlet that is any angle other than 90° to the main run or axis.

A	Area Square Feet; Ampere	Calc	Calculated	D.H.	Double Hung
ABS	Acrylonitrile Butadiene Styrene;	Cap.	Capacity	DHW	Domestic Hot Water
	Asbestos Bonded Steel	Carp.	Carpenter	Diag.	Diagonal
A.C.	Alternating Current ;	C.B.	Circuit Breaker	Diam.	Diameter
	Air Conditioning;	C.C.F.	Hundred Cubic Feet	Distrib.	Distribution
	Asbestos Cement	cd	Candela	Dk.	Deck
A.C.I.	American Concrete Institute	cd/sf	Candela per Square Foot	D.L.	Dead Load; Diesel
Addit.	Additional	CD	Grade of Plywood Face & Back	Do.	Ditto
Adj.	Adjustable	CDX	Plywood, grade C&D, exterior glue	Dp.	Depth
af	Audio-frenquency	Cefi.	Cement Finisher	D.P.S.T.	Double Pole, Single Throw
A.G.A.	American Gas Association	Cem.	Cement	Dr.	Driver
Agg.	Aggregate	CF	Hundred Feet	Drink.	Drinking
A.H.	Ampere Hours	C.F.	Cubic Feet	D.S.	Double Strength
A hr	Ampere-hour	CFM	Cubic Feet per Minute	D.S.A.	Double Strength A Grade
A.I.A.	American Institute of Architects	c.g.	Center of Gravity	D.S.B.	Double Strength B Grade
AIC	Ampere Interrupting Capacity	CHW	Commercial Hot Water	Dty.	Duty
Allow.	Allowance	C.I.	Cast Iron	DWV	Drain Waste Vent
alt.	Altitude	C.I.P.	Cast in Place	DX	Deluxe White, Direct Expansion
Alum.	Aluminum	Circ.	Circuit	dyn	Dyne
a.m.	ante meridiem	C.L.	Carload Lot	e	Eccentricity
Amp.	Ampere	Clab.	Common Laborer	E	Equipment Only; East
Approx.	Approximate	C.L.F.	Hundred Linear Feet	Ea.	Each
Apt.	Apartment	CLF	Current Limiting Fuse	Econ.	Economy
Asb.	Asbestos	CLP	Cross Linked Polyethylene	EDP	Electronic Data Processing
A.S.B.C.	American Standard Building Code	cm	Centimeter	E.D.R.	Equiv. Direct Radiation
Asbe.	Asbestos Worker	CMP	Corr. Metal Pipe	Eq.	Equation
A.S.H.R.A.E.	American Society of Heating,	C.M.U.	Concrete Masonry Unit	Elec.	Electrician; Electrical
	Refrig. & AC Engineers	Col.	Column	Elev.	Elevator; Elevating
A.S.M.E.	American Society of	CO_2	Carbon Dioxide	EMT	Electrical Metallic Conduit;
	Mechanical Engineers	Comb.	Combination		Thin Wall Conduit
A.S.T.M.	American Society for	Compr.	Compressor	Eng.	Engine
	Testing and Materials	Conc.	Concrete	EPDM	Ethylene Propylene
Attchmt.	Attachment	Cont.	Continuous;		Diene Monomer
Avg.	Average		Continued	Eqhv.	Equip. Oper., heavy
Bbl.	Barrel	Corr.	Corrugated	Eqlt.	Equip. Oper., light
B.&B.	Grade B and Better;	Cos	Cosine	Eqmd.	Equip. Oper., medium
	Balled & Burlapped	Cot	Cotangent	Eqmm.	Equip. Oper., Master Mechanic
B.&S.	Bell and Spigot	Cov.	Cover	Eqol.	Equip. Oper., oilers
B.&W.	Black and White	CPA	Control Point Adjustment	Equip.	Equipment
b.c.c.	Body-centered Cubic	Cplg.	Coupling	ERW	Electric Resistance Welded
B.F.	Board Feet	C.P.M.	Critical Path Method	Est.	Estimated
Bg. Cem.	Bag of Cement	CPVC	Chlorinated Polyvinyl Chloride	esu	Electrostatic Units
BHP	Brake Horsepower	C. Pr.	Hundred Pair	E.W.	Each Way
B.I.	Black Iron	CRC	Cold Rolled Channel	EWT	Entering Water Temperature
Bit.;		Creos.	Creosote	Excav.	Excavation
Bitum.	Bituminous	Crpt.	Carpet & Linoleum Layer	Exp.	Expansion
Bk.	Backed	CRT	Cathode-ray Tube	Ext.	Exterior
Bkrs.	Breakers	CS	Carbon Steel	Extru.	Extrusion
Bldg.	Building	Csc	Cosecant	f.	Fiber stress
Blk.	Block	C.S.F.	Hundred Square Feet	F	Fahrenheit; Female; Fill
Bm.	Beam	C.S.I.	Construction Specifications	Fab.	Fabricated
Boil.	Boilermaker		Institute	FBGS	Fiberglass
B.P.M.	Blows per Minute	C.T.	Current Transformer	F.C.	Footcandles
BR	Bedroom	CTS	Copper Tube Size	f.c.c.	Face-centered Cubic
Brg.	Bearing	Cu	Cubic	f'c.	Compressive Stress in Concrete;
Brhe.	Bricklayer Helper	Cu. Ft.	Cubic Foot		Extreme Compressive Stress
Bric.	Bricklayer	cw	Continuous Wave	F.E.	Front End
Brk.	Brick	C.W.	Cool White	FEP	Fluorinated Ethylene
Brng.	Bearing	Cwt.	100 Pounds		Propylene (Teflon)
Brs.	Brass	C.W.X.	Cool White Deluxe	F.G.	Flat Grain
Brz.	Bronze	C.Y.	Cubic Yard (27 cubic feet)	F.H.A.	Federal Housing Administration
Bsn.	Basin	C.Y./Hr.	Cubic Yard per Hour	Fig.	Figure
Btr.	Better	Cyl.	Cylinder	Fin.	Finished
BTU	British Thermal Unit	d	Penny (nail size)	Fixt.	Fixture
BTUH	BTU per Hour	D	Deep; Depth; Discharge	Fl. Oz.	Fluid Ounces
BX	Interlocked Armored Cable	Dis.;		Flr.	Floor
c	Conductivity	Disch.	Discharge	F.M.	Frequency Modulation;
C	Hundred;	Db.	Decibel		Factory Mutual
	Centigrade	Dbl.	Double	Fmg.	Framing
C/C	Center to Center	DC	Direct Current	Fndtn.	Foundation
Cab.	Cabinet	Demob.	Demobilization	Fori.	Foreman, inside
Cair.	Air Tool Laborer	d.f.u.	Drainage Fixture Units	Foro.	Foreman, outside

Fount.	Fountain	I.P.S.	Iron Pipe Size	M.C.M.	Thousand Circular Mils
FPM	Feet per Minute	I.P.T.	Iron Pipe Threaded	M.C.P.	Motor Circuit Protector
FPT	Female Pipe Thread	J	Joule	MD	Medium Duty
Fr.	Frame	J.I.C.	Joint Industrial Council	M.D.O.	Medium Density Overlaid
F.R.	Fire Rating	K.	Thousand; Thousand Pounds	Med.	Medium
FRK	Foil Reinforced Kraft	K.D.A.T.	Kiln Dried After Treatment	MF	Thousand Feet
FRP	Fiberglass Reinforced Plastic	kg	Kilogram	M.F.B.M.	Thousand Feet Board Measure
FS	Forged Steel	kG	Kilogauss	Mfg.	Manufacturing
FSC	Cast Body; Cast Switch Box	kgf	Kilogram force	Mfrs.	Manufacturers
Ft.	Foot; Feet	kHz	Kilohertz	mg	Milligram
Ftng.	Fitting	Kip.	1000 Pounds	MGD	Million Gallons per Day
Ftg.	Footing	KJ	Kiljoule	MGPH	Thousand Gallons per Hour
Ft. Lb.	Foot Pound	K.L.	Effective Length Factor	MH	Manhole; Metal Halide; Man Hour
Furn.	Furniture	Km	Kilometer	MHz	Megahertz
FVNR	Full Voltage Non Reversing	K.L.F.	Kips per Linear Foot	Mi.	Mile
FXM	Female by Male	K.S.F.	Kips per Square Foot	MI	Malleable Iron; Mineral Insulated
Fy.	Minimum Yield Stress of Steel	K.S.I.	Kips per Square Inch	mm	Millimeter
g	Gram	K.V.	Kilovolt	Mill.	Millwright
G	Gauss	K.V.A.	Kilovolt Ampere	Min.	Minimum
Ga.	Gauge	K.V.A.R.	Kilovar (Reactance)	Misc.	Miscellaneous
Gal.	Gallon	KW	Kilowatt	ml	Milliliter
Gal./Min.	Gallon Per Minute	KWh	Kilowatt-hour	M.L.F.	Thousand Linear Feet
Galv.	Galvanized	L	Labor Only; Length; Long	Mo.	Month
Gen.	General	Lab.	Labor	Mobil.	Mobilization
Glaz.	Glazier	lat	Latitude	Mog.	Mogul Base
GPD	Gallons per Day	Lath.	Lather	MPH	Miles per Hour
GPH	Gallons per Hour	Lav.	Lavatory	MPT	Male Pipe Thread
GPM	Gallons per Minute	lb.; #	Pound	MRT	Mile Round Trip
GR	Grade	L.B.	Load Bearing; L Conduit Body	ms	millisecond
Gran.	Granular	L. & E.	Labor & Equipment	M.S.F.	Thousand Square Feet
Grnd.	Ground	lb./hr.	Pounds per Hour	Mstz.	Mosaic & Terrazzo Worker
H	High; High Strength Bar Joist;	lb./L.F.	Pounds per Linear Foot	M.S.Y.	Thousand Square Yards
	Henry	lbf/sq in.	Pound-force per Square Inch	Mtd.	Mounted
H.C.	High Capacity	L.C.L.	Less than Carload Lot	Mthe.	Mosaic & Terrazzo Helper
H.D.	Heavy Duty; High Density	Ld.	Load	Mtng.	Mounting
H.D.O.	High Density Overlaid	L.F.	Linear Foot	Mult.	Multi; Multiply
Hdr.	Header	Lg.	Long; Length; Large	MVAR	Million Volt Amp Reactance
Hdwe.	Hardware	L. & H.	Light and Heat	MV	Megavolt
Help.	Helper average	L.H.	Long Span High Strength Bar Joist	MW	Megawatt
HEPA	High Efficiency Particulate Air Filter	L.J.	Long Span Standard Strength	MXM	Male by Male
Hg	Mercury		Bar Joist	MYD	Thousand yards
H.O.	High Output	L.L.	Live Load	N	Natural; North
Horiz.	Horizontal	L.L.D.	Lamp Lumen Depreciation	nA	nanoampere
H.P.	Horsepower; High Pressure	lm	Lumen	NA	Not Available; Not Applicable
H.P.F.	High Power Factor	lm/sf	Lumen per Square Foot	N.B.C.	National Building Code
Hr.	Hour	lm/W	Lumen Per Watt	NC	Normally Closed
Hrs./Day	Hours Per Day	L.O.A.	Length Over All	N.E.M.A.	National Electrical
HSC	High Short Circuit	log	Logarithm		Manufacturers Association
Ht.	Height	L.P.	Liquefied Petroleum;	NEHB	Bolted Circuit Breaker to 600V.
Htg.	Heating		Low Pressure	N.L.B.	Non-Load-Bearing
Htrs.	Heaters	L.P.F.	Low Power Factor	nm	nanometer
HVAC	Heating, Ventilating &	Lt.	Light	No.	Number
	Air Conditioning	Lt. Ga.	Light Gauge	NO	Normally Open
Hvy.	Heavy	L.T.L.	Less than Truckload Lot	N.O.C.	Not Otherwise Classified
HW	Hot Water	Lt. Wt.	Lightweight	Nose.	Nosing
Hyd.;		L.V.	Low Voltage	N.P.T.	National Pipe Thread
Hydr.	Hydraulic	M	Thousand; Material; Male;	NQOB	Bolted Circuit Breaker to 240V.
Hz.	Hertz (cycles)		Light Wall Copper	N.R.C.	Noise Reduction Coefficient
I.	Moment of Inertia	m/hr	Man-hour	N.R.S.	Non Rising Stem
I.C.	Interrupting Capacity	mA	Milliampere	ns	nanosecond
ID	Inside Diameter	Mach.	Machine	nW	nanowatt
I.D.	Inside Dimension;	Mag. Str.	Magnetic Starter	OB	Opposing Blade
	Identification	Maint.	Maintenance	OC	On Center
I.F.	Inside Frosted	Marb.	Marble Setter	OD	Outside Diameter
I.M.C.	Intermediate Metal Conduit	Mat.	Material	O.D.	Outside Dimension
In.	Inch	Mat'l.	Material	ODS	Overhead Distribution System
Incan.	Incandescent	Max.	Maximum	O & P	Overhead and Profit
Incl.	Included; Including	MBF	Thousand Board Feet	Oper.	Operator
Int.	Interior	MBH	Thousand BTU's per hr.	Opng.	Opening
Inst.	Installation	M.C.F.	Thousand Cubic Feet	Orna.	Ornamental
Insul.	Insulation	M.C.F.M.	Thousand Cubic Feet	O.S.&Y.	Outside Screw and Yoke
I.P.	Iron Pipe		per Minute	Ovhd	Overhead

Oz.	Ounce	S.	Suction; Single Entrance; South	T.S.	Trigger Start
P.	Pole; Applied Load; Projection	Scaf.	Scaffold	Tr.	Trade
p.	Page	Sch.;		Transf.	Transformer
Pape.	Paperhanger	Sched.	Schedule	Trhv.	Truck Driver, Heavy
PAR	Weatherproof Reflector	S.C.R.	Modular Brick	Trlr.	Trailer
Pc.	Piece	S.D.R.	Standard Dimension Ratio	Trlt.	Truck Driver, Light
P.C.	Portland Cement; Power Connector	S.E.	Surfaced Edge	TV	Television
P.C.F.	Pounds per Cubic Foot	S.E.R.;		T.W.	Thermoplastic Water Resistant Wire
P.E.	Professional Engineer; Porcelain Enamel; Polyethylene; Plain End	S.E.U.	Service Entrance Cable	UCI	Uniform Construction Index
		S.F.	Square Foot	UF	Underground Feeder
Perf.	Perforated	S.F.C.A.	Square Foot Contact Area	U.H.F.	Ultra High Frequency
Ph.	Phase	S.F.G.	Square Foot of Ground	U.L.	Underwriters Laboratory
P.I.	Pressure Injected	S.F. Hor.	Square Foot Horizontal	Unfin.	Unfinished
Pile.	Pile Driver	S.F.R.	Square Feet of Radiation	URD	Underground Residential Distribution
Pkg.	Package	S.F.Shlf.	Square Foot of Shelf		
Pl.	Plate	S4S	Surface 4 Sides	V	Volt
Plah.	Plasterer Helper	Shee.	Sheet Metal Worker	VA	Volt/amp
Plas.	Plasterer	Sin.	Sine	V.C.T.	Vinyl Composition Tile
Pluh.	Plumbers Helper	Skwk.	Skilled Worker	VAV	Variable Air Volume
Plum.	Plumber	SL	Saran Lined	Vent.	Ventilating
Ply.	Plywood	S.L.	Slimline	Vert.	Vertical
p.m.	Post Meridiem	Sldr.	Solder	V.G.	Vertical Grain
Pord.	Painter, Ordinary	S.N.	Solid Neutral	V.H.F.	Very High Frequency
pp	Pages	S.P.	Static Pressure; Single Pole; Self Propelled	VHO	Very High Output
PP; PPL	Polypropylene			Vib.	Vibrating
P.P.M.	Parts per Million	Spri.	Sprinkler Installer	V.L.F.	Vertical Linear Foot
Pr.	Pair	Sq.	Square; 100 square feet	Vol.	Volume
Prefab.	Prefabricated	S.P.D.T.	Single Pole, Double Throw	W	Wire; Watt; Wide; West
Prefin.	Prefinished	S.P.S.T.	Single Pole, Single Throw	w/	With
Prop.	Propelled	SPT	Standard Pipe Thread	W.C.	Water Column; Water Closet
PSF; psf	Pounds per Square Foot	Sq. Hd.	Square Head	W.F.	Wide Flange
PSI; psi	Pounds per Square Inch	S.S.	Single Strength; Stainless Steel	W.G.	Water Gauge
PSIG	Pounds per Square Inch Gauge	S.S.B.	Single Strength B Grade	Wldg.	Welding
PSP	Plastic Sewer Pipe	Sswk.	Structural Steel Worker	Wrck.	Wrecker
Pspr.	Painter, Spray	Sswl.	Structural Steel Welder	W.S.P.	Water, Steam, Petroleum
Psst.	Painter, Structural Steel	St.; Stl.	Steel	WT, Wt.	Weight
P.T.	Potential Transformer	S.T.C.	Sound Transmission Coefficient	WWF	Welded Wire Fabric
P. & T.	Pressure & Temperature	Std.	Standard	XFMR	Transformer
Ptd.	Painted	STP	Standard Temperature & Pressure	XHD	Extra Heavy Duty
Ptns.	Partitions	Stpi.	Steamfitter, Pipefitter	Y	Wye
Pu	Ultimate Load	Str.	Strength; Starter; Straight	yd	Yard
PVC	Polyvinyl Chloride	Strd.	Stranded	yr	Year
Pvmt.	Pavement	Struct.	Structural	Δ	Delta
Pwr.	Power	Sty.	Story	%	Percent
Q	Quantity Heat Flow	Subj.	Subject	~	Approximately
Quan.; Qty.	Quantity	Subs.	Subcontractors	Ø	Phase
Q.C.	Quick Coupling	Surf.	Surface	@	At
r	Radius of Gyration	Sw.	Switch	#	Pound; Number
R	Resistance	Swbd.	Switchboard	<	Less Than
R.C.P.	Reinforced Concrete Pipe	S.Y.	Square Yard	>	Greater Than
Rect.	Rectangle	Syn.	Synthetic		
Reg.	Regular	Sys.	System		
Reinf.	Reinforced	t.	Thickness		
Req'd.	Required	T	Temperature; Ton		
Resi	Residential	Tan	Tangent		
Rgh.	Rough	T.C.	Terra Cotta		
R.H.W.	Rubber, Heat & Water Resistant Residential Hot Water	T.D.	Temperature Difference		
		TFE	Tetrafluoroethylene (Teflon)		
rms	Root Mean Square	T. & G.	Tongue & Groove; Tar & Gravel		
Rnd.	Round				
Rodm.	Rodman	Th.; Thk.	Thick		
Rofc.	Roofer, Composition	Thn.	Thin		
Rofp.	Roofer, Precast	Thrded	Threaded		
Rohe.	Roofer Helpers (Composition)	Tilf.	Tile Layer Floor		
Rots.	Roofer, Tile & Slate	Tilh.	Tile Layer Helper		
R.O.W.	Right of Way	THW.	Insulated Strand Wire		
RPM	Revolutions per Minute	THWN;			
R.R.	Direct Burial Feeder Conduit	THHN	Nylon Jacketed Wire		
R.S.	Rapid Start	T.L.	Truckload		
RT	Round Trip	Tot.	Total		

Plumbing Fixture Symbols

Baths					Dishwasher
	Corner			DW	
	Recessed				Sinks
					Single Basin
	Angle				Twin Basin
					Single Drainboard
	Whirlpool				Double Drainboard
Showers					
	Stall			DF	Floor — Drinking Fountains
				DF	Recessed
	Corner Stall			DF	Semi-Recessed
	Wall Gang			LT	Single — Laundry Trays
Water Closets					
	Tank			L T	Double
	Flush Valve			SS	Wall — Service Sinks
	Bidet			SS	Floor
Urinals					
	Wall			WF	Circular — Wash Fountains
	Stall			WF	Semi-Circular
	Trough			WH	Heater — Hot Water
Lavatories					
	Counter			HWT	Tank
	Wall			G	Gas — Separators
	Corner				
	Pedestal				Oil

Piping Symbols

Valves, Fittings & Specialties					Valves, Fittings & Specialties (Cont.)
Gate			Up/Dn	Pipe Pitch Up or Down	
Globe				Expansion Joint	
Check				Expansion Loop	
Butterfly				Flexible Connection	
Solenoid			T	Thermostat	
Lock Shield				Thermostatic Trap	
2-Way Automatic Control			F&T	Float and Thermostatic Trap	
3-Way Automatic Control				Thermometer	
Gas Cock				Pressure Gauge	
Plug Cock			F S	Flow Switch	
Flanged Joint			B	Pressure Switch	
Union				Pressure Reducing Valve	
Cap			P	Humidistat	
Strainer			A	Aquastat	
Concentric Reducer				Air Vent	
Eccentric Reducer			M	Meter	
Pipe Guide				Elbow	
Pipe Anchor				Tee	
Elbow Looking Up					
Elbow Looking Down					
Flow Direction					

Piping Symbols (Cont.)

Plumbing

Floor Drain	⊡
Indirect Waste	——— W ———
Storm Drain	——— SD ———
Combination Waste & Vent	——— CWV ———
Acid Waste	——— AW ———
Acid Vent	– – – AV – – –
Cold Water	——— CW ———
Hot Water	——— HW ———
Drinking Water Supply	——— DWS ———
Drinking Water Return	——— DWR ———
Gas-Low Pressure	——— G ———
Gas-Medium Pressure	——— MG ———
Compressed Air	——— A ———
Vacuum	——— V ———
Vacuum Cleaning	——— VC ———
Oxygen	——— O ———
Liquid Oxygen	——— LOX ———
Liquid Petroleum Gas	——— LPG ———

HVAC

Hot Water Heating Supply	——— HWS ———
Hot Water Heating Return	——— HWR ———
Chilled Water Supply	——— CHWS ———
Chilled Water Return	——— CHWR ———
Drain Line	——— D ———
City Water	——— CW ———
Fuel Oil Supply	——— FOS ———
Fuel Oil Return	——— FOR ———
Fuel Oil Vent	——— FOV ———

HVAC (Cont.)

——— FOG ———	Fuel Oil Gauge Line
——o— PD —o——	Pump Discharge
– – – – – –	Low Pressure Condensate Return
——— LPS ———	Low Pressure Steam
——— MPS ———	Medium Pressure Steam
——— HPS ———	High Pressure Steam
——— BD ———	Boiler Blow-Down

Fire Protection

——— F ———	Fire Protection Water Supply
——— WSP ———	Wet Standpipe
——— DSP ———	Dry Standpipe
——— CSP ———	Combination Standpipe
——— SP ———	Automatic Fire Sprinkler
——⊖———⊖——	Upright Fire Sprinkler Heads
——●———●——	Pendent Fire Sprinkler Heads
⊲	Fire Hydrant
⊢<	Wall Fire Dept. Connection
⊖<	Sidewalk Fire Dept. Connection
FHR ⊖╫╫╫	Fire Hose Rack
FHC	Surface Mounted Fire Hose Cabinet
FHC	Recessed Fire Hose Cabinet

HVAC Ductwork Symbols

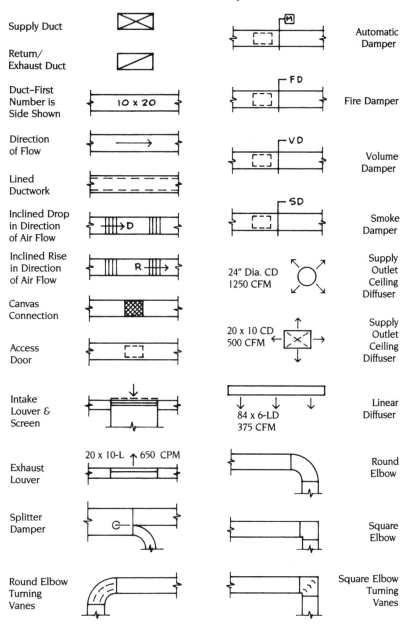

Supply Duct	Automatic Damper
Return/ Exhaust Duct	Fire Damper (FD)
Duct–First Number is Side Shown (10 x 20)	Volume Damper (VD)
Direction of Flow	Smoke Damper (SD)
Lined Ductwork	Supply Outlet Ceiling Diffuser (24" Dia. CD 1250 CFM)
Inclined Drop in Direction of Air Flow (D)	Supply Outlet Ceiling Diffuser (20 x 10 CD 500 CFM)
Inclined Rise in Direction of Air Flow (R)	Linear Diffuser (84 x 6-LD 375 CFM)
Canvas Connection	Round Elbow
Access Door	Square Elbow
Intake Louver & Screen	
Exhaust Louver (20 x 10-L 650 CPM)	
Splitter Damper	
Round Elbow Turning Vanes	Square Elbow Turning Vanes

Double Duct Air System

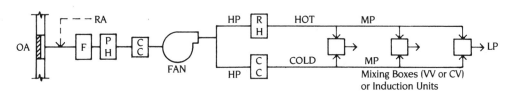

OA = Outside Air
RA = Return Air
F = Filter
PH = Preheat Coil

CC = Cooling Coil
RH = Reheat Coil
HP = High Pressure Duct
MP = Medium Pressure Duct

LP = Low Pressure Duct
VV = Variable Volume
CV = Constant Volume

INDEX